Developments in Paleoenvironmental Research

Volume 23

Series Editor

John P. Smol, Department of Biology, Queen's University, KINGSTON, Canada

Paleoenvironmental research continues to enjoy tremendous interest and progress in the scientific community. The overall aims and scope of the *Developments in Paleoenvironmental Research* book series is to capture this excitement and document these developments. Volumes related to any aspect of paleoenvironmental research, encompassing any time period, are within the scope of the series. For example, relevant topics include studies focused on terrestrial, peatland, lacustrine, riverine, estuarine, and marine systems, ice cores, cave deposits, palynology, isotopes, geochemistry, sedimentology, paleontology, etc. Methodological and taxonomic volumes relevant to paleoenvironmental research are also encouraged. The series will include edited volumes on a particular subject, geographic region, or time period, conference and workshop proceedings, as well as monographs.

Prospective authors and/or editors should consult the **Series Editor John P. Smol for more details. Any comments or suggestions for future volumes are welcomed;**

Hilary H. Birks • Carl D. Sayer
Richard E. Walton

Seeds and Fruits of Aquatic Plants of Northern Europe - A Handbook

Springer

Hilary H. Birks
Department of Biological Sciences
University of Bergen
Bergen, Norway

Carl D. Sayer
Department of Geography
University College London
London, UK

Richard E. Walton
School of Geography and
Environmental Science
University of Southampton
Southampton, UK

ISSN 1571-5299 ISSN 2215-1672 (electronic)
Developments in Paleoenvironmental Research
ISBN 978-3-032-04169-2 ISBN 978-3-032-04170-8 (eBook)
https://doi.org/10.1007/978-3-032-04170-8

This Springer imprint is published by the registered company Springer Nature Switzerland AG
The registered company address is: Gewerbestrasse 11, 6330 Cham, Switzerland

Acknowledgements

We are most grateful to the following people and institutions for the gift or the loan of plant material.

The University of Bergen Herbarium and Astrid Botnen, The University of Helsinki Herbarium and Minna Väliranta, The Czech Academy of Sciences at Trebon and Lenká Moravcová, and Anna Šolcová, Mike Field, Pim van der Knapp, Lubomir Adamec, Astrid Biddle, and Richard Unsworth. We gratefully acknowledge that the Kew Millennium Seed Bank provided us with the following material: *Alopecurus aequalis*, 0187750, *Berula erecta*, 0678333, *Bidens cernua*, 0078221, *Callitriche obtusangula*, 939265, *C. stagnalis*, 0134473, *C. truncata*, 0135377, *Damasonium alisma*, 164627, *Eriocaulon aquaticum*, 143606, *Glyceria declinata*, 0130512, *Hypericum elodes*, 0136525, *Isolepis cernua*, 0135632, *I. setacea*, 0143226, *Juncus subnodulosus*, 0136776, *Mentha aquatica*, 0478315, *Myriophyllum alterniflorum*, 0135241, *Nasturtium microphyllum*, 128609, *Ranunculus circinatus*, 0127163, *R. peltatus*, 0133627, *R. peltatus* ssp. *baudotii*, 0127174, *R. omiophyllus*, 0134803, *Rorippa amphibia*, 0143215, *Stachys palustris*, 0137692, *Valeriana dioica*, 134989, *Veronica catenata*, 0423187.

We thank all those who have given advice and shared their knowledge with us, in particular Richard Lansdown, Gerhard Weigleb, Mariusz Gałka, Lubomir Adamec, and John Birks.

We acknowledge the use of photographs taken by Eftesum Eftesum, B. Brandon Curry, and Ben Goldsmith. All the other photographs were taken by the authors.

We would not have been able to complete our project without the invaluable help of Arild Breistøl, Lene Halvorsen, and Linn Krüger in the University of Bergen Palaeoecology Laboratory, Ambroise Baker of Teesside University, and Cath D'Alton of University College London Department of Geography.

We thank John Smol for inviting us to contribute our book to the Springer DPER (Developments in Paleoenvironmental Research) Series and for help and support. We also thank Aaron Schiller and Tina Silva at Springer Publishers for their help and advice.

We acknowledge funding from the Mead Fund, University College London to Richard Walton to visit Bergen, and from the European Research Council (ERC) under the European Union's Horizon 2020 research and innovation programme (grant agreement no. 741413) HOPE: Humans On Planet Earth - Long-term impacts on biosphere dynamics, awarded to H.J.B. Birks, which funded repairs and updates to the stereo-photomicroscope in the University of Bergen Palaeoecology Laboratory.

Contents

1 Introduction 1
1.1 The Aims of this Handbook 1
1.2 What Is an Aquatic Plant, or Macrophyte? 2
1.3 Reasons for Identifying Seeds and Fruits of Aquatic Plants 4

2 Aquatic Ecosystems and Their Aquatic Plants 7
2.1 Aquatic Ecosystems 7
2.2 Wetland Types 7
2.3 Ecosystem Services and Benefits of Aquatic Plants 10
2.4 Threats to Aquatic Ecosystems and Their Plants 11
2.5 Mitigation and Restoration 12

3 Palaeolimnology 15
3.1 Palaeolimnology 15
3.2 Multiproxy Studies 17
3.3 Palaeolimnological Proxies Used in Conjunction with Plant Macrofossil Analysis 17
3.3.1 Pollen and Macrofossils 17
3.3.2 Other Biological Proxies 18
3.3.3 Some Non-biological Proxies Used in Palaeolimnology 19
3.4 Applications of Plant Macrofossils in Palaeolimnology 20
3.4.1 Aquatic Ecosystem History 20
3.4.2 Water-level Reconstructions 22
3.4.3 Eutrophication 22
3.4.4 The Application of Macrofossils to Aquatic Conservation 23

4 Plant Macrofossil Analysis 25
4.1 What Is a Plant Macrofossil? 25
4.2 What Is a Fruit and What Is a Seed? 26
4.3 Study Design and Obtaining Sediment for Macrofossil Analysis 27
4.4 Method of Macrofossil Analysis 28

4.5 Sources of Error 30
4.6 Interpretation of Macrofossil Data 32
4.6.1 Macrofossil Production and Taphonomy 32
4.6.2 Dispersal 34
4.7 Applications of Plant Macrofossil Analysis 36

5 Species Accounts 37
5.1 Introduction 37
5.2 Species Accounts 38
5.2.1 *Acorus calamus* L. Acoraceae 38
5.2.2 *Alisma gramineum* Lej. Alismataceae 39
5.2.3 *Alisma lanceolatum* With. Alismataceae 40
5.2.4 *Alisma plantago-aquatica* L. Alismataceae 41
5.2.5 *Baldellia ranunculoides* (L.) Parl. *sensu lato* Alismataceae 42
5.2.6 *Damasonium alisma* Mill. Alismataceae 44
5.2.7 *Luronium natans* (L.) Raf. Alismataceae 45
5.2.8 *Sagittaria sagittifolia* L. Alismataceae 47
5.2.9 *Berula erecta* (Huds.) Coville Apiaceae 48
5.2.10 *Cicuta virosa* L. Apiaceae 49
5.2.11 *Helosciadium inundatum* (L.) W.D.J.Koch (Syn. *Apium inundatum* (L.) Rchb.fil.) Apiaceae 50
5.2.12 *Helosciadium nodiflorum* (L.) W.J.D.Koch (Syn. *Apium nodiflorum* (L.) Lag.) Apiaceae 51
5.2.13 *Oenanthe crocata* L. Apiaceae 52
5.2.14 *Oenanthe fistulosa* L. Apiaceae 53
5.2.15 *Oenanthe lachenalii* C.C.Gmel. Apiaceae 54
5.2.16 *Peucedanum ostruthium* (L.) Koch (Syn. *Imperatoria ostruthium* L.) Apiaceae 55
5.2.17 *Peucedanum palustre* (L.) Moench. (Syn. *Thysselinum palustre* (L.) Hoffm) Apiaceae 56
5.2.18 *Sium latifolium* L. Apiaceae 57
5.2.19 *Calla palustris* L. Araceae 58
5.2.20 *Lemna* L., *Spirodela* (L.) Schleid., *Wolffia* Horkel ex Schleid. Araceae (Lemnaceae) 59
5.2.21 *Hydrocotyle vulgaris* L. Araliaceae (Hydrocotylaceae) 64
5.2.22 *Warnstorfia fluitans* (Hedwig.) Loeske, *Warnstorfia exannulata* (Schimp.) Loeske Calliergonaceae *Drepanocladus aduncus* (Hedwig.) Warnst. Amblystegiaceae (Synonyms: *Drepanocladus fluitans* (Hedw.) Warnst., *Drepanocladus exannulatus* (Schimp.) Warnst., *Sarmenthypnum exannulatum* (Schimp.) Hedenäs 65
5.2.23 *Thelypteris palustris* (Salisb.) Schott Aspleniaceae 67
5.2.24 *Bidens cernua* L. Asteraceae 68
5.2.25 *Bidens radiata* Thuill. Asteraceae 69

5.2.26 *Bidens tripartita* L. Asteraceae 70
5.2.27 *Cirsium palustre* (L.) Scop. Asteraceae 71
5.2.28 *Eupatorium cannabinum* L. Asteraceae. 72
5.2.29 *Jacobaea aquatica* (Hill) G.Gaertn., B.Mey. & Scherb. (Syn. *Senecio aquaticus* Hill) Asteraceae 73
5.2.30 *Petasites hybridus* (L.) G.Gaertn., B.Mey. & Scherb. Asteraceae 74
5.2.31 *Sonchus palustris* L. Asteraceae 75
5.2.32 *Impatiens glandulifera* Royle Balsaminaceae 76
5.2.33 *Alnus glutinosa* (L.) Gaertn. Betulaceae 77
5.2.34 *Alnus incana* (L.) Moench. Betulaceae 79
5.2.35 *Myosotis scorpioides* L. Boraginaceae 80
5.2.36 *Nasturtium microphyllum* (Boenn. ex Rchb.) Rchb. Brassicaceae. 81
5.2.37 *Nasturtium officinale* W.T.Aiton (Syn. *Rorippa nasturtium-aquaticum*) Brassicaceae. 82
5.2.38 *Rorippa amphibia* (L.) Besser Brassicaceae 83
5.2.39 *Rorippa islandica* (Oeder) Borbás and *Rorippa palustris* (L.) Besser Brassicaceae. 84
5.2.40 *Rorippa sylvestris* (L.) Besser Brassicaceae 85
5.2.41 *Subularia aquatica* L. Brassicaceae. 86
5.2.42 *Butomus umbellatus* L. Butomaceae 87
5.2.43 *Lobelia dortmanna* L. Campanulaceae 88
5.2.44 *Valeriana officinalis* L., *Valeriana excelsa* ssp. *sambucifolia* (J.C.Mikan ex Pohl) Holub., *Valeriana dioica* L. Caprifoliaceae (formerly Valerianaceae). 89
5.2.45 *Silene flos-cuculi* L. (Syn. *Lychnis flos-cuculi*) Caryophyllaceae. 91
5.2.46 *Ceratophyllum demersum* L. and *Ceratophyllum submersum* L. Ceratophyllaceae 92
5.2.47 Characeae S.F.Gray 94
5.2.48 *Crassula aquatica* (L.) Schönl. Crassulaceae 100
5.2.49 *Bolboschoenus maritimus* (L.) Palla (Syn. *Scirpus maritimus* L.) Cyperaceae 101
5.2.50 *Carex* L. Cyperaceae 102
5.2.51 *Carex acuta* L. Cyperaceae 104
5.2.52 *Carex appropinquata* Schumach. Cyperaceae. 106
5.2.53 *Carex aquatilis* Wahlenb. Cyperaceae 107
5.2.54 *Carex chordorrhiza* L.f. Cyperaceae 108
5.2.55 *Carex demissa* Hornem. (Syn. *Carex viridula* subsp. *oedocarpa*) Cyperaceae 109
5.2.56 *Carex diandra* Schrank Cyperaceae. 110
5.2.57 *Carex dioica* L. Cyperaceae. 111
5.2.58 *Carex elata* All. Cyperaceae 112
5.2.59 *Carex hispida* Willd. ex Schkuhr (Syn. *Carex acutiformis* Ehrh.) Cyperaceae. 113

5.2.60 *Carex lasiocarpa* Ehrh. Cyperaceae 114
5.2.61 *Carex nigra* (L.) Reichard Cyperaceae 115
5.2.62 *Carex paniculata* L. Cyperaceae 116
5.2.63 *Carex pseudocyperus* L. Cyperaceae.................. 117
5.2.64 *Carex riparia* Curt. Cyperaceae....................... 118
5.2.65 *Carex rostrata* Stokes Cyperaceae.................... 119
5.2.66 *Carex vesicaria* L. Cyperaceae 120
5.2.67 *Cladium mariscus* (L.) Pohl. Cyperaceae 121
5.2.68 *Cyperus fuscus* L. Cyperaceae........................ 123
5.2.69 *Cyperus longus* L. Cyperaceae 124
5.2.70 *Eleocharis acicularis* (L.) Roem. & Schult. Cyperaceae... 125
5.2.71 *Eleocharis mamillata* (H.Lindb.) H.Lindb. Cyperaceae ... 126
5.2.72 *Eleocharis multicaulis* (Smith) Desvaux Cyperaceae 127
5.2.73 *Eleocharis palustris* (L.) Roem. & Schult. Cyperaceae.... 128
5.2.74 *Eleocharis quinqueflora* (Hartmann) O.Schwarz Cyperaceae....................................... 129
5.2.75 *Eleocharis uniglumis* Schult. & Zinserl. Cyperaceae 130
5.2.76 *Isolepis cernua* (Vahl) Roem. & Schult. Cyperaceae 131
5.2.77 *Isolepis fluitans* (L.) R.Br. (Synonyms: *Eleogiton fluitans*; *Scirpus fluitans*) Cyperaceae 132
5.2.78 *Isolepis setacea* (L.) R.Br. Cyperaceae 133
5.2.79 *Scirpoides holochoenus* (L.) Soják Cyperaceae.......... 134
5.2.80 *Scirpus sylvaticus* L. Cyperaceae..................... 135
5.2.81 *Schoenoplectus* (L.) Palla (Syn. *Scirpus lacustris*) Cyperaceae....................................... 136
5.2.82 *Schoenoplectus tabernaemontani* (C.C.Gmel.) Palla Cyperaceae....................................... 137
5.2.83 *Schoenoplectus triqueter* (L.) Palla Cyperaceae 138
5.2.84 *Aldrovanda vesiculosa* L. Droseraceae 139
5.2.85 *Elatine* L. Elatinaceae 140
5.2.86 *Equisetum* L. Equisetaceae 144
5.2.87 *Eriocaulon aquaticum* Druce Eriocaulaceae 145
5.2.88 *Lathyrus palustris* L. Fabaceae 146
5.2.89 *Fontinalis antipyretica* Hedw. Fontinalaceae............ 147
5.2.90 *Myriophyllum alterniflorum* DC. Haloragaceae 148
5.2.91 *Myriophyllum spicatum* L. Haloragaceae 150
5.2.92 *Myriophyllum verticillatum* L. Haloragaceae 152
5.2.93 *Elodea canadensis* Michx. and *Elodea nuttallii* (Planch) H.St.John Hydrocharitaceae 153
5.2.94 *Hydrocharis morsus-ranae* L. Hydrocharitaceae......... 154
5.2.95 *Lagarosiphon major* (Ridl.) Moss Hydrocharitaceae 155
5.2.96 *Najas flexilis* (Willd.) Rostk. & W.L.E.Schmidt Hydrocharitaceae (Najadaceae)...................... 156
5.2.97 *Najas marina* L. Hydrocharitaceae (Najadaceae) 159

5.2.98 *Najas minor* All. Hydrocharitaceae (Najadaceae) 161
5.2.99 *Najas tenuissima* (A.Braun ex Magnus) Magnus Hydrocharitaceae (Najadaceae) 162
5.2.100 *Stratiotes aloides* L. Hydrocharitaceae 163
5.2.101 *Hypericum elodes* L. Hypericaceae 165
5.2.102 *Iris pseudacorus* L. Iridaceae. 166
5.2.103 *Isoetes lacustris* L. and *Isoetes echinospora* Durieu Isoetaceae 167
5.2.104 *Juncus* L. Juncaceae. 169
5.2.105 *Juncus acutiflorus* Ehrh. ex Hoffm. Juncaceae 170
5.2.106 *Juncus articulatus* L. Juncaceae. 171
5.2.107 *Juncus bufonius* L. Juncaceae 172
5.2.108 *Juncus bulbosus* L. Juncaceae 173
5.2.109 *Juncus conglomeratus* L. and *Juncus effusus* L. Juncacaeae 174
5.2.110 *Juncus filiformis* L. Juncaceae 176
5.2.111 *Juncus subnodulosus* Schrank Juncaceae 177
5.2.112 *Lycopus europaeus* L. Lamiaceae 178
5.2.113 *Mentha aquatica* L. Lamiaceae 179
5.2.114 *Stachys palustris* L. Lamiaceae 180
5.2.115 *Utricularia intermedia* Hayne Lentibulariaceae 181
5.2.116 *Utricularia minor* L. Lentibulariaceae. 183
5.2.117 *Utricularia vulgaris* L. Lentibulariaceae 184
5.2.118 *Lythrum hyssopifolia* L. Lythraceae. 185
5.2.119 *Lythrum portula* (L.) D.A.Webb Lythraceae 186
5.2.120 *Lythrum salicaria* L. Lythraceae 187
5.2.121 *Trapa natans* L. Lythraceae 188
5.2.122 *Menyanthes trifoliata* L. Menyanthaceae 190
5.2.123 *Nymphoides peltata* (S.G.Gmel.) Kuntze Menyanthaceae. 191
5.2.124 *Pilularia globulifera* L. Marsileaceae 192
5.2.125 *Montia fontana* L. Montiaceae. 193
5.2.126 *Nuphar lutea* (L.) Sm. Nymphaeaceae 194
5.2.127 *Nuphar pumila* (Timm.) DC. Nymphaeaceae 195
5.2.128 *Nymphaea alba* L. Nymphaeaceae 196
5.2.129 *Nymphaea tetragona* Georgi Nymphaeaceae 198
5.2.130 *Epilobium hirsutum* L. Onagraceae 199
5.2.131 *Epilobium palustre* L. Onagraceae. 200
5.2.132 *Epilobium parviflorum* Schreb. Onagraceae 201
5.2.133 *Erythranthe guttata* (DC.) G.L.Nesom (Syn. *Mimulus guttatus* Fisch. ex DC.) Phrymaceae (Scrophulariaceae). 202
5.2.134 *Pedicularis palustris* L. Orobanchaceae 203
5.2.135 *Callitriche* L. Plantaginaceae (Callitrichaceae). 204

5.2.136 *Callitriche brutia* ssp *brutia* Petagna and *Callitriche brutia* ssp *hamulata* (Kütz. ex W.D.J.Koch) O.Bolòs & Vigo Plantaginaceae (Callitrichaceae)....... 206
5.2.137 *Callitriche cophocarpa* Sendtn. Plantaginaceae (Callitrichaceae)................................ 208
5.2.138 *Callitriche hermaphroditica* L. Plantaginaceae (Callitrichaceae)................................ 209
5.2.139 *Callitriche obtusangula* Le Gall Plantaginaceae (Callitrichaceae)................................ 210
5.2.140 *Callitriche palustris* L. Plantaginaceae (Callitrichaceae) 211
5.2.141 *Callitriche platycarpa* Kütz. Plantaginaceae (Callitrichaceae)................................ 212
5.2.142 *Callitriche stagnalis* Scop. Plantaginaceae (Callitrichaceae)................................ 213
5.2.143 *Callitriche truncata* ssp. *occidentalis* (Rouy) Schotsman Plantaginaceae (Callitrichaceae)........... 214
5.2.144 *Gratiola officinalis* L. Plantaginaceae 215
5.2.145 *Hippuris vulgaris* L. Plantaginaceae (Hippuridaceae).... 216
5.2.146 *Littorella uniflora* (L.) Asch. Plantaginaceae........... 217
5.2.147 *Veronica* L. Plantaginaceae 219
5.2.148 Poaceae L. (Syn. Gramineae)...................... 221
5.2.149 *Alopecurus aequalis* Sobol., and *Alopecurus geniculatus* L. Poaceae........................... 223
5.2.150 *Calamagrostis canescens* (Weber) Roth Poaceae 225
5.2.151 *Calamagrostis epigejos* (L.) Roth. Poaceae............ 226
5.2.152 *Calamagrostis purpurea* (Trin.) Trin. Poaceae 227
5.2.153 *Calamagrostis stricta* (Timm) Koeler Poaceae 228
5.2.154 *Catabrosa aquatica* (L.) P.Beauv. Poaceae 229
5.2.155 *Glyceria declinata* Bréb Poaceae.................... 230
5.2.156 *Glyceria fluitans* (L.) R.Br. Poaceae 231
5.2.157 *Glyceria lithuanica* (Gorski) Gorski Poaceae 233
5.2.158 *Glyceria maxima* (Hartm.) Holmb. (Syn. *Glyceria aquatica* (L.) Wahlenb.) Poaceae.................... 234
5.2.159 *Glyceria notata* Chevall. (Syn. *Glyceria plicata* (Fries) Fries) Poaceae 235
5.2.160 *Phalaris arundinacea* L. Poaceae 236
5.2.161 *Phragmites australis* ssp. *australis* (Cav.) Trin. ex Steud. Poaceae.............................. 237
5.2.162 *Poa palustris* L. Poaceae 239
5.2.163 *Persicaria amphibia* (L.) Delarbre (Syn. *Polygonum amphibium* L.) Polygonaceae 240
5.2.164 *Persicaria hydropiper* (L.) Delarbre (Syn. *Polygonum hydropiper* L.) Polygonaceae 241
5.2.165 *Rumex aquaticus* L. Polygonaceae 242

5.2.166 *Rumex hydrolapathum* Huds. Polygonaceae 243
5.2.167 *Rumex palustris* Sm. Polygonaceae 244
5.2.168 *Groenlandia densa* (L.) Fourr. Potamogetonaceae 245
5.2.169 *Potamogeton* L. Potamogetonaceae 247
5.2.170 *Potamogeton acutifolius* Link ex Roem. & Schult. Potamogetonaceae 252
5.2.171 *Potamogeton alpinus* Balb. Potamogetonaceae 254
5.2.172 *Potamogeton x angustifolius* J.Presl. (Syn. *Potamogeton x zizii* W.D.J.Koch ex Roth) Potamogetonaceae 256
5.2.173 *Potamogeton berchtoldii* Fieber Potamogetonaceae 257
5.2.174 *Potamogeton coloratus* Hornem. Potamogetonaceae 259
5.2.175 *Potamogeton compressus* L. Potamogetonaceae 260
5.2.176 *Potamogeton crispus* L. Potamogetonaceae. 262
5.2.177 *Potamogeton epihydrus* Raf. Potamogetonaceae. 264
5.2.178 *Potamogeton friesii* Rupr. Potamogetonaceae 265
5.2.179 *Potamogeton gramineus* L. Potamogetonaceae 267
5.2.180 *Potamogeton lucens* L. Potamogetonaceae 269
5.2.181 *Potamogeton natans* L. Potamogetonaceae 270
5.2.182 *Potamogeton nodosus* Poir. Potamogetonaceae. 272
5.2.183 *Potamogeton obtusifolius* Mert. & Koch Potamogetonaceae 273
5.2.184 *Potamogeton perfoliatus* L. Potamogetonaceae. 275
5.2.185 *Potamogeton polygonifolius* Pourr. Potamogetonaceae. . . 276
5.2.186 *Potamogeton pusillus* L. (Syn. *P. panormitanus* Biv.) Potamogetonaceae 278
5.2.187 *Potamogeton rutilus* Wolfg. Potamogetonaceae 280
5.2.188 *Potamogeton trichoides* Cham. & Schltdl. Potamogetonaceae 281
5.2.189 *Stuckenia filiformis* (Pers.) Börner (Syn. *Potamogeton filiformis* Pers.) Potamogetonaceae 283
5.2.190 *Stuckenia pectinata* (L.) Börner (Syn. *Potamogeton pectinatus* L.) Potamogetonaceae 285
5.2.191 *Stuckenia vaginata* (Magnin) Holub (Syn. *Potamogeton vaginatus* Turcz.) Potamogetonaceae. 287
5.2.192 *Hottonia palustris* L. Primulaceae. 288
5.2.193 *Lysimachia thyrsiflora* L. Primulaceae 289
5.2.194 *Lysimachia vulgaris* L. Primulaceae 290
5.2.195 *Caltha palustris* L. Ranunculaceae 291
5.2.196 *Ranunculus* L. Section *Batrachium* (DC.) Gray Ranunculaceae 292
5.2.197 *Ranunculus flammula* L. Ranunculaceae. 298
5.2.198 *Ranunculus hyperboreus* Rottb. Ranunculaceae 299
5.2.199 *Ranunculus lingua* L. Ranunculaceae 300
5.2.200 *Ranunculus reptans* L. Ranunculaceae 301

5.2.201 *Ranunculus sceleratus* L. Ranunculaceae 302
5.2.202 *Comarum palustre* L. Rosaceae...................... 303
5.2.203 *Filipendula ulmaria* (L.) Maxim. Rosaceae............ 304
5.2.204 *Geum rivale* L. Rosaceae........................... 305
5.2.205 *Galium palustre* L. Rubiaceae 306
5.2.206 *Galium uliginosum* L. Rubiaceae..................... 307
5.2.207 *Ruppia cirrhosa* (Petagna) Grande and *Ruppia maritima* L. Ruppiaceae........................... 308
5.2.208 *Salix* L. spp. Salicaceae 310
5.2.209 *Azolla filiculoides* Lam. Salviniaceae 312
5.2.210 *Limosella aquatica* L. Scrophulariaceae 314
5.2.211 *Scrophularia auriculata* L. Scrophulariaceae 315
5.2.212 *Solanum dulcamara* L. Solanaceae 316
5.2.213 *Sparganium angustifolium* Michx. Typhaceae (Sparganiaceae) 317
5.2.214 *Sparganium emersum* Rehmann Typhaceae (Sparganiaceae) 318
5.2.215 *Sparganium erectum* L. Typhaceae (Sparganiaceae)..... 320
5.2.216 *Sparganium glomeratum* (Laest. ex Beurl.) Beurl. Typhaceae (Sparganiacaeae) 322
5.2.217 *Sparganium gramineum* Georgi Typhaceae (Sparganiaceae) 323
5.2.218 *Sparganium hyperboreum* Laest. ex Burl. Typhaceae (Sparganiaceae) 324
5.2.219 *Sparganium natans* L. Typhacaeae (Sparganiaceae) 325
5.2.220 *Typha latifolia* L., *Typha angustifolia* L., *Typha domingensis* Pers. Typhaceae....................... 327
5.2.221 *Sphagnum* L. spp. Sphagnaceae..................... 329
5.2.222 *Viola palustris* L. Violaceae........................ 330
5.2.223 *Zannichellia palustris* L. Zannichelliaceae 331
5.2.224 *Zostera marina* L. Zosteraceae 332
5.2.225 *Zostera noltii* Hornem. Zosteraceae.................. 334

Appendices.. 335

Glossary 341

References 345

Index.. 369

Chapter 1
Introduction

1.1 The Aims of this Handbook

Aquatic plants are key ecological engineers in aquatic ecosystems, driving both structure and function. As important primary producers, aquatic plants contribute a large proportion of the organic matter that accumulates and is preserved in the sediments of still waters. The remains of aquatic macrophytes in lake, pond, and other wetland sediments play a fundamental role in the study of past ontogenetic and human-driven environmental changes via the fields of palaeolimnology and palaeoecology. Macrophyte macrofossil remains are an essential component of multiproxy studies of the past development of aquatic habitats that may involve pollen and spores, diatoms, and remains of invertebrates and fish, as well as sediment characteristics.

Without easy access to a sizeable physical reference collection, which takes many years to accumulate, the identification of aquatic plant seeds, fruits, and vegetative parts is highly challenging, as suitable pictorial references are scattered and often incomplete. Seed atlases and identification guides usually include all plant diaspores from an area, both terrestrial and aquatic, but many of these are currently out of print. This book aims to fill this long-standing and substantial gap. A list of some seed atlases, as well as useful general papers on macrofossil analysis are provided in Appendix 1. Because the majority of macrofossils in lake, pond, and wetland sediments are derived from local macrophyte stands, we concentrate our handbook on these aquatic species. We focus on northern Europe, but many aquatic macrophytes have much wider distributions and therefore their fruits and seeds, or those of their close relatives, may be identified for other parts of the world using this guide. It must be emphasised that any illustrated handbook (as here) is a supplement

H. H. Birks et al., *Seeds and Fruits of Aquatic Plants of Northern Europe - A Handbook*, Developments in Paleoenvironmental Research 23,
https://doi.org/10.1007/978-3-032-04170-8_1

to a good physical reference collection. Hence this book affords a guide for a tentative identification that should then be checked against reference material wherever possible.

In this book we provide photographs and some illustrations of the fruits and seeds, and occasionally the distinctive often preserved vegetative parts, of aquatic macrophytes of northern Europe, as well as a few other plants such as bryophytes and macro-algae. The species are arranged alphabetically by Family, Genus, and Species. It is a guide to aid identification, both in the fresh form and when aquatic plant macro-remains are retrieved from sediments as subfossils. We provide notes on seed, fruit, leaf, and spore morphology that are intended to assist identification, and we typically include illustrations of more than one specimen to show different aspects and some idea of variability. In addition, we provide brief notes on the life form and ecology of all the plant species that we cover with the intention of informing environmental reconstructions. We reference the global distribution of the species using the maps in The Global Biodiversity Information Facility (GBIF) and from species-specific literature. If known, we also provide information on the Quaternary history of each species, concentrating on the Late Glacial and Holocene (last 15,000 years). The latter is not meant to be exhaustive and is largely based upon Godwin's (1975) History of the British Flora together with much other literature. This book will be an essential reference for palaeolimnologists and palaeoecologists and is aimed at encouraging and supporting the use of plant macrofossils. It will also be invaluable for archaeobotanists who come across aquatic plant remains, as well as for scientists concerned with seed banks, the diets of wetland birds and mammals, and aquatic plant dispersal ecology. More generally, this book will be an important reference for scientists, aquatic ecosystem managers, students, and naturalists working on, or interested in aquatic plants, as seeds and fruits quite often can be diagnostic.

1.2 What Is an Aquatic Plant, or Macrophyte?

An aquatic macrophyte is an "aquatic photosynthetic organism, large enough to see with the naked eye, that actively grows permanently or periodically submerged below, floating on, or up through the water surface" of inland freshwater or brackish waterbodies (Chambers et al. 2008). As the interface of land and water is constantly shifting in space and time, many aquatic plants have developed adaptations that allow them to cope with wetting and drying. Some species require exposure to set seed and favour temporary or ephemeral water bodies that frequently dry up in

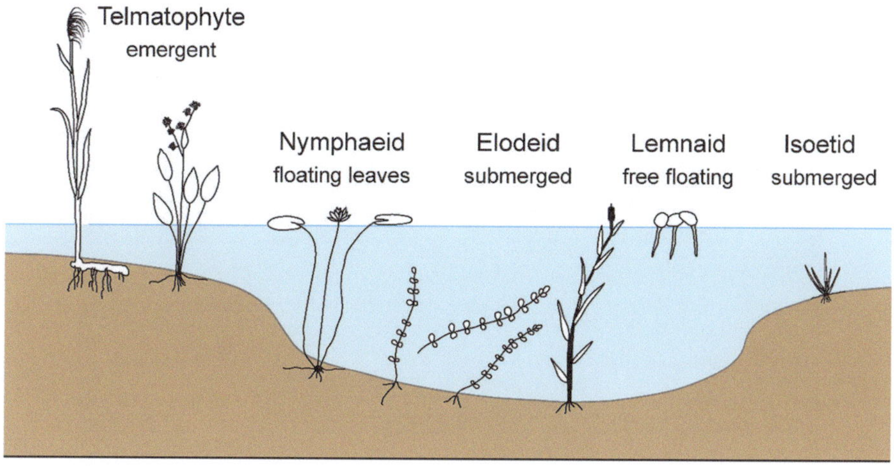

Fig. 1.1 Life forms of aquatic macrophytes

summer. By contrast, many other species are ideally permanently submerged but can nevertheless cope with periodic shallow water and drying out events. Due to this blurring of the aquatic-terrestrial boundary it is difficult to strictly define an aquatic plant. In this book, we consider submerged, floating, floating-leaved, and some emergent plants found in or on the shores of water bodies in northern Europe.

Our book considers the key aquatic-plant life forms (Fig. 1.1). Emergent telmatophytes are rooted in the sediment below the water with the main body of the plant located above the water. They usually grow in shallow water, for example at the margins of water bodies and often form a marginal reedswamp or fen vegetation. Floating-leaved nymphaeids, are rooted in the sediment and have leaves that float on the water surface. Free-floating lemnaids and other plants (e.g. *Stratiotes aloides*, *Hydrocharis morsus-ranae*) have no connection with the sediment. Submerged elodeids have long stems that are either rootless or rooted in the sediment, with no floating leaves, although they may produce flowering stems above the water. Isoetids are stemless tufted plants with rosettes that grow submerged in shallow water, often forming a short turf at lake margins. Water depth is an important controlling factor determining the zonation of macrophytes out from a lake shore. Within the overall zonation of these life forms there is also much mingling of species, and aquatic plants have a strong sociology, with for example, dense emergent plant beds often sheltering non-rooted free-floating and submerged plants (Fig. 1.2).

Fig. 1.2 The sheltered edge of a Northern Irish shallow lake where free-floating plants *(Stratiotes aloides* and *Spirodela polyrhiza*) are protected by rooted floating and emergent plant beds (Photo, Ben Goldsmith)

1.3 Reasons for Identifying Seeds and Fruits of Aquatic Plants

There is much interest in the environmental and ecological history of lakes, ponds, and other wetlands. How did they originate? How were they colonised? How have their aquatic ecosystems developed? How have aquatic macrophyte communities responded to human activities? The sediments of lakes, ponds, and wetlands often contain a temporal archive of these changes. Macrofossils of macrophytes in a lake sediment sequence provide information about past environmental conditions and changes over time. They reflect climate, water chemistry, water depth, sediment type, and wider aquatic ecosystem structure. Natural and human-induced changes in environmental and ecological characteristics of aquatic habitats and their catchments can be reconstructed and in turn the factors driving these changes inferred.

There is increasing interest in conserving still-water habitats and wetlands to maintain the crucial biodiversity and ecosystem services that they provide. Many are currently threatened by human modifications, including drainage, water diversion, or damming of inflows and outflows, climate change, invasive species, and pollution. These pressures have detrimental effects on aquatic plants and associated species, and so major restoration efforts are in progress, especially for lakes. To be successful and to judge success, it is important to know what the causes of ecosystem degradation were as well as the nature of pre-disturbance plant communities that can be used as restoration targets. Palaeolimnological studies utilising macrofossils of aquatic macrophytes are excellent at providing this evidence (Sayer et al. 2012; Bennion et al. 2018).

Aquatic plant macrofossils play important roles in many aspects of palaeolimnology and archaeology. Among many palaeoecological applications they can be used to infer the effects of eutrophication on submerged and floating vegetation and associated ecological interactions (e.g. Sayer et al. 2016), They can also track acidification, either natural acidification due to nutrient depletion (Birks 2000) or through bog development (e.g. Birks and Mathewes 1978), or due to the recent addition of strong acids through acid rain (Renberg and Wallin 1985). They can illustrate the timing of immigration and local extirpation of species and contribute to historical biogeography. Over long timescales, plant macrofossils may provide evidence of lake infilling (e.g. Bradshaw et al. 2005), as well as changes in water levels which can in turn be used to infer climate change impacts (Hannon and Gaillard 1997). Macro-remains of aquatic plants can provide data to assist archaeologists in reconstructing wetland environments where humans have built dwellings (e.g. Peregro et al. 2025) and in the direct use of aquatic plants as food (Buckley et al. 2023). In many cases aquatic plant remains can provide material for radiocarbon dating, although, due to 'hardwater effects', care must be taken to utilise only those plants that use atmospheric CO_2 in photosynthesis (see Chap. 3).

Chapter 2
Aquatic Ecosystems and Their Aquatic Plants

2.1 Aquatic Ecosystems

Aquatic ecosystems, the habitats for aquatic plants, cover a myriad of freshwater bodies including lakes (both deep and shallow), ponds, ditches, streams, rivers and shallow wetlands. In natural or semi-natural settings these freshwater habitats often form continua (Sayer 2014). Towards the coast, they transition into freshwater tidal marshes and shallow-water marine environments, including coastal lagoons and estuaries, which also support aquatic plants. Although aquatic ecosystems occupy a very small proportion of the earth's surface, they support disproportionally high biodiversity (Grizett and Poikane 2024; Keddy 2024).

2.2 Wetland Types

Distinctions between different wetland ecosystems are often hard to make, although many attempts have been made (Richardson et al. 2022; Keddy 2024). Just as different aquatic habitats are hard to delineate, so is the focus of this book—aquatic plants. This is because many plants are adapted to survive in areas where water level fluctuation renders the habitat rhythmically and/or periodically wet and dry. In turn, this means that some plants typically considered to be terrestrial in habit, particularly weedy species, can be found in temporarily dried out aquatic habitats (lake and pond shores, desiccated upper river sections). Equally, many terrestrial plants may at least temporarily exist underwater as water levels rise. All of this makes for a blurred boundary between aquatic and terrestrial plants. From the perspective of this book, it is important to note that the presence/absence and the type of aquatic vegetation present is important in delineating different aquatic and wetland habitats.

H. H. Birks et al., *Seeds and Fruits of Aquatic Plants of Northern Europe - A Handbook*, Developments in Paleoenvironmental Research 23,
https://doi.org/10.1007/978-3-032-04170-8_2

Fig. 2.1 Artistic depiction of vegetation zonation in a lake. 1. *Schoenoplectus lacustris*, 2. *Berula erecta*, 3. *Sagittaria sagittifolia*, 4. *Sparganium emersum*, 5. *Nuphar lutea*, 6. *Ceratophyllum demersum*, 7. *Elodea canadensis*, 8. *Potamogeton crispus*, 9. *Potamogeton lucens*. (Source unknown)

Aquatic vegetation grows in shallow water near the shore in lakes, ponds, streams, ditches, and slow-flowing rivers where there is sufficient light for photosynthesis. A lake, for example, may host several wetland types in a typical zonation (Figs. 2.1 and 2.2). With increasing water depth, wet woods, fen, or marsh vegetation gives way to floating-leaved and often isoetid species on stony shores. With increased depth, elodeids may dominate. Charophyte species can be important components of the submerged vegetation of lakes, ponds and occasionally rivers and can grow in both shallow and deep water. In deep lakes, plants gradually peter out as water depth depletes the light. In lakes with fluctuating water levels, temporarily exposed mud flats can be rapidly colonised by annual plants from a long-lived seed bank as increased light-levels and temperatures stimulate germination (Poschlod and Rosbakh 2018). Seeds that die can be recovered as macrofossils from the sediments and bear witness to past lowered lake levels.

Where a pond ends and a small lake begins is a topic of much debate, but it is well known that, despite their small size, ponds are major contributors to aquatic biodiversity (Law et al. 2024). Ditches are human-constructed channels used to convey water through the landscape. Often ditches effectively operate as linear ponds

Fig. 2.2 Zonation of aquatic macrophytes at the edge of Corraharra Lough, County Fermanagh, Northern Ireland (Photo, Ben Goldsmith)

and like ponds, and where water quality is good, they can support highly species-rich plant communities (Hinojosa-Garro et al. 2008). Ecological succession can often happen very quickly in ponds and ditches due to a greater influence of surrounding terrestrial (in terms of organic matter inputs) and emergent wetland vegetation (in terms of plant invasion of open water), and sometimes due to high sediment influx in agricultural landscapes. Rivers are also important habitats for aquatic plants and while in fast flowing streams aquatic vegetation may be confined to margins and islands, under low flows in larger slower rivers, similar to shallow lakes, a species-rich marginal aquatic vegetation can develop (Dawson 1988). River-floodplains, including mosaics of beaver ponds, oxbow lakes, temporary floodplain lakes and other wetland features, can also support extremely high aquatic plant diversity.

Shallow freshwater wetland habitats fall into six basic types (see Keddy 2024). All wetland habitats are characterized by anaerobic water-logged soils that are under water or prone to flooding, but there are intermediates between the habitat types and they grade into each other. A *swamp* is dominated by trees, and in temperate regions includes 'carr', a habitat that is usually dominated by *Alnus* or *Salix* spp. trees. A *wet meadow* is dominated by herbaceous plants on damp soils that are periodically flooded, such as a water meadow on a floodplain. A *marsh* is a wetland dominated by herbaceous plants emergent above the water and rooted in submerged soils, such as a reed marsh often dominated by *Phragmites, Typha, Carex* spp., and other Cyperaceae. A *fen* consists of grasses and sedges growing in shallow water subjected to water movement and can occur on alkaline (rich fen) or acidic (poor fen) soils. Fen peat accumulates below the vegetation of marshes and fens. A *bog* is characterized by *Sphagnum* moss accompanied by Cyperaceae species such as *Eriophorum* and *Carex* spp. and ericaceous shrubs and sometimes trees on acid substrates, and it accumulates distinctive bog peat. Topography is an important factor controlling the distribution of wetlands within a landscape.

Estuaries, lagoons, and shallow marine bays are brackish and saline habitats where macroalgae (seaweeds) frequently dominate. A few vascular plants tolerate brackish water, however, particularly sea grasses (including *Zostera* and *Ruppia* spp.) which can form extensive, highly productive meadows. Salt marsh vegetation often develops on exposed mud flats in the intertidal zone.

2.3 Ecosystem Services and Benefits of Aquatic Plants

As well as being critical to biodiversity conservation, aquatic plants are food sources for herbivores, and many are regarded as keystone species in aquatic systems. Through oxygen production and the manufacture of complex structured habitat, aquatic plants afford living, prey-refuge, egg-laying and hunting habitat for a vast array of other species (Jeppesen et al. 1998). They form the basis of complex aquatic food webs and are critical in mediating species interactions. Macrophytes that are rooted in the substrate can stabilise sediments. In some circumstances they may also act as ecosystem engineers which can physically modify habitats. For example, *Sparganium erectum,* due to its strong root anchorage and overall toughness, has been shown to affect channel form and position in rivers (Gurnell et al. 2013).

Aquatic plants can buffer the negative effects of flooding by extreme rainfall events by slowing water movement though the landscape and they can also help to retain water in the landscape during dry periods, for example by the reduction of water evaporation by floating leaves. Rooted plants recycle nutrients such as phosphorus and nitrogen that have been sequestered in the sediments and also beneficially absorb nutrients from the water during eutrophication (e.g. Li et al. 2021). Other key regulating ecosystem services provided by macrophytes include carbon storage, erosion regulation, and pest control (Thomaz 2023).

Some aquatic macrophytes benefit humans directly by acting as sources of food such as *Oryza sativa* (rice), *Typha* spp. rhizomes, and *Trapa natans* nuts, as well as providing materials used in buildings (e.g. thatching, shelter construction) and for textiles, mats and baskets. In addition, wetlands and their aquatic plants have high aesthetic value, improving mental health and adding to the attractiveness of sites used for human leisure activities such as swimming, boating, and fishing. They play an important role in ecosystem and historical science, and in education, as especially promoted in this book. The seeds and other remains derived from wetland plants that are deposited in sediments form a temporal archive of ecosystem history which can be exploited to reconstruct past environments and hence inform conservation. Thus, it is important, for a whole host of reasons, that society and conservation activities place high priority on aquatic plants. Keddy (2024) provides extensive additional information on the ecosystem services provided by wetlands and their associated aquatic plants.

2.4 Threats to Aquatic Ecosystems and Their Plants

Aquatic ecosystems and their associated vegetation are naturally dynamic. Immigration of species and succession alters the macrophyte vegetation and associated animal populations over time (e.g. Birks 2000; Birks et al. 2000). Natural successional processes (the hydrosere) modify aquatic vegetation structure over long timescales, leading to the gradual accumulation of sediment which eventually fills in waterbodies resulting in the formation of fens and bogs (see Chap. 3). Sequential changes in plant communities may also occur as a response to climate change, mainly through temperature changes and an excess or lack of precipitation, resulting in effects on water levels.

Over recent centuries aquatic systems have been heavily impacted by human activity and anthropogenic drivers of aquatic vegetation change have taken over from natural climatic and successional influences. A major cause of plant decline has been drainage activities and often the conversion of wetlands to arable land and human settlements. It has been estimated that some 21% of global wetland area has been lost since 1700 due to cropland replacing wetland (Fluet-Chouinard et al. 2023). The loss of natural wetlands has not been matched by human-led wetland construction and is hence a key cause of freshwater species decline.

One of the most pervasive threats to aquatic ecosystems is eutrophication, which has occurred in the majority of lowland European lakes over the last century. Nutrient enrichment, the excessive addition of primarily nitrogen and phosphorus derived from sewage, animal waste, and agricultural fertilisers, to aquatic systems has been shown to have profound negative effects on aquatic vegetation abundance, composition and diversity. Initially, phytoplankton blooms decrease water transparency thereby reducing macrophyte colonization depth in lakes resulting in a loss of deep-water charophytes and isoetids (Sand-Jensen et al. 2000; Salgado et al. 2010). A typical second stage is the expansion of aggressive canopy-forming submerged elodeid macrophytes and/or water lilies often at the expense of short-stature isoetid, characean, elodeid and bryophyte species (Sand-Jensen et al. 2000, 2008). With time and increasing nutrient concentrations, aquatic plant species richness is steadily reduced, ultimately with a loss of almost all aquatic vegetation and the prevalence of phytoplankton (Davidson et al. 2023). Fewer studies have been undertaken on eutrophication effects on ponds compared to lakes, but they are similarly profound, with major reductions in species richness, losses of rare species (Rosset et al. 2013) and an often-observed transition to floating plant (e.g. Lemnaceae) dominance (Feuchtmayr et al. 2009; Emson et al. 2018). In rivers aquatic plant responses to eutrophication are less fully understood. Nonetheless, clear negative impacts are evident, with Hilton et al. (2006) showing that both slow and fast-flowing river sections tend to lose their plants as eutrophication progresses.

There are many other anthropogenic causes of aquatic macrophyte degradation in aquatic habitats. Acid deposition in oligotrophic soft-water lakes decreases pH and causes the loss of many aquatic plants and their replacement by a few acid-tolerant species such as *Sphagnum* spp. and *Juncus bulbosus* (Brouwer et al. 2002). Human modification of natural hydrological regimes through water withdrawal, drainage, and damming can destroy aquatic habitats, and also cause a loss of plants that require water level changes to thrive, especially sensitive drawdown species. In addition, the introduction of invasive aquatic plant species can eliminate natives and detrimentally change plant habitats. Indeed, several of the worst invasive weeds in the world are aquatic plants. In northern Europe these include *Azolla filiculoides, Crassula helmsii*, and *Elodea* spp. (Hofstra et al. 2020) which can shade out native vegetation, decrease habitat diversity for aquatic animals and, in the case of *A. filiculoides*, reduce oxygen diffusion into the water.

2.5 Mitigation and Restoration

The importance of wetlands and their aquatic plants is now widely recognised, and efforts are being made to restore them and their associated ecosystem services. Restoration of desirable ecological conditions in waterbodies is of utmost concern given the ecosystem services that humans derive from them. This involves political, economic, and scientific considerations. Driven by legislation (such as the EU Water Framework Directive), much money is spent every year to try and reduce nutrient concentrations in European rivers and lakes and to mitigate the impacts of nutrient enrichment on drinking water sources and important fisheries. Evidence for successful restoration efforts is limited, however. A great many restoration studies have been undertaken for shallow lakes with the conclusion that even partial recovery takes decades to manifest (e.g. Søndergaard et al. 2007; McCrackin et al. 2017; Hilt et al. 2018). There is some hope, however, as recent studies show that aquatic plant communities can be rather rapidly restored at degraded terrestrialised farm ponds by the removal of overhanging woody vegetation and sediment (Sayer et al. 2012; Walton et al. 2021b). Local biodiversity is increased, and the increase in pollinators (Walton et al. 2021a) benefits the wider agricultural landscape as well as insect-pollinated food crops. Macrophyte propagules in the sediments of lakes and ponds may contribute to restoring past macrophyte diversity (Verhofstad et al. 2017). Seed banks of many aquatic plants can be long-lived (10s to 100s of years), as shown for ponds that have long been filled in (ghost ponds) and then re-excavated (Alderton et al. 2017).

A historical baseline of aquatic habitats and their plants is needed to inform restoration. Past conditions can be inferred from palaeoecological evidence provided by seeds, fruits and other macro-remains in the sediment (see Chap. 3). This approach can assist with targeting restoration activities in many freshwater and coastal habitats (Dick et al. 2011; Bennion et al. 2018; Walton et al. 2021b).

Chapter 3
Palaeolimnology

3.1 Palaeolimnology

Palaeolimnology, or historical limnology, is a multidisciplinary science that uses the physical, chemical, and biological information preserved in sedimentary profiles (proxy data) to reconstruct past environmental conditions in inland aquatic systems, especially for lakes and ponds (Smol 2008). A sediment core can be analysed for the remains of plants and animals in sequential samples to reconstruct past ecological and environmental changes through time. These changes can be dated by radiochronological and other methods, permitting an understanding of the timing and rates of key environmental changes, spanning recent decades to hundreds or thousands of years. An excellent introduction to palaeolimnology is given by Smol (2023).

Palaeolimnology asks questions about the history and dynamics of environmental and biological changes such as: How did present aquatic vegetation and animal communities develop? What were the drivers of ecosystem changes? When did taxa arrive or become extinct, and why? How did aquatic vegetation initiation proceed, for example after deglaciation or new lake formation and how did it develop as a natural hydrosere, perhaps infilling the basin completely over time? When and how fast did changes occur through time? What internal lake processes influenced the aquatic ecosystem? How did the catchment influence and interact with the development of the aquatic ecosystem? What was the influence of climatic changes? Were there lake-level changes, and what caused them? In recent millennia, human activities have severely impacted aquatic ecosystems. Palaeolimnology can address questions on the timing and nature of environmental and ecological degradation as well as the structure of pre-disturbance biological communities (especially including aquatic plant structure) that can help to direct conservation and restoration actions.

To answer such questions, we look back through time by analysing remains of organisms and organic and inorganic sediment properties in sedimentary sequences. The data that are acquired represent the past environment and are known as proxy

H. H. Birks et al., *Seeds and Fruits of Aquatic Plants of Northern Europe - A Handbook*, Developments in Paleoenvironmental Research 23,
https://doi.org/10.1007/978-3-032-04170-8_3

data. Whole organisms are seldom preserved so the analyst must identify and record the remains that are present. In the case of aquatic plants, fruits and seeds are typical examples of proxy representatives of aquatic plants.

A lake ecosystem is influenced by its catchment, and the wider region. A diagrammatic lake ecosystem and the processes acting upon it is depicted in Fig. 3.1. The proxy data preserved in sediments can be interpreted in terms of all of the processes acting upon the lake.

The sediment in a lake comes from two main sources. Internal sources within the lake include macrophytes, algae, invertebrates and chemical precipitates. External sources include material derived from a lake's marginal vegetation, the catchment, and beyond it. The catchment supplies plant material, soil and minerogenic particles, charcoal, pollen, and more recently, microplastics, transported by surface water run-off and streams. Other components can be derived from beyond the catchment, such as airborne pollen, dust, contaminants, and volcanic tephra, and chemicals dissolved in rain. Analyses of these proxies can be used to reconstruct the

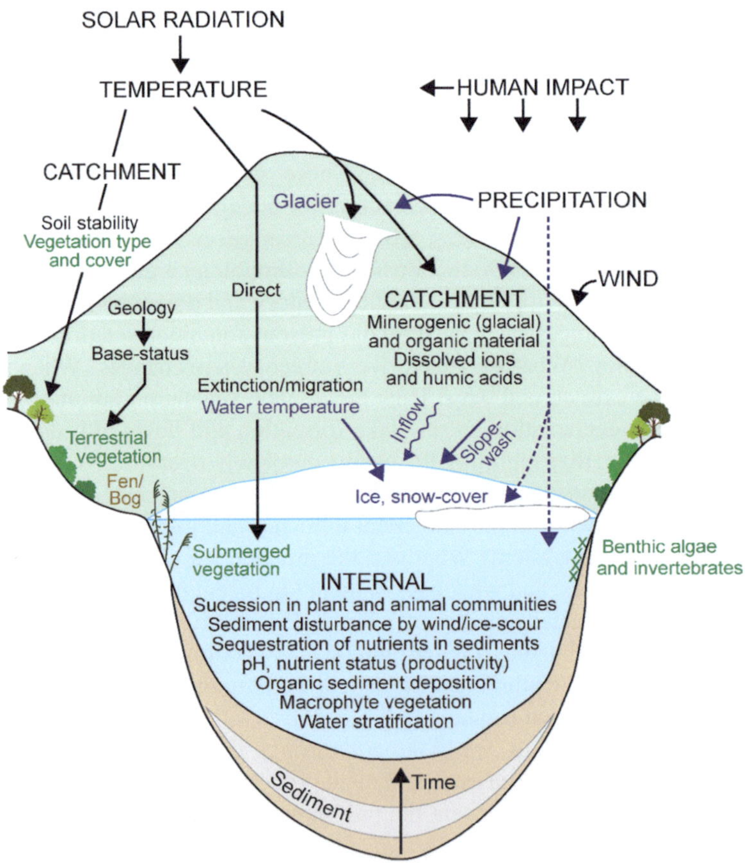

Fig. 3.1 Diagrammatic lake ecosystem, with the processes (upper case) occurring within the lake (internal), in the catchment, and beyond the catchment that are all reflected in the sediment. Adapted from Birks et al. (2000)

history of a water body, as well as of the wider landscape. For this process, good taxonomic, ecological, and limnological knowledge is necessary (Smol 2023).

3.2 Multiproxy Studies

Lake sediments are a time-stratified archive of a wide array of biological and inorganic proxies. Most palaeolimnological studies use more than one proxy source of evidence, as a single proxy provides only a limited amount of information about past conditions. In recent decades paleolimnological studies have included several proxies, spanning both organisms and sediment characteristics, in so-called multiproxy studies. The interpretation of each biological proxy record requires expert knowledge of their present-day tolerances and requirements. This means that a large multidisciplinary study needs a coordinated team of specialists with a common goal, as typified by the Kråkenes Lake Project in western Norway (Birks and Wright 2000). As each of the several proxies analysed in a sequence provides its own sphere of evidence, they can be combined to provide a more complete interpretation of past conditions than any single proxy (Smol 2008; Birks and Birks 2006). The evidence from seeds, fruits, spores and sometimes vegetative parts of aquatic plants has often been invaluable. Indeed, many multiproxy studies have included plant macrofossils as a central component, as inferences on changing plant structure are critical to the understanding of changes in other species groups. In a conservation context, it is vital that robust environmental and ecological reconstructions are communicated, in a meaningful way, to the public, politicians, and policy makers (Smol 2023; Siggery et al. 2025). Having plant macrofossils in a palaeoecological study and hence the names of well-known aquatic plants, including rare species, is often a great help in this respect.

3.3 Palaeolimnological Proxies Used in Conjunction with Plant Macrofossil Analysis

Detailed accounts of the many proxies that have been used in combination with plant macrofossil analysis can be found in the 'Tracking Environmental Changes Using Lake Sediments' (Springer) series. For physical proxies, see Last and Smol (2001), for terrestrial, algal, and siliceous indicators, see Smol et al. (2001a) and for zoological indicators, see Smol et al. (2001b).

3.3.1 Pollen and Macrofossils

The most common fossil proxy used alongside plant macrofossils is pollen. Compared with macrofossils, pollen can often only be identified to genus or family, resulting in reduced taxonomic precision. A striking example is Cyperaceae pollen, that includes *Carex* spp. and many other genera, most of whose fruits can be identified to species (See Family 20 in Chap. 5). Other pollen types with poor taxonomic resolution include Poaceae, subgenera of Asteraceae, *Potamogeton,* and *Sparganium* spp., and many Apiaceae and Brassicaceae. The fruits and seeds of these taxa can quite often be identified to species-level. The pollen of many macrophytes is thin-walled and often not well preserved (e.g. *Callitriche, Juncus, Najas, Potamogeton, Ceratophyllum* spp., and Lemnaceae). In contrast some macrophyte species have distinctive pollen such as *Typha latifolia, Myriophyllum* spp., *Cladium mariscus, Littorella uniflora, Eriocaulon aquaticum,* and *Menyanthes trifoliata*, as well as the genera *Nuphar, Nymphaea*, and *Ruppia*. Some aquatic taxa are only recorded as pollen and macrofossils are rarely found, as for *Lemna* spp. and *L. uniflora.* The opposite also applies. Non-vascular plants do not produce pollen. For Characeae, for example, it is only possible to find oospores (female fruiting bodies) and for mosses leaf and stem remains, except for *Sphagnum* spp. whose spores can be identified on pollen slides.

The composition of a macrofossil assemblage depends on taphonomic processes, including diaspore production, dispersal, deposition, and preservation. Because macrofossils are heavier than pollen, they are usually deposited closer to their source (e.g. Birks 1973; Birks and Bjune 2010 and see Chap. 4), whereas pollen may be locally deposited but can often be dispersed over large distances by wind or water. Therefore, macrofossils give a good representation of local vegetation within a lake or swamp (Birks 1973; Zhao et al. 2006) whereas pollen reflects a wider area, falling as a homogenous 'pollen rain' derived from terrestrial as well as aquatic sources. The two proxies are complementary. Macrofossil records can be used to validate and expand on pollen records leading to increased confidence in reconstructions of past vegetation structure (Birks and Birks 2000; Jackson and Booth 2007).

Pollen is usually produced in large quantities, and a small sample of sediment (ca. 1 cm^3) is often sufficient to obtain a representative fossil assemblage. Macrofossils are usually produced in much smaller numbers and require a larger sediment sample (ca. 30–100 cm^3). This can be acquired using a wide-diameter sediment corer (e.g. Patmore et al. 2014), or from an open section. Because of the abundance of pollen, pollen counts can profitably be converted to percentages and a pollen diagram drawn. This is usually impractical for a macrofossil assemblage, however, as fossil production and representation are highly variable (see Chap. 4). Therefore, counts are usually converted to concentrations per unit volume before drawing a stratigraphic diagram.

3.3.2 Other Biological Proxies

The analysis of algal fossils including diatoms and chrysophytes (see Battarbee et al. and Zeeb and Smol in Smol et al. 2001b) can provide valuable information on water chemistry conditions (nutrients and acidity) in palaeolimnological studies, as well as being useful in the reconstruction of past habitat structure. Animal remains are frequent in lake sediments, derived from Chironomidae, Trichoptera, Cladocera, Oribatidae, Bryozoa, Ostracoda, Mollusca and fish. Frey (1964) illustrates and discusses many animal remains. Courtney-Mustaphi et al. (2023) present an exhaustive illustrated compilation of invertebrate mandibles and sclerotized remains. For information on Coleoptera and Trichoptera see Elias (2001), for oribatid mites see Solhøy (2001), and for Bryozoan statoblasts see Francis (2001). Other proxies are covered in Smol et al. (2001b). Many animal fossils are found during plant macrofossil analysis and can contribute to many palaeoecological studies, even being used, in the case of the fish-leach *Piscicola geometra* cocoons, to help to infer past plant abundances (Odgaard and Rasmussen 2001).

Analysis of DNA preserved in sediments (sedDNA) can record more taxa at a higher taxonomic resolution than other palaeoecological proxies and is successful at detecting aquatic plants (Alsos et al. 2018; Stoof-Leichsenring et al. 2022). However, sedDNA provides only a semi-quantitative assessment of species' abundance, depending on the number of reads of a species in at least 8 analyses. SedDNA can provide complementary information to plant macrofossil data (e.g. Alsos et al. 2016) and is a valuable approach when macrofossils are scarce, as in cores from deep lakes (Clarke et al. 2020). Alsos et al. (2018) found good representation of the modern flora by sedDNA in surface lake sediments in northern Norway, and in Siberia and Tibet, where Stoof-Leichsenring et al. (2022) used surface sample data in downcore reconstructions of macrophyte diversity and environmental changes. Applications of sedDNA analysis in palaeolimnology and palaeoecology are provided by Revéret et al. (2023) and Garcés-Pastor et al. (2023).

3.3.3 Some Non-biological Proxies Used in Palaeolimnology

3.3.3.1 Sediments

Non-biological proxies are covered by Last and Smol (2001) and are the physical structure and chemical composition of the sediment. Physical properties include colour, grain size, and magnetic properties. Common chemical proxies include stable isotopes of H, O, C, and N, organic and carbonate content, and various organic and inorganic pollutants. Analyses of nitrogen and phosphorus can be useful in studies of eutrophication, and the C/N ratio is also often used in this respect. Stable isotopes, such as $\delta^{13}C$ and $\delta^{15}N$, can indicate whether sediment is derived from aquatic or terrestrial sources. Elemental analyses can be rapidly carried out with an XRF core scanner.

Percentage loss-on-ignition (or LOI) is a very useful index of the organic content of sediment (Heiri et al. 2001). Weighed samples are dried at 105 °C to remove free water, re-weighed, ignited at 500 °C, and re-weighed. Much of the organic matter in lake sediments may be derived from biogenic production in the lake catchment, whereas that in bog or fen sediments is derived from local plant growth (Birks and Birks 2006). The remainder is the minerogenic component. Further heating to 950 °C will drive off carbonate. The origin of minerogenic material is usually from inwash eroded from the catchment by, for example, streams, slope-wash, and glaciers. Less commonly, significant amounts of minerogenic material can originate from biogenic silica produced by diatoms, or volcanic fallout (tephra). For more information on other inorganic proxies and their roles in palaeolimnology see Last and Smol (2001).

3.3.3.2 Sediment Dating

Most sediments containing organic carbon younger than about 50,000 years can be dated by radiocarbon dating. In photosynthetic organisms that utilise atmospheric CO_2 the proportion of ^{14}C in their tissues reflects the proportion of ^{14}C in the atmosphere. After death, the amount of ^{14}C decays at a known rate. The ratio of ^{14}C to ^{12}C in a sediment sample therefore allows its age to be estimated to within a few decades. Presently most samples for ^{14}C dating are measured by Accelerator Mass Spectrometry (AMS) which requires much smaller samples than gas proportional counting. It is important to date terrestrial plant material picked out from sediments. This is because most lake sediments and aquatic plant remains contain a 'reservoir effect', which gives an anomalously old radiocarbon age. This can be due to calcium carbonate in the sediments, or the incorporation of 'old' carbon by macrophytes that utilise CO_2 from the sediment, dissolved CO_2, or HCO_3 from the water during photosynthesis. MacDonald et al. (1991) compared dates of terrestrial material, bulk lake sediment, *Pisidium* shells, and remains of *Drepanocladus* sp. moss. All the limnic components showed a reservoir effect compared with the dates derived from terrestrial material. To obtain a reliable radiocarbon date, plant material of terrestrial origin and of aquatic plants that only used contemporary atmospheric CO_2 should be utilised for dating.

Recent sediments can be dated using the short-lived isotope of lead (^{210}Pb) and peaks of radioactive caesium and strontium isotopes can be used as markers for 1960s nuclear bomb testing. The appearance and expansion of carbonaceous particles produced by coal and oil burning may also be used as recent time markers (Rose 2015). In some cases, sediments may be annually laminated (varved), with distinct winter and summer layers. If these can be counted, precise age estimates can be made, permitting detailed high-resolution reconstructions of past ecological events (e.g. Peglar et al. 1984). A protocol for the preparation of material for radiocarbon dating is given in Appendix 2.

3.4 Applications of Plant Macrofossils in Palaeolimnology

3.4.1 Aquatic Ecosystem History

Ecosystem history is a fundamental aspect of palaeolimnology. New lakes may be formed after deglaciation of an area, by landscape disturbance such as landslips or subsidence, as oxbow lakes by river cut-off, or by human means, such as reservoirs and ornamental lakes. The development of all of these ecosystems can be tracked by multiproxy analyses of their sediments.

As temperatures rose after the end of deglaciation, newly formed lakes provided suitable habitats for aquatic species. Aquatic plants are excellent colonisers and spread very rapidly, often becoming established in a few decades, making them better indicators of post-glacial temperature rise than, for example, trees that have a long generation time and relatively slow spread (Iversen 1954). A ready supply of minerals from freshly deglaciated soils and decreasing turbidity as soils stablised in the catchment allowed aquatic plants to become abundant in the Bølling and Allerød warm phases ca. 14,600–12,800 years ago at the end of the last glacial period in northern Europe, and again in the Early Holocene after the end of the Younger Dryas cold reversal around 11,700 years ago (Iversen 1954; Birks 2000). A succession (hydrosere) then proceeded, controlled by internal in-lake factors and inputs from the catchment. In Linsley Pond, Connecticut, USA, organic and microfossil content increased logarithmically in the sediment at the start of the Holocene - the so-called sigmoidal growth phase (Livingstone 1957). However, after the initial abundance, a slow decline in productivity followed, likely due to phosphorus being bound up in minerogenic sediments. A similar pattern was seen at Kråkenes Lake, western Norway, where availability of phosphorus decreased as it was leached from the catchment and sequestered into the lake sediments resulting in a massive decline in plants and invertebrates and a concomitant decrease in pH. This was only reversed once deep-rooted macrophytes such as *Nymphaea alba* and *Potamogeton* spp. colonized the lake and recycled phosphorus from the lake sediments (Birks 2000). Competition through shading from these tall-growing canopy-forming macrophytes then reduced the biodiversity of low stature species in the lake, such as isoetids.

A decline in aquatic plant abundance in the Early Holocene is found in many lake sequences and usually explained as a response to reduced nutrient availability. In some cases, where both sexual and vegetative remains are recorded, it can be demonstrated that macrophytes persist but do not fruit. For example, *Ceratophyllum* fruits are rare after the Early Holocene, but vegetative remains may continue to be recorded (Watts 1978). *Myriophyllum alterniflorum* fruits occur in the Early Holocene in Kråkenes Lake, Norway but disappear thereafter, although continued presence of *Myriophyllum* sp. is shown by leaf fragments (Birks 2000). The balance between fruiting and vegetative persistence is not well understood (Arbor 1920). Sexual reproduction maintains genetic variability which is important in pioneer situations. Later, as macrophyte competition increases, vegetative reproduction of a successful genotype might be advantageous.

Over centuries and millennia sediment accumulates at the bottom of a lake, composed of organic and minerogenic material. Eventually, as the water becomes shallower, telmatophytes can advance into the lake and they start to fill the basin with reedswamp and fen peat. As the peat accumulates, it becomes dry enough for colonisation by shrubs and trees such as *Salix* and *Alnus* spp. and eventually woodland will develop (Fig. 3.2). In acid situations, the final stage will be a bog vegetation rather than woodland. Hydrological changes may affect the speed and direction of hydroseral succession. Palaeolimnological records, and in particular plant macrofossils, are the only way of tracking long-term hydroseral processes in lake ecosystems and of elucidating the history underlying their present status.

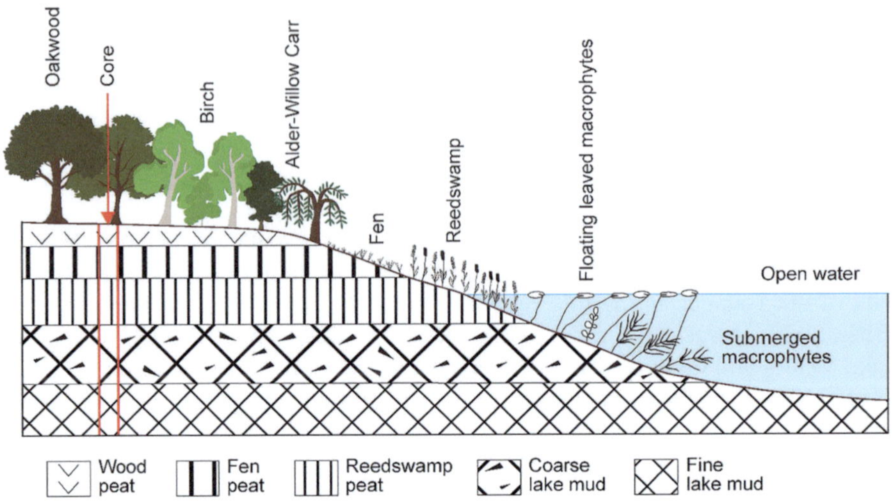

Fig. 3.2 Generalised hydrosere and the sediments produced by each vegetational stage. If the processes have been continuous, a core under the oakwood (arrowed) should produce a sequence of sediments laid down in time which is the same as the sequence being laid down in space at the present day

3.4.2 Water-level Reconstructions

Using the known water-depth tolerances of macrophytes, plant macrofossils can be used to indicate water-level fluctuations (Hannon and Gaillard 1997; Birks and Gaillard 2007). Lowered water-levels may result from drought or drainage. Subsequent erosion from the exposed shore often deposits a layer of minerogenic material in the lake. Exposed shores can be colonised by annual 'drawdown' species. This is well demonstrated in macrofossil sequences from the warm dry mid-Holocene period of the North American prairie region, such as at Pickerel Lake, South Dakota (Watts and Bright 1968). Similar evidence of a less extreme 'prairie period' was demonstrated by macrofossils at Kirchner Marsh, in the deciduous forest zone of Minnesota (Watts and Winter 1966). As the climate became wetter in the later Holocene, water levels rose and macrophyte vegetation recolonized the drawdown mud. Shallower water accelerated marginal reedswamp growth and the Kirchner basin became filled by marsh or swamp.

3.4.3 Eutrophication

Many studies have used plant macrofossils, often combined with animal macrofossils, Cladocera, and diatoms, to track the development of eutrophication in lakes and occasionally in ponds. An early eutrophication study was commissioned by authorities in St Clair County, North Dakota because an amenity lake, Lake Sallie, was becoming intolerably polluted (Birks et al. 1976). The progress of eutrophication was tracked by multiproxy analyses of the recent sediment, including plant macrofossils, and related to changes in the catchment, mainly the stages of expansion of human settlement over the previous 150 years. *Ceratophyllum demersum*, *Potamogeton* spp. and *Ruppia* became dominant macrophytes in this lake as eutrophication proceeded. The nutrient-enrichment of Edku Lake in the Nile Delta, initiated by the building of the Aswan High Dam, resulted in expansion of *Azolla filicoides, Lemna* sp., *Potamogeton pectinatus,* and *Typha* sp. (Birks et al. 2001). Plant macrofossils have been used to record the timing and nature of transitions from species-rich oligotrophic and mesotrophic floras to eutrophic plant communities and quite often complete aquatic plant loss. At Groby Pool, a small, shallow lake in the English Midlands, it was possible to compare aquatic plant macrofossil sequences with 250 years of detailed historically documented plant data (back to 1747). A remarkable overlap was found between these two data sources (Davidson et al. 2005). The study showed clear-cut sequential changes in the aquatic vegetation of the lake as a response to eutrophication with the conclusion that macrofossil remains afford a reliable indication of temporal change in the dominant structural components of the lake's aquatic vegetation along a long gradient of eutrophication. Similar conclusions have been reached in other studies (e.g. Zhao et al. 2006; Levi et al. 2014).

3.4.4 The Application of Macrofossils to Aquatic Conservation

Macrofossil analysis has become an important tool for informing aquatic conservation and restoration. Establishing past aquatic macrophyte communities through macrofossil analysis has many applications in aquatic conservation management. Reconstructions of past macrophyte communities can be used to set restoration targets for lakes and ponds and in turn assess restoration success (Sayer et al. 2012; Bennion et al. 2018; Walton et al. 2021b; Dendievel et al. 2024). In addition, macrofossils can be used to help diagnose the causes of aquatic ecological change, helping to determine appropriate restoration and management strategies. For example, at Upton Great Broad, a shallow eastern English lake, there was debate over the best way to manage the site for its rare plant *Najas marina.* Should sediment be removed to arrest shallowing and terrestrialisation? A macrofossil-based study (Ayres et al. 2008) showed a succession from *Chara* spp. to a more mixed vegetation and finally to *N. marina* dominance after around 1968, synchronous with a transition from marl to a fluid green sediment (Fig. 3.3). It was concluded that *N. marina* was rare until the 1940s–50s when eutrophication allowed its expansion and that, therefore, sediment removal should not be undertaken as it would remove the very factor that had permitted its dominance. In other studies macrofossil analysis has been used to identify the development of macrophyte communities indicating water quality deterioration and hence the need for managers to intervene to stop further ecological deterioration (e.g. Bennion et al. 2018; Zhang et al. 2025).

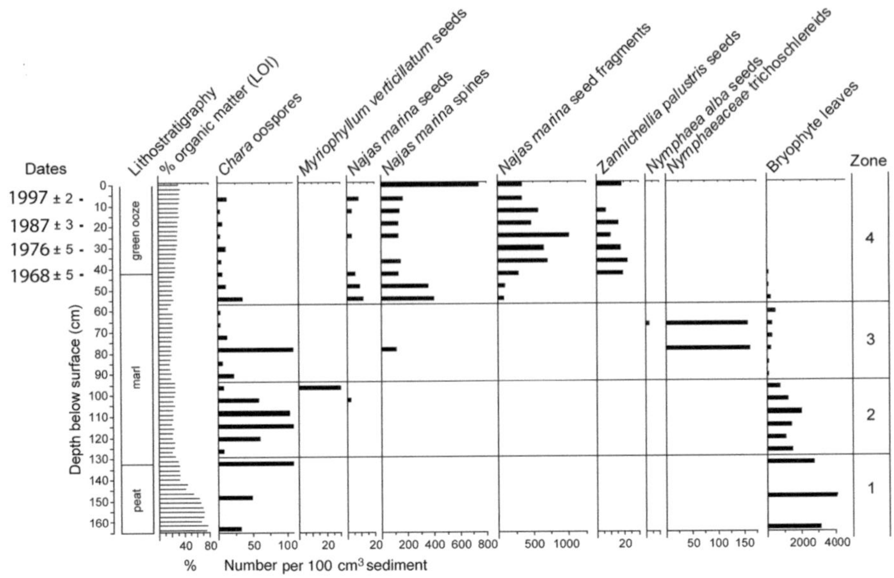

Fig. 3.3 Plant macrofossil sequence from Upton Great Broad, eastern England (modified from Ayres et al. 2008)

Chapter 4
Plant Macrofossil Analysis

4.1 What Is a Plant Macrofossil?

A plant macrofossil can be considered to be the remains of a plant in sediment that is visible to the naked eye and that can be manipulated by hand. Plant macrofossils in Quaternary sediments are seldom true stony fossils where the organic material is replaced by inorganic minerals. They are usually the original organic material and thus are often termed subfossils. The sizes of macrofossils range from 0.5 mm to enormous tree trunks. The median size is 0.5–2.0 mm. Macrofossils are usually studied using a stereomicroscope at magnifications of between 10 and 40x. Greater magnification (e.g. 100x) using a high-power optical microscope with overhead lighting can be employed if necessary. Photographs can be taken under a stereomicroscope, an optical microscope, or a scanning electron microscope, depending on requirements.

Plant remains encountered in lake sediments are not only fruits and seeds, but also vegetative parts: leaves, cuticles, buds, bud scales, anthers, flower parts, and rhizomes (Fig. 4.1). Some of these are distinctive (see e.g. Mauquoy and van Geel 2007; Katz et al. 1977). Sometimes, whole leaves are preserved, as for *Salix* spp., or parts of leaves may be identifiable, as for many *Potamogeton*, species as well as for *Ceratophyllum demersum* and *Najas marina*, which produce distinctive leaf spines. Several monocot cuticle remnants can be identified with care and with adequate reference material. Sometimes, bark fragments are distinctive. If they are large enough, wood fragments can be identified using their anatomy. *Sphagnum* moss leaves are frequently found but are difficult to identify to species. Remains of other mosses, both aquatic and terrestrial (e.g. Jonsgard and Birks 1995) may be identified using appropriate manuals. Vegetative parts can supplement the record of fruits and seeds and can also provide evidence when diaspores are absent. Vegetative parts are generally not discrete entities and have to be recorded on a relative abundance scale.

H. H. Birks et al., *Seeds and Fruits of Aquatic Plants of Northern Europe - A Handbook*, Developments in Paleoenvironmental Research 23,
https://doi.org/10.1007/978-3-032-04170-8_4

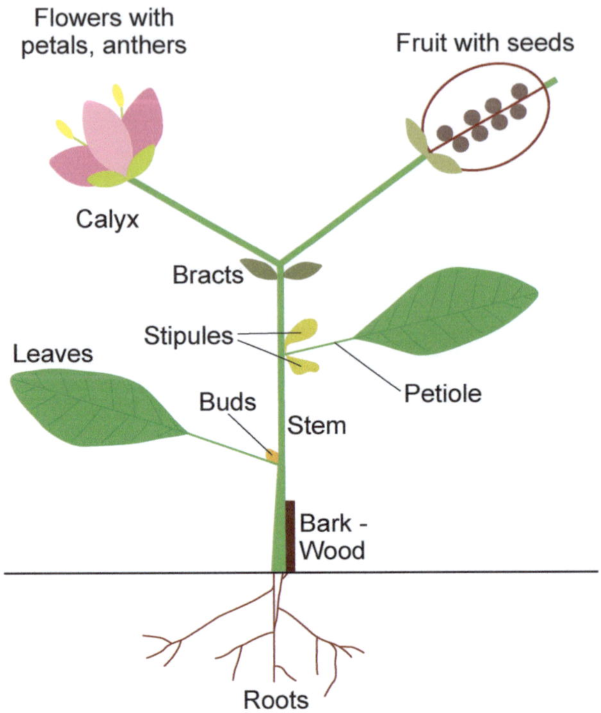

Fig. 4.1 Parts of a plant that may be preserved in sediment

Remains of lower plants found as macrofossils include pteridophyte sporangia and megaspores, moss leaves (but seldom liverworts, as they are usually not preserved), and oospores of Characeae. We do not consider these in this book, apart from charophyte oospores, some pteridophyte megaspores, and three examples of mosses.

4.2 What Is a Fruit and What Is a Seed?

A fruit is the fleshy or dry ripened ovary of a flowering plant, enclosing the seed or seeds. A fruit has two points of attachment, the stalk or receptacle and the style or stigma (Fig. 4.2). The receptacle may support or enclose the fruit and become fleshy. The ovary wall may be modified to become dry, succulent, woody, dehiscent, or indehiscent. Different layers can have different structures. For example, *Potamogeton* and *Sparganium* spp. have fleshy outer exocarps and tough inner endocarps which are not digested by ducks and which protect the embryo. They are often well preserved in lake sediments. Dehiscent fruits release the seeds for dispersal. Fleshy

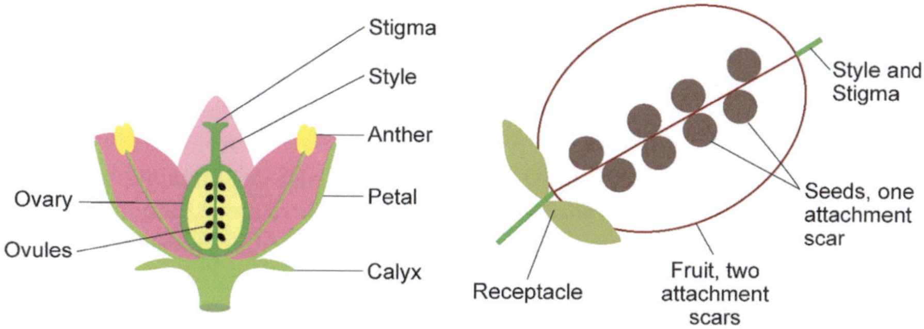

Fig. 4.2 Idealised flower and fruit

fruits are designed to be eaten by animals and the seeds are spat out or excreted. A single-seeded fruit may be dispersed as a unit, protecting the seed inside. A dispersal unit is termed a diaspore.

A seed is the product of sexual reproduction in angiosperms (flowering plants) and gymnosperms (e.g. conifers, cycads, and ginkgo). The seed is a fertilized ovule and contains the embryo and usually a store of food, either within the embryo or around it as endosperm. It is surrounded by a protective coat, the testa. A seed has one point of attachment, the micropyle, where it was attached to the ovary wall by the funicle (Fig. 4.2). Seeds perform a variety of functions: multiplication, perennation (surviving seasons of stress such as winter or drought), dormancy (a state of arrested development), and dispersal. The arrangement and structure of seeds and fruits are very important in plant taxonomy. In general, plant families and genera have similar seeds and fruits that have a characteristic morphology.

It is desirable for the macrofossil analyst to have a good knowledge of the morphology of aquatic plants and their seeds and fruits. Atlases of seeds and fruits are helpful for preliminary identifications, which should then be checked by comparison with reference material.

4.3 Study Design and Obtaining Sediment for Macrofossil Analysis

Plant macrofossil analysis is a time-consuming and therefore expensive procedure. Therefore, it is essential to start with a clear question and some well-defined aims for a project and to choose an appropriate study site. A suitable site will depend on these aims. For macrofossil analyses in temperate regions, Watts (1978) proposed that "an ideal site for macrofossil analysis is a small, deep lake with sloping shores and no marginal reedswamp that would trap macrofossils, no inflow and outflow streams to create currents and sorting, and a stable water level to prevent re-deposition from shallow-water sediments." Of course, it is hard to find an ideal site, so site deficiencies and their effects on taphonomy should be considered during the interpretation of the results. Although the ideal situation is seldom encountered, the sediment from most small lakes and ponds can be utilised. Because seeds and fruits are not usually dispersed far from their sources, a sediment core from near the shore will provide the best information about local lake development. However, in multiproxy studies, cores are usually taken from the deepest part of the lake. In Arctic or tree-less situations, a small lake with steep slopes and small inflowing streams is preferable. Such lakes often have a sparse macrophyte vegetation and the catchment provides an important source of information on lake development (e.g. Birks 1991, 2000). Macrofossils can also be preserved in the organic deposits of ox-bow lakes in river-floodplain systems, in peat deposits, and within the anaerobic or charred deposits of archaeological sites. Unusually, vegetation can also be buried directly by tephra deposition (e.g. Goetcheus and Birks 2001).

Once a project has been defined and a site chosen, sediment for analysis must be obtained. Cores may be collected from open water using boats and rafts, or from ice, or a marsh/bog surface. Commonly used corers include the hand-operated, side-sampling 'Russian' corer (Jowsey 1966), which is especially useful for exploring the distribution of sediments in a site, and the 'Livingstone' piston corer, especially with the square rod modification (Livingstone 1955; Wright et al. 1984) and a large (10 cm) diameter. Various wide-diameter piston corers (e.g. Patmore et al. 2014) are available that provide sufficient material for macrofossil analysis at a fine temporal resolution. Large sediment volumes can also be sequentially obtained from open sections that can be sampled at high resolution. Large spot samples can be obtained from archaeological sites.

The position of the core site needs to be considered carefully in any macrofossil study and tailored to the question being addressed. In contrast to many palaeoecological proxies, the deepest point in the centre of a lake is often not the best place to find macrofossils, as they are usually deposited more abundantly in shallower water close to vegetated lake shores (e.g. Birks 1973). Because of this local nature, one core from shallow water sediments will not necessarily be representative for the whole lake.

4.4 Method of Macrofossil Analysis

1. Describe the sediments; colour, type and size of minerogenic particles and type of organic material. A good guide is by Troels-Smith (1955) and the symbols of this system can be drawn on macrofossil diagrams.

2. Take subsamples of the sediment of known volume. A suitable volume is usually about 50 cm^3, but good results can often be obtained with 20–30 cm^3, depending on the concentration of remains. Sediment volume is typically measured by the displacement of water in a measuring cylinder.
3. Store the samples in the cold and dark in a refrigerator or freeze them until needed.
4. If necessary, disaggregate the sediment by adding sodium pyrophosphate ($NaPO_4O_7$ + $10H_2O$) or 'Calgon' water softener, or in the case of peats by adding dilute sodium hydroxide. In the latter case, do not use NaOH for long periods of time, or the sediment and the macrofossils in it will dissolve.
5. Wash the sediment through a 125 μm mesh. This will retain the smallest macrofossils. Use coarser sieve(s) in addition if necessary.
6. Transfer the residue to a fully labelled storage bottle and keep it cool until the sample is examined.
7. Suspend a small quantity of the residue in water in a convenient dish (e.g. a petri dish with a grid or lines marked on it). Examine the sample systematically under a stereomicroscope at about 12x magnification.
8. Pick out the remains of interest using entomological (soft) forceps or a brush. If the remains of a taxon are very numerous, they can be counted but not picked out. If an item is not immediately recognisable, it should be picked out to be examined later. "If in doubt, pick it out!".
9. Sort the remains and identify them. Seed atlases and handbooks, such as this one, will be useful to aid identification. All identifications based on photographs and diagrams should ideally be checked by comparison with reference material.
10. Record and count each taxon. Some remains are not units like seeds or fruits but are vegetative parts like leaf fragments and mosses. These can be recorded on an ordinal scale (e.g. abundant, frequent, occasional, rare) or on a numerical abundance scale.
11. Choose a suitable standard volume for expressing the results, such as 50 cm^3 or 100 cm^3 and calculate the number of fossils that would be found in that volume for all the samples in the sequence.
12. Store the identified material in small vials containing glycerol with a small amount of fungicide added. If the material is to be used for radiocarbon dating, rinse and dry it following the protocol in Appendix 2.
13. Tabulate the results and convert any ordinal counts to a number convenient for plotting. Presence/absence records can be incorporated. Counts may be converted to percentages, but this is usually not useful for plant macrofossils due to the variable representation of remains and the difficulty of defining a calculation sum. A macrofossil concentration diagram (number of macrofossils per unit volume) can be produced (see Fig. 3.3) and if sediment accumulation rates are known, macrofossil influx (number of macrofossils per unit area per year) can also be calculated.

14. Draw a macrofossil diagram (see Fig. 3.3) using a software package such as TGView, (Grimm 1990, 2004) C2, or rioja (Juggins 2022, 2023, 2024).
15. Numerical analysis of macrofossil sequences can be undertaken but this is difficult because of the diverse qualitative and quantitative nature of macrofossils (point 10 above). The abundance of macrofossil taxa in an assemblage can vary hugely within a sample and between samples and there are usually many zero records through a sequence. Birks (2014) discusses these difficulties and suggests numerical procedures that help to exploit the full potential of macrofossil stratigraphic data and their contribution to palaeoecology.

4.5 Sources of Error

1. A plant macrofossil may be wrongly identified. Or new knowledge may come to light to enable a more detailed identification. It is important to retain analysed material so it can be re-examined at a later date.
2. The results may be wrongly interpreted. This is a matter for discussion.
3. A macrofossil sample may suffer from contamination, which can lead to a faulty interpretation. Contamination may occur in various ways both in the field and in the laboratory. A sample may have a mixed origin, due to long-distance dispersal from other vegetation zones. This may be most common in mountainous districts. The sample may contain redeposited material from earlier strata, for example interglacial sediments may be incorporated into glacial deposits, or material may be reworked from shallow water during or after a low-water stand. The sample may be sorted during the processes of deposition, for example by streams, lake processes such as overturn, and 'filtering' of remains by marginal reed-swamp vegetation.
4. The significance of the presence and absence of a taxon in a fossil assemblage is discussed by Nowak et al. (2000). The true presence of a taxon occurs when it was present in the vegetation. A false presence may occur when the taxon was recorded in the sediment but was absent in the vegetation. The true absence of a macrofossil in an assemblage occurs when it was absent in the vegetation. A false absence may be due to poor representation, through low production, weak dispersal, and bad preservation.

The processes of plant macrofossil analysis, including potential sources of error may be summarised as a flowchart (Fig. 4.3).

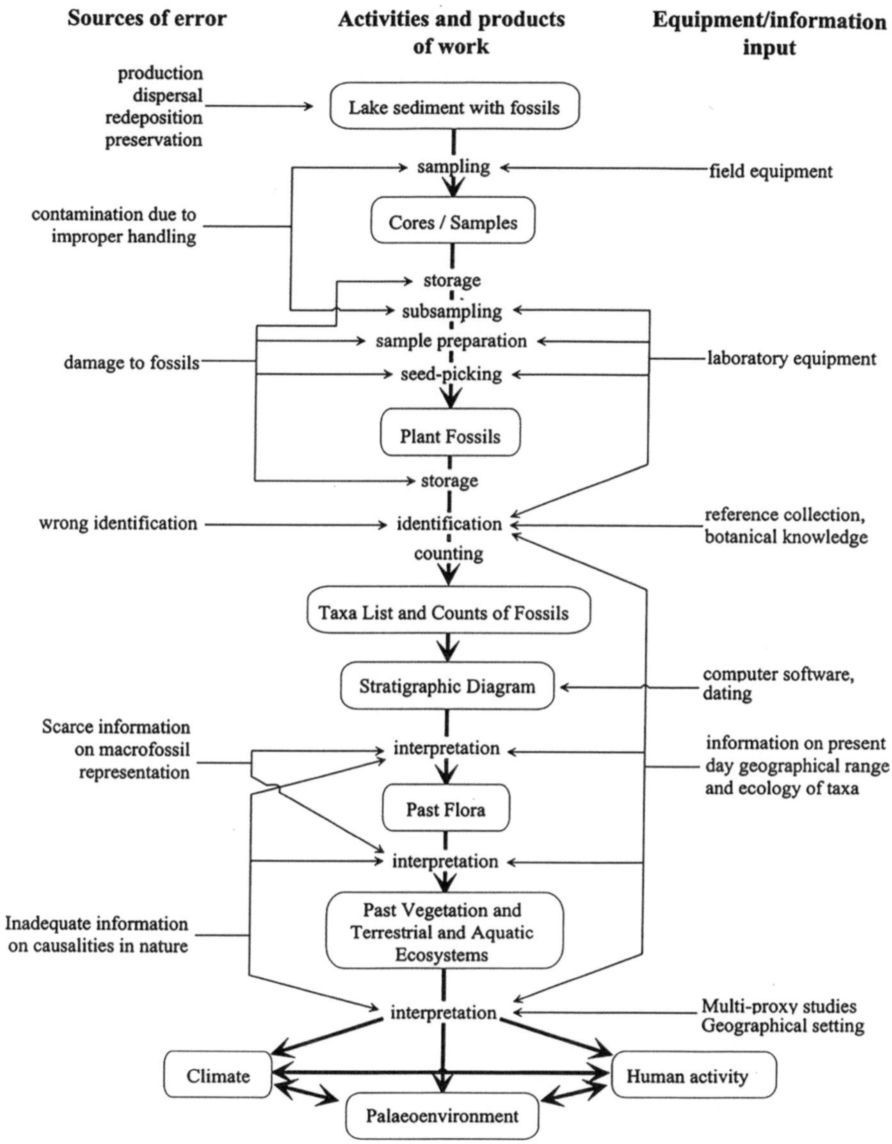

Fig. 4.3 The processes of plant macrofossil analysis summarised in a flowchart (Birks 2007; see also Birks 2001)

4.6 Interpretation of Macrofossil Data

All phases from seed production to deposition as a fossil as well as fossil preservation need to be considered when interpreting sediment assemblages of aquatic plants.

4.6.1 Macrofossil Production and Taphonomy

To interpret a fossil assemblage correctly, knowledge of the differential production of seeds and fruits (diaspores) and other identifiable preserved plants parts is needed. The fruits, seeds, and vegetative structures that eventually form sediment assemblages are produced in different amounts by different species. For example, large amounts of charophyte oospores and seeds of some annual aquatic plants (e.g. *Zannichellia palustris, Najas flexilis*) are produced, compared to *Potamogeton* and Nymphaeaceae species (Birks 1973). This means that even very low concentrations of *Potamogeton* sp. seeds (5–10 per 100 cm^3) may potentially represent abundant *Potamogeton* in the vegetation, whereas high concentrations of charophyte oospores (300–400 per 100 cm^3) are needed to infer dense local charophyte meadows (Zhao et al. 2006). Surface samples taken in relation to the vegetation provide a qualitative representation of the diaspores produced by a certain density of the parent species (e.g Birks 1973; Dieffenbacher-Krall and Halteman 2000) (Fig. 4.4).

Taphonomy is the process whereby remains produced by plants (or animals) are transferred to the sediment. The taphonomic processes for plant macrofossils are visualized in Fig. 4.4. They depend upon the sedimentary environment, ranging

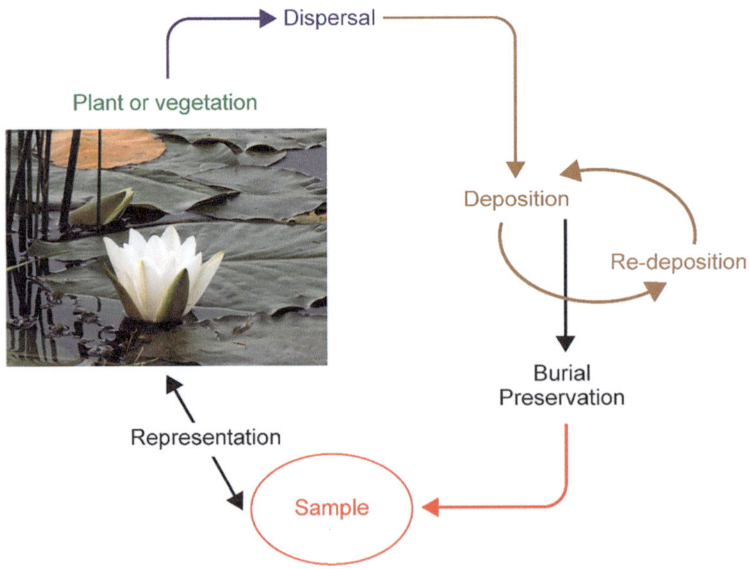

Fig. 4.4 Taphonomic processes leading to the formation of a macrofossil sample

from glacio-fluvial deposits and organic lenses in gravels, to lake sediments, marsh or bog sediments, backwaters of slow rivers or oxbow lakes, and archaeological contexts.

As shown in Fig. 4.4, plant macro-remains are produced by the plant or vegetation. These are dispersed differentially by various means from the plant or vegetation before being deposited in a sedimentary environment. Redeposition may occur before the remains are finally buried. On the way, they may be destroyed by decay, being eaten, or erosion and thus not reach the sediment sample, or they may be poorly preserved in the sediment.

The representation of former aquatic plant communities by sediment assemblages is the net result of all of the taphonomic processes. This may be investigated in a surface sediment sample taken beside the vegetation. By comparison with the parent vegetation, an estimate can be made of the representation of a species by its macrofossils. Surface-sample studies have shown that most macrofossils of aquatic plants are deposited near the lake shore, usually within the vegetated zone (e.g. Birks 1973; Dieffenbacher-Krall and Halteman 2000), never very far from the parent plants. Those that produce a large number of diaspores are usually more abundant in a fossil assemblage than those with a low production, but the relationship is not proportional, so the abundance of the taxon in the parent vegetation can only be estimated by using this general knowledge. The depth tolerance and dispersal characteristics of diaspores may be used to estimate water depth and water-level changes in the past (Hannon and Gaillard 1997; Birks and Gaillard 2007). However, because of the variability in dispersal and representation of diaspores, water-level changes need to be in the range of at least 2 m to be positively inferred.

The contrasting taphonomic processes operating in a bog environment and a lake are shown in Fig. 4.5. In a marsh or bog, the remains are produced by the vegetation growing on the surface and are largely autochthonous. There may be some atmospheric input or water input from streams or flooding. In a lake, there may be local sources of macrofossils, but there are also other possible allochthonous sources

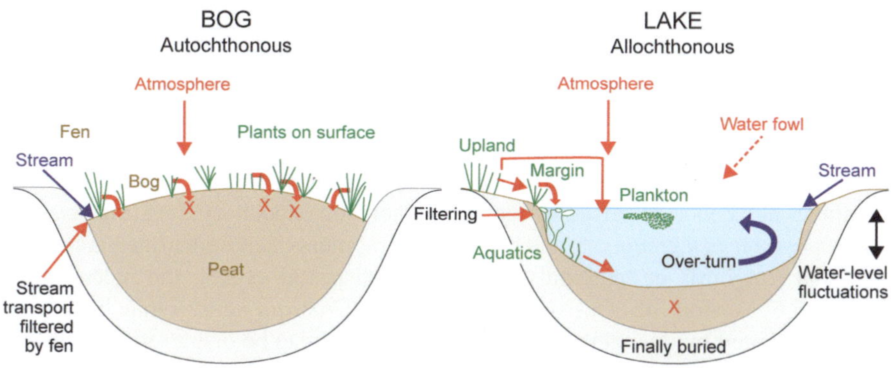

Fig. 4.5 A comparison of the taphonomic processes shaping macrofossil assemblages in a bog/marsh and in a lake

from the catchment and other parts of the lake. Therefore, lake-sediment assemblages reflect the basin itself and its catchment, whereas marsh or bog sediments reflect the local vegetation that formed the peat.

4.6.2 Dispersal

Dispersal agents are various, involving wind, floating, animal dispersal, and water currents. Wind is a common dispersal mechanism for seeds or fruits of telmatophytes that produce tiny 'balloon' seeds (e.g. *Gratiola officinalis*), or seeds with plumes (e.g. *Epilobium* and *Typha* spp.), wings, or parachutes (pappus) (e.g. Asteraceae). Floating on water (hydrochory) is also a common means of dispersal for many aquatic macrophytes. Many aquatic and wetland plants have seeds or fruits that can float for long periods of time, months, or even indefinitely (Praeger 1913). Exceptionally, streams and rivers may transport diaspores far from their origin. Thus distant parts of a lake or stream can be colonised, as the diaspores will be deposited in a suitable wet or damp habitat. The outer wall of a fruit may be composed of aerenchyma (tissue containing air) or of corky tissue with negative buoyancy that aids floating. Small seeds may have a reticulum which traps air and delays sinking. Diaspores of submerged plants often sink quickly, or if they are small, are dispersed within the water body by water currents (e.g. Characeae oospores) (Dieffenbacher-Krall and Halteman 2000). The seeds of *Nymphoides peltata* have a hydrophobic surface. Plumed seeds also usually float well.

Many macrophyte seeds and fruits are adapted for animal dispersal which can result in a larger dispersal range, including inter-lake transport. For fleshy fruits, this usually involves internal endozoochory with seeds being eaten and passed out in faeces or spat out before ingestion. Water birds (ducks, geese, and swans) are very important dispersal agents in this respect (Green et al. 2022). Most often seeds and fruits are eaten and pass through the digestive tract (e.g. *Potamogeton* fruits) after which they may be deposited locally or occasionally dispersed over long distances as birds migrate (e.g. Wilkinson 2023). Navarro-Ramos et al. (2024) found that digestive processes negatively affect seed buoyancy, leading to more rapid sinking and deposition to the sediment surface after egestion. Such long-distance events are important for the colonisation of distant habitats and the movement of plants following climate change, both in the past and at the present day.

Fruits may also bear hooks barbs, or bristles that allow them to become attached to animals and hence dispersed by fur and feathers (ectozoochory) (e.g. *Bidens* spp., *Ceratophyllum demersum*). Small vegetative fragments can sometimes adhere to the bodies or birds or mammals resulting in dispersal events; the recent rapid European expansions of invasive *Elodea* spp. and *Crassula helmsii* may have been partly facilitated by this method. Aquatic plants may also be dispersed by humans. This often results in the establishment of non-native and sometimes invasive species, for example by the emptying of aquaria into water bodies or the stocking of garden ponds. Several modes of dispersal can play a role in a species' dispersal and can interact.

4.6.2.1 Representation in Lake Sediments

What do the numbers of macrofossils in a fossil assemblage mean in terms of the plants and vegetation that produced the assemblage? Some of the processes leading to the formation of a macrofossil assemblage in a lake once plant remains have been produced are shown in Fig. 4.6. Many plant macro-remains are relatively heavy and are generally not dispersed far from parent plants, but greater dispersal distances are possible when floating occurs (Praeger 1913). Strong currents and lake turnover may result in redeposition. On the way to the sediments, diaspores may also be lost to a fossil assemblage by germination or by predation and finally, when permanently buried in sediment aquatic plant remains may suffer from decay and damage. A relatively small proportion of the total output of aquatic plant remains may reach the sediment and be preserved and supply long-term evidence of the former vegetation and environment. In most cases aquatic plant diaspores are better preserved than vegetative material, but leaf remains can often be abundant in sediments, including lcaf spincs, lcaf tips, buds and turion structurcs, all of which can and havc been used in palaeoecological reconstructions.

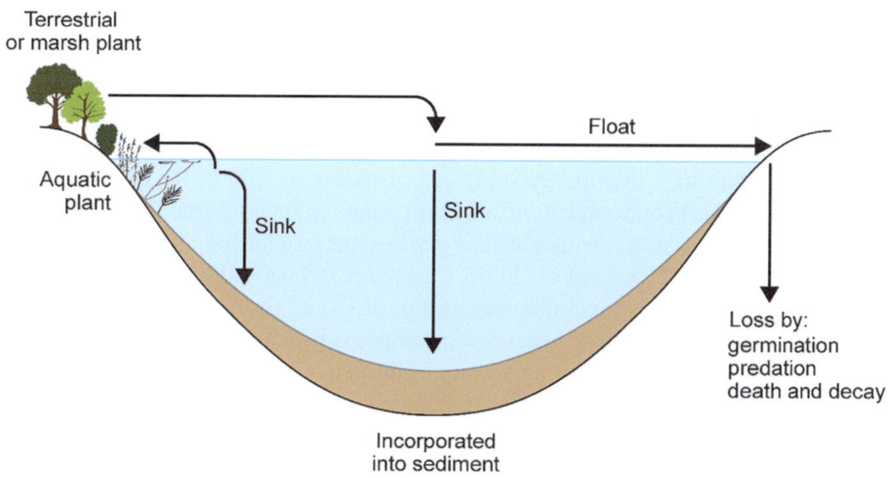

Fig. 4.6 The processes leading to the formation of a macrofossil assemblage in lake sediments

The relationship of a macrofossil assemblage to the parent plant or vegetation is summarised by the term 'representation'. Representation can be investigated directly by analysing surface sediment samples and comparing the composition of the assemblage with the vegetation that produced them. This aids the inference of past vegetation from fossil assemblages. Indicator species that grow in precise habitats are also useful when interpreting a fossil assemblage and are commonly used in macrofossil analysis. A number of studies on the representation of living plant communities in sediment assemblages have concluded that macrofossils can provide robust reconstructions of changing aquatic vegetation structure, especially in the case of dominant plants (Birks 1973; Davidson et al. 2005; Zhao et al. 2006). Because of their local deposition in general, a macrofossil assemblage best represents local, patch-scale vegetation structure rather than whole-lake environment (Birks 1973; Zhao et al. 2006), allowing past local vegetation to be reconstructed in some detail.

Those aquatic plants that produce a large number of diaspores are usually more abundant in a fossil assemblage than those with low production, but the relationship is not proportional, so the abundance of the taxon in the parent vegetation can only be estimated. It is possible to down-weight the bias towards plants that produce abundant and persistent remains by converting macrofossil concentration data to frequency scales (1–4) with the frequency of each fossil standardized against its maximum abundance in all of the samples under study (Odgaard and Rasmussen 2001). Combining plant macrofossil and pollen analysis is also advantageous in enhancing interpretations of aquatic vegetation change from sediment sequences (see Chap. 3).

4.7 Applications of Plant Macrofossil Analysis

Most macrofossil studies are aimed at the reconstruction of the past ecosystem, environment, and climate, both at the site of deposition and at the catchment scale. They usually form part of a multiproxy study (e.g. Birks et al. 2000). Reconstruction of recent changes due to eutrophication or human intervention are particularly important for restoration and management of aquatic ecosystems (e.g. Flower 2001; Sayer 2014; Bennion et al. 2018; Perego et al. 2025). Diaspores in lake sediment that have been buried, for example, by agriculture, may still be alive and form a seedbank (Alderton et al. 2017). Reconstruction of water-level changes is important for inferring precipitation: evaporation ratios, reflecting periods of drought and flooding (e.g. Hannon and Gaillard 1997). Seeds and fruits in the digestive systems of aquatic animals, such as wildfowl, are evidence of the diets of these animals and to what extent they rely on seeds and fruits of aquatic plants. Macrofossils may be preserved in wet environments related to human occupation, for example in wells and latrines. They provide excellent evidence of the past environment, food, and uses of plant material and are widely used in environmental archaeology (Jacomet 2007). Macrofossils of terrestrial plants and certain aquatic plants can provide reliable material for radiocarbon dating.

We hope that this book, by providing guidance on how to identify and interpret aquatic plant macrofossils, will greatly enhance many future palaeolimnological, archaeological, and seed bank studies.

Chapter 5
Species Accounts

5.1 Introduction

This chapter contains written and photographic descriptions and accounts of seeds, fruits, and some vegetative remains of individual aquatic macrophytes species that are found in northern Europe and that are or could potentially be recovered as macrofossils from sediments. They are arranged alphabetically by Family, and then Genus and Species within the family. Plant nomenclature follows the World Checklist of Vascular Plants (WCVP). This is embedded in Plants of the World Online (POWO) https://powo.science.kew.org/.

The accounts show photographs or diagrams of the relevant parts of the taxon and notes on their morphology. They contain a short description of the life form and ecology of the species, and its distribution in northern Europe and beyond, illustrated by the distribution map in GBIF (Global Biodiversity Information Facility). If relevant, the record of fossil remains of the taxon in Quaternary sediments, especially in the Late Glacial and the Early Holocene is noted, based on Godwin (1975) and other literature. This is not an exhaustive literature survey.

H. H. Birks et al., *Seeds and Fruits of Aquatic Plants of Northern Europe - A Handbook*, Developments in Paleoenvironmental Research 23,
https://doi.org/10.1007/978-3-032-04170-8_5

5.2 Species Accounts

5.2.1 *Acorus calamus* L. Acoraceae

Morphology The shiny pale-brown fruits are 4–5 mm long. They are irregularly wedge-shaped and four-angled, depending on how they were packed into the fruiting spike. The stigma is indented. Each three-locular fruit contains 1–9 (usually 5–7) narrowly ovoid, angled (as a result of crowding inside the fruit) seeds that are 3–4 mm long according to Buell (1935). They are encased in dense mucus that swells and bursts the fruit on contact with water. *A. calamus* is sterile in much of Europe and Eastern North America, where it is triploid and has probably been introduced. The diploid form is probably native in central North America and is fertile (Buell 1935).

Life form and ecology *Acorus calamus* is a tall, bright-green, *Iris*-like, rhizomatous perennial typically forming a narrow fringe along the shallow margins of mesotrophic-eutrophic lakes, rivers, and canals. It is a plant with several medicinal uses (Imam et al. 2013; Mittal et al. 2015) and when crushed or trampled it gives off a distinctive odour of citrus orange.

Distribution The native range of *A. calamus* is probably Asia and North America but there is uncertainty in this respect. *A. calamus* is widely naturalised in Europe where it is believed to have been introduced in the mid-sixteenth century (Imam et al. 2013).

Acorus calamus L. (gbif.org)

5.2.2 *Alisma gramineum* Lej. Alismataceae

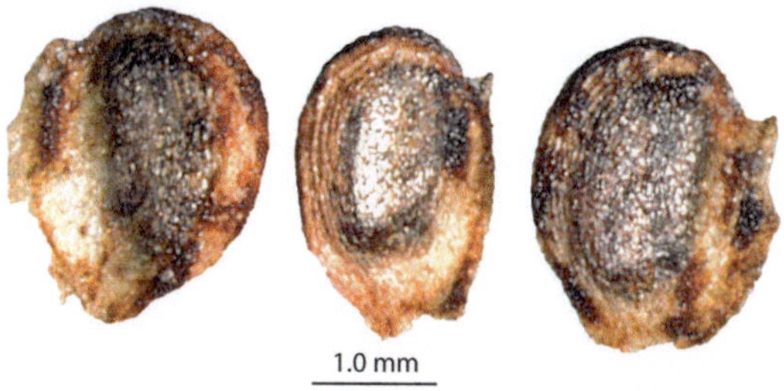

Morphology The dispersal unit is a single-seeded achene, 2.0–2.7 mm long, which floats on water. It has a rounded outline with a small vestige of the stigma. The sides are flattened due to the arrangement of the achenes on the receptacle. The fruit wall usually decays in sediments, but the seeds are robust. A seed is about 1.0 mm long with a characteristically curved embryo. The seed surface is obscurely reticulate.

Life form and ecology *Alisma gramineum* is a perennial herb up to 0.5 m high. The ribbon-like linear leaves are usually submerged. The aerial leaves are elliptic-lanceolate. It grows in shallow, naturally nutrient rich water at the edge of lakes, ponds, and rivers, and in marshes. Plants grow in small to medium-sized patches, but not in thick stands.

Distribution *Alisma gramineum* is widespread throughout the temperate Northern Hemisphere and reaches subarctic regions. It is considered critically endangered in the United Kingdom.

Quaternary history *Alisma gramineum* has been recorded occasionally in Holocene sediments (e.g. Matthews 1975; Pokorny et al. 2000). However, the seeds are hard to distinguish from other *Alisma* species.

Alisma gramineum Lej. (gbif.org)

5.2.3 *Alisma lanceolatum* With. Alismataceae

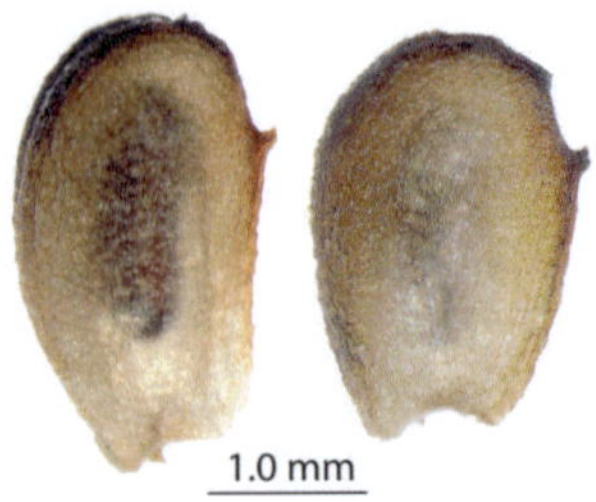

Morphology The single-seeded achenes of *Alisma lanceolatum* float in water and are dispersed by water currents or wild-fowl. They are oval in outline with flat sides and a vestigial stigma about one third of the way from the top, about 2.5–2.0 mm long and 1.6–2.0 mm wide. The fruit wall decays readily in sediments, but the seeds persist. The seeds are about 1.5 mm long. The embryo is curved, with a furrow where the cotyledons and radicle lie adjacent. The seed surface is coarsely ridged transversely.

Life form and ecology *Alisma lanceolatum* is a telmatic perennial herb with emergent lanceolate leaves and stems up to 1 m tall. It grows in shallow, usually calcareous water at the margins of lakes, slow-flowing rivers, ponds, and in fens, often with *Typha latifolia* and *Phragmites australis*. Plants occur in small to medium-sized patches, but not in extensive stands.

Distribution *Alisma lanceolatum* is widespread throughout temperate Europe and central Asia, but is rare in Scandinavia, Scotland, and Ireland. It reaches North Africa. It has been introduced in Australia and parts of North and South America where it can become a noxious weed in some places.

Alisma lanceolatum With. (gbif.org)

5.2.4 *Alisma plantago-aquatica* L. Alismataceae

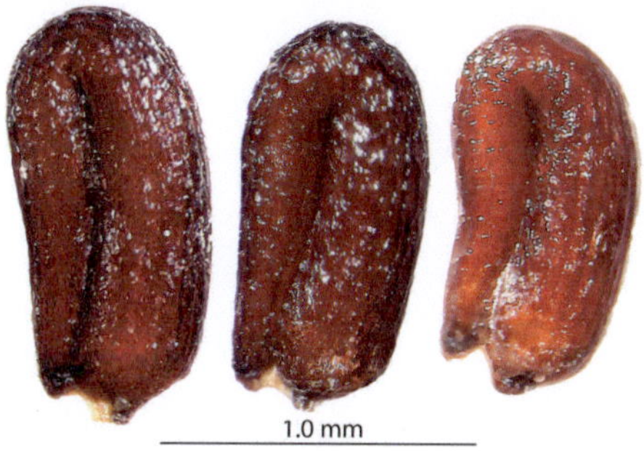

Morphology The single-seeded achenes of *Alisma plantago-aquatica* are composed of spongy tissue that allows them to float for long periods on water (Praeger 1913). They are oval in outline. The flattened sides reflect the arrangement of the achenes on the receptacle. The remains of the stigma are not obvious. The fruit wall decays readily in sediments, releasing the seeds that are resistant to decay. The seeds are about 1.3 mm long. The embryo is curved with the cotyledons closely fitted to the radical and covered by a continuous testa. The seed surface is obscurely reticulate.

Life form and ecology *Alisma plantago-aquatica* is a telmatic herb with emergent elliptical leaves on long petioles and stems up to 1 m high. It grows on exposed mud in fens or, more usually, in shallow water at the shores of ponds, lakes, and slow-flowing streams. It typically grows in clumps on nutrient-rich mud but does not form large stands. The fruits are primarily dispersed by water, and the seeds remain viable for several years (DiTamaso and Healy 2003).

Distribution *Alisma plantago-aquatica* occurs widely throughout Europe south of the Arctic Circle, and through temperate Asia (ssp. *orientale*), and is also native in north and east Africa. It is introduced and considered naturalised in large areas of North America, South America, southern Africa, Australia, and New Zealand.

Quaternary history *Alisma plantago-aquatica* has been recorded in Quaternary sediments (Wasylikowa 1967; Hannon and Gaillard 1997; Lapateva et al. 2023; Gałka et al. 2024) where it is recognised by its distinct seed shape and reticulate cell pattern.

Alisma plantago-aquatica L. (gbif.org)

5.2.5 ***Baldellia ranunculoides*** (L.) Parl. *sensu lato* Alismataceae

Morphology The seeds of *Baldellia ranunculoides* are about 1.0 mm long and 0.5 mm wide. They have the typical Alismataceae curved embryo. The cotyledons are wider and longer than the radical forming a pointed base. The enveloping testa has a well marked reticulate pattern with raised edges and sunken lumina. The fruits of *B. ranunculoides* ssp. *ranunculoides* are about 2.5 mm long and 1.0 mm wide, smooth, slightly curved towards the stigma base, and with a spongy pericarp and a distinct beak. Those of *B. ranunculoides* ssp. *repens* are about 2 mm long and papillose (Kozlowski et al. 2008). The seeds can form a long-lived seed bank. They can be dispersed by floating or by waterfowl.

Life form and ecology *Baldellia ranunculoides* forms a perennial rosette with elliptical to lanceolate aerial leaves some 20 cm long. When out of flower, it can resemble *Ranunculus flammula.* The submerged leaves are linear. *B. ranunculoides* ssp. *repens* produces new rosettes on stolons and is a smaller plant. *B. ranunculoides* ssp. r*anunculoides* usually grows in shallow, mildly to strongly calcareous or brackish mesotrophic water on a range of minerogenic and peaty substrates. It is found in shallow water at the margins of ponds and lakes, in ditches and canals, and in mildly saline marshes. It is intolerant of competition and needs its open habitat to be maintained by fluctuating water levels or cattle poaching. *B. ranunculoides* ssp. *repens* prefers weakly acidic and oligotrophic water. After a reservoir drainage in southwest Norway, *B. ranunculoides* ssp. *repens* germinated from its seed bank and covered the exposed sediment and flowered (Fadne 2023). Overall, the rapid, recent decline of both subspecies has been caused by habitat destruction, overgrowth by more competitive species, drought, and eutrophication. For a detailed account see Kozlowski et al. (2008).

Distribution *Baldellia ranunculoides s.l.* has a southern sub-oceanic distribution in northwest Europe, just reaching southern Sweden and southwest Norway. It occurs southwards from Scotland, Denmark, and the Low Countries, through western France to Iberia. From northern Italy it is spread across to Turkey and along the North African coast of Morocco, Algeria, and Tunisia, and in the Azores.

Baldellia ranunculoides (L.) Parl. (gbif.org)

5.2.6 *Damasonium alisma* Mill. Alismataceae

Morphology *Damasonium alisma* seeds are about 2 mm long and 1 mm wide. The curve of the embryo is obvious. The radicle and cotyledons are of equal length and joined by the testa. The testa bears prominent lateral ridges.

Life form and ecology *Damasonium alisma* is a tufted annual with basal, long-petiolate submerged, floating, or sometimes emergent ovate-cordate leaves. It grows on muddy shores of small acid ponds and the drawdown zones of lake shores. It requires a disturbed habitat, kept open by water fluctuations and grazing animals. It is usually flooded in winter but may be wet or dry in summer. The seeds persist in a long-lived seed bank.

Distribution *Damasonium alisma* has a western European distribution, particularly in France and Iberia. It is scarce and endangered in southern England, where conservation efforts are bringing it 'back from the brink'. It also occurs in North Africa, Israel and Jordan, and central Asia.

Damasonium alisma Mill. (gbif.org)

5.2.7 ***Luronium natans*** (L.) Raf. Alismataceae

Morphology The seeds of *Luronium natans* are about 1.5 mm long and 0.7 mm wide. In common with all Alismataceae, the embryo is curved. The cotyledons are wider than the radical and extend beyond it. The testa covering the embryo is smooth, with faintly discernible isodiametric cells. The fruits are distinctive, about 2 mm long. The fruit encloses a single seed. They have about 15 well marked longitudinal ribs and are slightly bent to the shape of the embryo.

Life form and ecology *Luronium natans* is a stoloniferous perennial with rosettes of linear submerged leaves in deeper water and rounded floating leaves in shallow water or on exposed mud where it flowers freely. Vegetatively, it resembles *Baldellia ranunculoides* and *Alisma plantago-aquatica* but apart from *B. ranunculoides* ssp. *repens*, these do not have stolons. Reproduction is largely by seed. Fruits and detached plantlets at the end of the stolons are dispersed by floating on water. *L. natans* grows on sand, gravel, silt, or peat substrates in clear water in mesotrophic or oligotrophic lakes, pools. and slow-flowing rivers, often with fluctuating water levels and a wide range of pH values, from 3.6–8.0. It is intolerant of competition and benefits from moderate disturbance. It has been lost from many of its lowland sites because of eutrophication, but not acidification. Lansdown and Wade (2003) provide an extensive account of *Luronium* ecology.

Distribution *Luronium natans* is endemic to northwest Europe. It is most common in France and the Low Countries. It extends south to Spain and north to southern Scandinavia and Scotland. In the UK, it is commonest in Wales and Shropshire.

Luronium natans (L.) Raf. (gbif.org)

5.2.8 *Sagittaria sagittifolia* L. Alismataceae

Morphology The seeds of *Sagittaria sagittifolia* have the typical cuved embryo of Alismataceae. At about 4.0 mm long, they are longer than most of the Alismataceae considered here. They are relatively narrow. The cotyledons are slightly wider than the radical, and extend a little beyond it. The testa has fairly distinct isodiametric cells. The fruits are about 6 mm long and 5 mm wide, with a spongy wall extended into poorly defined wings. The beak-shaped style base is prominent, giving the fruits a characteristic shape. The wall contains air and has a waxy surface, and the achenes float readily, often for long periods, until frost damages them and they sink (Hroudová et al. 1988). The seeds germinate the following spring on exposed mud.

Life form and ecology *Sagittaria sagittifolia* is a perennial emergent aquatic plant with a rosette of variable, usually hastate (arrow-shaped) aerial leaves and submerged linear leaves. Stolons grow downwards from the rosette and form overwintering tubers. It grows in mesotrophic to eutrophic, slow flowing or standing rivers, canals, or ditches and in shallow standing water up to about 1 m deep. However, it grows best on exposed very moist muddy loam at the margins of ponds and ditches where the seeds germinate when exposed by fluctuating water levels. A comprehensive account of its ecology and biology is given by Hroudová et al. (1988).

Distribution The main distribution of *Sagittaria sagittifolia* is across temperate Europe and western Asia, north of the Mediterranean area and Spain. It also occurs in India, southeast Asia, and Australia. It is introduced in New Zealand and is invasive in North America.

Quaternary history Seeds of *Sagittaria sagittifolia* have been recorded in the Kama valley, western Siberia (Lapateva et al. 2023).

Sagittaria sagittifolia L. (gbif.org)

5.2.9 *Berula erecta* (Huds.) Coville Apiaceae

Morphology The mericarps of *Berula erecta* are rounded, 1.5–2.0 mm long, with indistinct ribs.

Life form and ecology *Berula erecta* is a stoloniferous perennial growing submerged in rivers and fast-flowing streams, or emergent at lake and pond margins and in fens and marshes on seasonally flooded fine silt or mud, where it is 20–100 cm tall. The fruits are dispersed by water (Cappers 1993a, b) and waterbirds (Soons et al. 2008).

Distribution *Berula erecta* occurs widely in temperate Europe and North America south of the boreal forest. It extends to western Asia, southern Africa, and southeast Australia.

Quaternary history *Berula erecta* macrofossils have been recorded in interglacial, glacial, and Holocene sediments (e.g. Godwin 1975; Gibbard et al. 1977; Smith et al. 2001; Schlütz et al. 2021).

Berula erecta (Huds.) Coville (gbif.org)

5.2.10 *Cicuta virosa* L. Apiaceae

Morphology The mericarps of *Cicuta virosa* are rounded and about 2 mm long. The corky ridges are well defined, but hardly project from the rest of the mericarp.

Life form and ecology *Cicuta virosa* is a perennial herb 1– 2 m tall. It grows in shallow water at the margins of lakes, ponds, rivers, streams, ditches, and canals, or in deeper water on floating mats of vegetation. It also occurs in tall-herb fen, in marshy pasture, and on damp mud. It is poisonous. The fruits are primarily dispersed by water (Shin et al. 2013).

Distribution *Cicuta virosa* has a circumboreal distribution throughout the temperate and sub-Arctic Northern Hemisphere.

Quaternary history *Cicuta virosa* fruits can be distinguished by their shape and corky ridges and have been recorded in Quaternary sediments (Godwin 1975), especially during the Holocene (e.g. Hannon and Gaillard 1997; Bennike and Jensen 2013; Gałka and Tobolski 2013; Lapateva et al. 2023). An early Holocene record in northern Norway testifies to a climate at that time warmer than present (Birks et al. 2012).

Cicuta virosa L. (gbif.org)

5.2.11 *Helosciadium inundatum* (L.) W.D.J.Koch (Syn. *Apium inundatum* (L.) Rchb.fil.) Apiaceae

Morphology The mericarps of *Helosciadium inundatum* are much longer than wide: 2.5–3.0 mm long and 1.0 mm wide. This distinguishes them from other *Helosciadium* species. There are five prominent spongy ridges along the back which enable floating and dispersal by water (e.g. Cappers 1993a, b).

Life form and ecology *Helosciadium inundatum* is a low-growing, often prostrate, perennial herb 10–50 cm long with aerial and submerged leaves. It grows in permanent shallow water in streams, ditches, ponds, canals, and even in fast-flowing water, and in periodically desiccated sites, such as the margins of lakes, pools, reservoirs, and dune-slacks. It is confined to oligotrophic or mesotrophic water, and most sites are base-poor. It has suffered from habitat destruction in Britain and elsewhere.

Distribution *Helosciadium inundatum* has a European suboceanic temperate distribution. It is centred in western and northwest Europe, from Iberia to the British Isles. It reaches Denmark and southern Sweden in Scandinavia.

Quaternary history *Helosciadium inundatum* seeds have been recorded in some interglacial, glacial, and in late Holocene sediments (Godwin 1975; Aalto et al. 1984; Waller 2010).

Apium inundatum (L.) Rchb.fil. (gbif.org)

5.2.12 ***Helosciadium nodiflorum*** (L.) W.J.D.Koch (Syn. *Apium nodiflorum* (L.) Lag.) Apiaceae

Morphology The typical Apiaceae mericarps of *Helosciadium nodiflorum* are produced in joined pairs (schizocarps). The junction is straight, and the back is rounded (left photo) with prominent longitudinal ridges. The mericarps are 1.5–2.5 mm long. They are dispersed by water and birds.

Life form and ecology *Helosciadium nodiflorum* is a sprawling perennial herb 30–80 cm tall. It grows in damp, nutrient-rich ditches, marshes, ponds, and river sides, and in wet meadows.

Distribution *Helosciadium nodiflorum* has a southern-temperate Euro-Siberian distribution. It is most frequent in western Europe north to Scotland but absent in Scandinavia. It extends to western Asia, North Africa, and the Canary Isles. It is an introduced weed in Chile and New Zealand where it is considered a nuisance species (Groenteman et al. 2020).

Quaternary history *Helosciadium (Apium) nodiflorum* fruits have been identified in late Holocene sediments, often associated with human activity (Godwin 1975; Aalto et al. 1984; Carrott et al. 1996; Collins et al. 1996; Brown et al. 2018).

Apium nodiflorum (L.) Lag. (gbif.org)

5.2.13 *Oenanthe crocata* L. Apiaceae

Morphology The mericarps of *Oenanthe crocata* are 4–6 mm long. The mericarps are semi-cylindrical, with dorsal ridges that end in apical hooks. The ventral side has a deep central furrow. The remnant of a long stigma may extend from the apex.

Life form and ecology *Oenanthe crocata* is a perennial growing up to 1.5 m tall. It inhabits wet places including marshes, lake and stream margins, and other wetlands. It prefers slow-moving water and can survive long periods of shallow inundation. The fruits are primarily dispersed by water.

Distribution *Oenanthe crocata* occurs in oceanic western Europe south of Scandinavia, and in the west and central Mediterranean basin, especially Corsica and Sardinia.

Quaternary history There are a few records of the distinctive fruit of *Oenanthe crocata* in Holocene sediments (Godwin 1975).

Oenanthe crocata L. (gbif.org)

5.2.14 *Oenanthe fistulosa* L. Apiaceae

Morphology The mericarps of *Oenanthe fistulosa* are about 4 mm long and widest at the top. The dorsal side has many ridges and a spongy crest bearing sepal teeth that facilitates dispersal by floating on water. The two resin canals on the ventral side are separated by a central ridge. The remnant of a long style extends from the apex.

Life form and ecology *Oenanthe fistulosa* is a slender perennial herb up to 80 cm tall. It grows in shallow water at the edges of ponds, lakes, streams, and in seasonally flooded wetlands with some disturbance. It grows on muddy nutrient-rich soils with neutral pH. Germination is best with some ground disturbance (Stroh 2015). Plants occur singly or in dense stands.

Distribution *Oenanthe fistulosa* has a sub-atlantic distribution in Europe south of Fennoscandia. It is infrequent in Scotland and Ireland and southwest England.

Quaternary history *Oenanthe fistulosa* has been occasionally recorded in Quaternary sediments (Buurman et al. 1995; Godwin 1975).

Oenanthe fistulosa L. (gbif.org)

5.2.15 ***Oenanthe lachenalii*** C.C.Gmel. Apiaceae

Morphology The mericarps of *Oenanthe lachenalii* are 2.0–2.5 mm long and semi-cylindrical. The dorsal side has three apical spines at the top of spongy longitudinal ridges. The ventral side is flat and smooth with two resin canals. The remnants of the style may extend from the top of the mericarp.

Life form and ecology *Oenanthe lachenalii* is a wetland perennial herb up to 1 m tall. It inhabits the upper regions of saltmarshes and tidal marshes with brackish water. Further inland it prefers base-rich habitats including marshes and fen-meadows with periodically flooded soils. The fruits are primarily dispersed by water.

Distribution *Oenanthe lachenalii* is found throughout Spain, France, and western Germany and in coastal regions in northwest Europe as far north as Scotland and Denmark.

Quaternary history *Oenanthe lachenalii* has been rarely recorded in coastal Quaternary sediments (e.g. Schepers et al. 2013; Godwin 1975).

Oenanthe lachenalii C.C.Gmel. (gbif.org)

5.2.16 ***Peucedanum ostruthium*** (L.) Koch (Syn. *Imperatoria ostruthium* L.) Apiaceae

Morphology The large mericarps of *Peucedanum ostruthium* are almost round, about 4–5 mm in diameter. The body is obovoid, widest above the middle, with three marked ribs on the dorsal side. The two lateral ribs are extended into wings, which aid dispersal. The ventral surface has marked resin canals.

Life form and ecology *Peucedanum ostruthium* is a perennial herb 30–100 cm tall. It grows in fertile moist meadows and pastures and on stream and river sides. It has been cultivated for medicinal purposes in the past and is often associated with dwellings and farms. Fossil fruits show that it was established in Ireland by 1000 CE (Godwin 1975).

Distribution *Peucedanum ostruthium* is native to the Pyrenees and the Alpine chain and in central Europe. It is naturalised in northern Britain, northeast Ireland, southern Scandinavia, and the USA.

Peucedanum ostruthium (L.) W.D.J.Koch (gbif.org)

5.2.17 ***Peucedanum palustre*** (L.) Moench. (Syn. *Thysselinum palustre* (L.) Hoffm) Apiaceae

Morphology The mericarps of *Peucedanum palustre* are large, 4–6 mm long and about 4 mm wide. They are broadly ellipsoid, with prominent dorsal ribs, the outer ones being slightly winged. The ventral side is smooth and flat, concealing the resin canals.

Life form and ecology *Peucedanum palustre* is a biennial, 30–100 cm tall. It grows on permanently damp peat of pH 5.0–6.5, often flooded in winter. It is characteristically found in tall-herb fen, occurring in both cut and uncut stands. It can survive in reed swamp, fen scrub, and alder-carr, and it also occurs rarely in marshes and damp pasture (Plant Atlas 2020). The fruits can be propelled about 75 cm by release of tension in the fruit stalk after disturbance by wind or birds (Meredith and Grubb 1993).

Distribution *Peucedanum palustre* has a Euro-Siberian boreo-temperate distribution, with a continental distribution in western Europe. In Britain, it is rare, almost confined to the Broads of East Anglia.

Peucedanum palustre (L.) Moench (gbif.org)

5.2.18 *Sium latifolium* L. Apiaceae

Morphology The robust mericarps of *Sium latifolium* are about 3 mm long and 1.5 mm wide. They have five strong light-brown dorsal ribs. The ventral side is almost flat and the resin canals are hidden. The fruits tend to be curved in lateral view.

Life form and ecology *Sium latifolium* is a robust, lowland, perennial herb 1.5–2.0 m tall. It grows on the banks of alkaline, eutrophic, still water and in reed swamps and fens. It has declined over the last 200 years due to the drainage of wetlands and loss of habitats.

Distribution *Sium latifolium* occurs in Europe and western Asia. It avoids oceanic climates and is therefore rare in Ireland, western UK, France, and Iberia. In Fennoscandia it is more or less confined to southern Sweden. It avoids the Arctic and is rare in the Mediterranean region.

Sium latifolium L. (gbif.org)

5.2.19 *Calla palustris* L. Araceae

Morphology The dispersal unit is a berry that contains several seeds. The seeds are oblong, with a longitudinal furrow (raphe). The seed surface is ridged near the base and pock-marked in the upper part. Seeds are approximately 3 mm long.

Life form and ecology *Calla palustris* is a rhizomatous perennial up to 30 cm tall. It grows in marshy ground as found near small ponds and in bogs and swamps. It prefers light shade and acidic conditions. The seeds are primarily dispersed by animals consuming the berries or by floating (Murray 1987).

Distribution *Calla palustris* is native in temperate and boreal Europe, especially central Europe, southern Sweden, and Finland. It extends across central Asia and boreal North America, where its main distribution is in eastern Canada and USA. It is introduced in Britain.

Quaternary history *Calla palustris* has been recorded in Quaternary sediments (Matthews 1974; Novenko et al. 2009) and is readily identifiable.

Calla palustris L. (gbif.org)

5.2.20 *Lemna* L., *Spirodela* (L.) Schleid., *Wolffia* Horkel ex Schleid. Araceae (Lemnaceae)

Lemnaceous plants (duckweeds) are the smallest, simplest, and fastest growing angiosperms (Fourounjian et al. 2021). They have efficient vegetative reproduction and dispersal. They seldom reproduce by seed. *Spirodela polyrhiza* very rarely flowers throughout its range (Preston and Croft 1997). The seeds of *Lemna* and *Spirodela* are easily recognisable by their whitish colour and obvious ribs. They are ovoid with a dark, pointed operculum at the top and rows of longitudinal cells forming 8–22 ribs with horizontally aligned cells between (Landolt 1986). They vary in the number of ribs and the size of the seed. Scanning electron micrographs of seeds of *Lemna* spp., *Spirodela*, and *Wolffia arrhiza* are shown in Landolt (1986). *Spirodela polyrhiza* seeds are very rarely produced. They are similar to those of *L. trisulca* but slightly larger.

Lemna seeds are produced in one- or few-seeded fruits which can be dispersed entire and release the seeds when they split. The seeds sink since they do not contain aerenchyma (air spaces). *Lemna minuta, L. aequinoctialis, L. turionifera,* and *L. valdiviana* are introduced in northwest Europe. Comprehensive accounts of *Lemna* and its relatives in northwest Europe are given by Schou et al. (2023).

Quaternary history *Lemna* seeds have been occasionally recorded alongside fossil pollen in British Holocene and interglacial deposits (Godwin 1975). They are rarely determined to species. Święta-Musznicka et al. (2011) reported *L. gibba* and *L. trisculca* seeds from Gdansk, Poland, and Gałka et al. (2020) recorded *L. trisulca* seeds in the Late Glacial of central Poland. *L. trisculca* seeds were recorded in the Early Holocene in the Kama valley, western Siberia (Lapateva et al. 2023) and in oxbow lake sediments in the Drava valley between Hungary and Croatia at around 1100 cal BP just before the infilling of the lake (Gałka et al. 2024). *Lemna trisulca* and *L. gibba/minor* seeds have been found in pond sediments in Norfolk UK (C.D. Sayer personal data). However, *Lemna* is very under-represented by seeds in sediments (Zhao et al. 2006). *Lemna minor* plants were recorded in recent lagoon sediments in Egypt by Birks et al. (2001).

5.2.20.1 *Lemna gibba* L.

Lemna gibba seeds

SEM photograph (modified from Fu et al. (2017))

Lemna gibba/minor fossil seed from Czech Republic

Morphology The winged fruit of *Lemna gibba* contains 1–5 seeds. The seeds are ovoid with 8–15 ribs (Sell and Murrell 1996–2018) and a dark pointed operculum at the top. The base can also be dark. The scanning electron micrographs by Landolt (1986) and Fu et al. (2017) show the ribs with longitudinal cells and horizontal cells between. The seeds are about 0.5 mm long. They cannot be differentiated from seeds of *L. minor*.

Life form and ecology *Lemna gibba* is a very small, (1–8 mm) free-floating duckweed. It reproduces vegetatively by budding from the plant body and can quickly cover the water surface. It grows in shallow ponds, canals, lakes, and ditches, preferring eutrophic, often brackish water of pH 7.4–9.8 and high conductivity and base status (Landolt 1975). Dispersal is primarily by waterfowl (Green et al. 2016) or human activities.

Distribution *Lemna gibba* has a wide distribution throughout temperate Europe, Africa, Asia, and the Americas, being absent only from polar regions and the tropics. It is widespread throughout Britain, except in much of Scotland and Shetland.

5.2.20.2 *Lemna minor* L.

Morphology The seeds of *Lemna minor* are very similar to those of *L gibba.*

Life form and ecology *Lemna minor* is a very small floating duckweed growing in still or slowly flowing mesotrophic to eutrophic water. It prefers water with a lower pH of 6.8–7.7 and lower conductivity and base status than *L. gibba* (Landolt 1975). A detailed account of its ecology and distribution is given by Landolt (1975) and Schou et al. (2023).

Distribution *Lemna minor* has a wide distribution throughout temperate Europe, Africa, Asia, and the Americas, but extends further north than *L. gibba* in Britain and reaches the Arctic Circle (Landolt 1986).

5.2.20.3 *Lemna trisulca* L.

Lemna trisulca seeds

Lemna trisulca seeds from recent sediments in Bonfire Pond) (left) and Guestwick Pond (right), both in Norfolk, UK

Morphology The whitish seeds of *Lemna trisulca* are about 1 mm long with 12–15 ribs and a dark operculum at the top. They are rather irregular in shape.

Life form and ecology *Lemna trisulca* fronds are usually submerged in ponds, ditches, and backwaters of slow-flowing rivers, only floating when in flower. It prefers clear mesotrophic water but can also sometimes occur in eutrophic or brackish situations. It suffers from competition from other floating macrophytes (Preston and Croft 1997).

Distribution *Lemna trisulca* has a circumboreal distribution in the Northern Hemisphere, and it also occurs in south and east Australia. Unlike *L. minor* it is almost absent from South America.

5.2.20.4 *Wolffia arrhiza* (L.) Horkel ex Wimm

Morphology The seeds of *Wolffia* and *Wolffiella* are less than 0.4 mm long and are smooth or slightly reticulate (Landolt 1986). The seeds of *Wolffia arrhiza* are minute and very rarely produced. Flowers have been found once in Europe (Schmitz and Kelm 2017).

Life form and ecology *Wolffia* spp. are a tiny, rootless perennial floating duckweeds which can reproduce vegetatively very rapidly. They cover open or partly shaded shallow base-rich and eutrophic water in lowland ditches and ponds. They are intolerant of desiccation and low temperatures. They appear to be pioneer species as they are adaptable to changing environmental conditions and can fix atmospheric nitrogen (Pietryka et al. 2023).

Distribution *Wolffia arrhiza* is widespread through temperate Europe but is always rare and often declining. It reaches east to the Caspian Sea and is sparse throughout Africa (Preston and Croft 1997).

W. australiana, W. columbiana, and *W. globosa* are introduced in northwest Europe (Schou et al. 2023).

W. columbiana has recently been recorded as spreading in England, whereas *W. arrhiza* is rarer than previously thought and is declining (Lansdown et al. 2022).

Lemna gibba L. (gbif.org)

Lemna minor L. (gbif.org)

Lemna trisulca L. (gbif.org)

Spirodela polyrhiza (L.) Schleid. (gbif.org)

Wolffia arrhiza (L.) Horkel ex Wimm. (gbif.org)

5.2.21 ***Hydrocotyle vulgaris*** L. Araliaceae (Hydrocotylaceae)

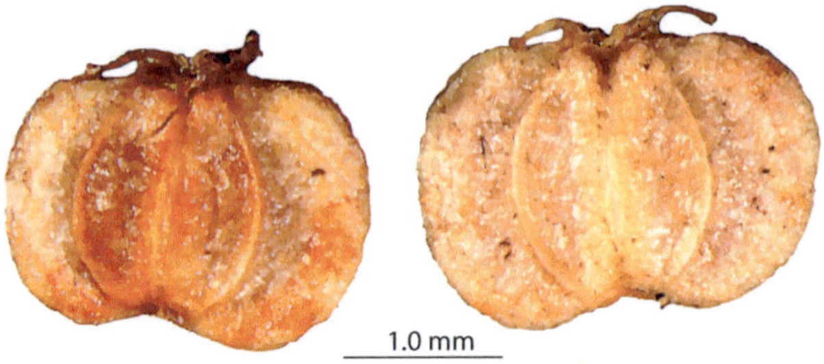

Morphology The schizocarp fruits of *Hydrocotyle vulgaris* are laterally flattened. In face view they are almost round with an indented base. The schizocarp may split between the two mericarps, each of which has a prominent rib in side view. The styles are usually persistent. The mericarps are about 1.5 mm long and 1.0 mm wide.

Life form and ecology *Hydrocotyle vulgaris* is a perennial creeping herb up to 40 cm tall. It grows in wet meadows, fens, and reed marshes, and beside lakes on non-calcareous clay soils.

Distribution *Hydrocotyle vulgaris* has a subatlantic distribution in northwest Europe. It occurs throughout the British Isles, but only reaches southernmost Sweden and to 60°N in coastal Norway. It is more scattered in southern Europe and occurs in Iceland and the Azores.

Quaternary history Fruits (and pollen) of *Hydrocotyle vulgaris* are recorded from Quaternary deposits in Britain, especially in the later Holocene (Godwin 1975), and also in the Early Holocene in Denmark (Bennike et al. 2020).

Hydrocotyle vulgaris L. (gbif.org)

5.2.22 ***Warnstorfia fluitans*** (Hedwig.) Loeske, ***Warnstorfia exannulata*** (Schimp.) Loeske Calliergonaceae ***Drepanocladus aduncus*** (Hedwig.) Warnst. Amblystegiaceae (Synonyms: ***Drepanocladus fluitans*** (Hedw.) Warnst., ***Drepanocladus exannulatus*** (Schimp.) Warnst., ***Sarmenthypnum exannulatum*** (Schimp.) Hedenäs

Warnstorfia fluitans/exannulata fossil specimens

cf. *Drepanocladus aduncus* fossil leaf

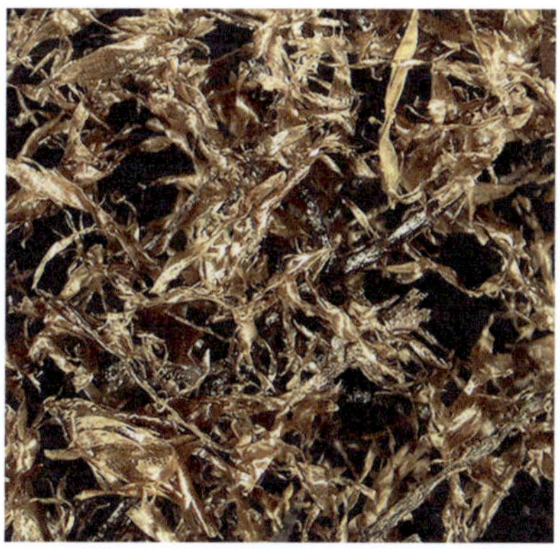

Fossil *Warnstorfia* layer in Higgins Lake, Michigan, sediments. Photo, B. Brandon Curry

Morphology *Warnstorfia* (syn. *Drepanocladus) exannulata* has leafy stems, with dense or sparse leaves with a distinct nerve that ends near the leaf apex. This species is not readily separable from W. *fluitans* in fossil material as the basal inflated cells are poorly preserved (Janssens 1983). *Drepanocladus aduncus* leaves can also be found in sediments. The leaves are often curved and have plicae (longitudinal folds). They have a strong nerve extending about half to two-thirds of the way up.

Life form and ecology *Warnstorfia/Drepanocladus* spp. are common aquatic creeping mosses that are frequently preserved in lake sediments. Species can be identified using a moss flora (e.g. Smith 1978) or Janssens (1983). *W. fluitans* and *W. exannulata* are the species most frequently found in acid or neutral situations. They both occur in base-poor pools and shallow lakes. *W. exannulata* is commonly found at higher altitudes and latitudes than the similar *W. fluitans.* If the water is clear, they can form a carpet on the lake floor down to a depth of at least 15 m (Light 1975; Light and Smith 1976). *D. aduncus* and especially *Scorpidium (Drepanocladus) revolvens* occur in more basic water, also sometimes with *Scopidium scorpioides* and *Calliergon s.l.* spp. The genus *Drepanocladus s.l.* has been split recently into several genera (Blockeel et al. 2021). *D. fluitans* is now *Warnstorfia fluitans* (Hedw.) Loeske and *D. exannulatus* is now *Sarmenthypnum exannulatum* (Schimp.) Hedenås. Likewise, *Calliergon* has been split into several genera (Blockeel et al. 2021). In general, most fossil mosses can be identified using a moss flora (e.g. Smith 1978). A large number of species can be identified from lake sediments (e.g. Birks and Mathewes 1978; Jonsgard and Birks 1995).

Quaternary history Bryophytes are infrequently identified from lake sediments and peats. A comprehensive account of fossil bryophytes of the British Isles is given by Dickson (1973). A detailed bryophyte record was used for palaeoecological reconstruction at Abernethy Forest, Scotland (Birks and Mathewes 1978) and at Kråkenes Lake, Norway (Jonsgard and Birks 1995). Many other records of mosses are scattered through the Quaternary literature. Liverworts are very seldom preserved. MacDonald et al. (1991) showed that *Drepanocladus crassicostatus* remains in North American lake sediment should not be used for radiocarbon dating, because they contain the same old-age reservoir effect as the lake sediment when compared with dates from terrestrial material at the same level.

Drepanocladus fluitans (Hedw.) Warnst.
Drepanocladus exannulatus (Schimp.) Warnst.
Drepanocladus aduncus (Hedw.) Warnst.

5.2.23 ***Thelypteris palustris*** (Salisb.) Schott Aspleniaceae

Morphology The detached pinnae of ferns all appear similar, and we illustrate *T. palustris* here as it is the most likely species to be preserved in fen peats. Well-preserved spores have characteristically echinate perines and have been reported widely. Fossil leaves strongly associated with *Alnus* remains have been recorded. Sometimes *Thelypteris* can form fen peat with its dense roots and rhizomes (Godwin 1975).

Life form and ecology *Thelypteris palustris* has long slender rhizomes bearing lanceolate to narrowly elliptic pale green leaves at intervals in sparse tufts. The sterile leaves are about 80 cm tall, and the fertile leaves are about 150 cm tall. It grows on mesotrophic wet organic soils in reed fens and marshes, often shaded by taller vegetation, and in alder and willow shrub carr. It is declining due to habitat destruction.

Distribution *Thelypteris palustris* occurs across Europe and Asia in the temperate zones. It is also widespread in eastern North America, but not west of the prairie.

Thelypteris palustris (Salisb.) Schott (gbif.org)

5.2.24 ***Bidens cernua*** L. Asteraceae

Morphology The achenes of *Bidens* spp. all have the same basic structure. The achene is ribbed. There are two or four spines extending from the main ribs. The ribs and spines bear downward pointing barbs. The achenes of *Bidens cernua* are long and narrow, tapering to the base. They are 4–6 mm long excluding the spines. The side and central ribs are thickened, giving the achenes a 4-angled cross section. The main ribs continue beyond the body as four spines, each bearing backward-pointing barbs. Similar barbs are present on the ribs.

Life form and ecology *Bidens cernua* is an annual herb up to 80 cm tall. It grows on often disturbed, damp soil beside ponds and streams where water levels are variable, particularly where there is standing water in winter, but the ground is dry during warmer months. The achenes are primarily dispersed by animals as the spines and barbs easily adhere to fur and feathers (Cox 2002). They also float well.

Distribution *Bidens cernua* is native in the temperate and subarctic Northern Hemisphere.

Quaternary history *Bidens cernua* has been recorded in Quaternary sediments in Britain (e.g. Godwin 1975), Europe (e.g. Gałka and Kasper 2011; Gaillard and Lemdahl 1994), and USA (Jackson et al. 1988; Grüger 1973). *Bidens* is easily identified due to the unique features of the achene, but the species may be difficult to differentiate.

Bidens cernua L. (gbif.org)

5.2.25 *Bidens radiata* Thuill. Asteraceae

Morphology The achenes of *Bidens radiata* are flattened and obovoid, about 6 mm long excluding the spines. The faces of the achene are smooth. The marginal ribs extend as two spines bearing backward-pointing barbs (Bojňanský and Fargašová 2007).

Life form and ecology *Bidens radiata* is an annual herb up to 60 cm high. It grows in wet places such as fens and ditches, lake and pond margins, and wet meadow margins (Mossberg and Stenberg 2018). The achenes are primarily dispersed by animals.

Distribution *Bidens radiata* is widespread in mainland central Europe, Sweden, and Finland, and across continental Eurasia.

Quaternary history *Bidens radiata* has been recorded in Quaternary interglacial sediments (Maul et al. 2013) but fossils may be hard to differentiate from other *Bidens* species.

Bidens radiata Thuill. (gbif.org)

5.2.26 *Bidens tripartita* L. Asteraceae

Morphology *Bidens tripartita* achenes are flattened, and narrow obovoid to oblong, tapering towards the base, and are about 6–8 (– 11) mm long excluding the spines. The edges and the central ribs bear downward-facing barbs. They continue into two, usually three, or occasionally four barbed spines. The achene faces are smooth.

Life form and ecology *Bidens tripartita* is an annual herb up 75 cm tall. It grows on damp or wet disturbed ground such as by ponds, lake margins, streams, and ditches, preferring neutral soils. The achenes are primarily dispersed by animals, but they also float well.

Distribution *Bidens tripartita* is native to the temperate Northern Hemisphere, extending in Europe from the Mediterranean to northernmost Norway. It is introduced in Australasia.

Quaternary history *Bidens tripartita* has been recorded in Quaternary sediments, in interglacials, and the Holocene in Europe (e.g. Godwin 1975; Stachowicz-Rybka et al. 2009; Field and Peglar 2010). Fossil achenes may be confused with other *Bidens* species.

Bidens tripartita L. (gbif.org)

5.2.27 *Cirsium palustre* (L.) Scop. Asteraceae

Morphology The achenes of *Cirsium palustre* are about 4 mm long and 1.5 mm wide, tapering towards the base. There is a distinct collar near the top where the pappus was attached. The remains of the style project from its centre. These features are sometimes eroded in fossil material, in which case the top appears rounded. The fruit surface is smooth with faint longitudinal ribs. *C. palustre* achenes cannot be identified to species as achenes of other *Cirsium* species are similar.

Life form and ecology *Cirsium palustre* is a tall, up to 2 m, sparsely branched biennial thistle. It grows on damp soil in mires, fens, marshes, damp grassland, rush-pastures, wet woodland, montane springs and flushes, and tall-herb vegetation on mountain ledges. The achenes have a pappus and are dispersed by wind. They can form a long-lived seedbank.

Distribution *Cirsium palustre* is widespread through Europe from northern Norway to southern Spain and extends east through western Asia. It is widely naturalised outside its native range in North America and New Zealand.

Cirsium palustre (L.) Scop. (gbif.org)

5.2.28 *Eupatorium cannabinum* L. Asteraceae

Morphology *Eupatorium cannabinum* achenes are black and oblong or fusiform, 2.5–3.0 mm long, with an angular cross section with 4–5 flat sides and projecting ribs between them. Fresh fruits bear transparent glands. The pappus at the top is not preserved. The pericarp is resistant to degradation, which means that fossil seeds, including small fragments, can be identified using the cell pattern of 'wavy rows of perforations' (Dickson 1970).

Life form and ecology *Eupatorium cannabinum* is a perennial herb up to 1.75 m tall. It grows in marshes, fens, ponds, stream banks and other damp habitats, and occasionally in dry grassland or wasteland. It can grow in shallow water but is usually found on saturated but unsubmerged soils. It prefers nutrient rich, stony, neutral to basic soils. The fruits are primarily dispersed by wind but the pappus enables them to float on water (e.g. Vogt et al. 2004).

Distribution *Eupatorium cannabinum* is widespread throughout Europe, extending to southern Fennoscandia and west-central Asia. It is introduced in some areas of North America and southeast Asia, where it is not invasive.

Quaternary history *Eupatorium cannabinum* achenes have been recorded in southern Hungary as part of the infilling succession of a late Holocene oxbow lake (Gałka et al. 2024).

Eupatorium cannabinum L. (gbif.org)

5.2.29 ***Jacobaea aquatica*** (Hill) G.Gaertn., B.Mey. & Scherb. (Syn. *Senecio aquaticus* Hill) Asteraceae

Morphology The cylindrical achenes of *Jacobaea aquatica* are 1.4–1.7 mm long and about 0.5 mm wide. They have longitudinal rather indistinct ribs. The surface is covered with short hairs. The attachment scar of the pappus is well marked. The achenes of *Jacobaea paludosa* (L.) G.Gaertn., B.Mey. & Scherb. (*Senecio paludosus*) are similar but larger at 2.5–3.2 mm long and glabrous. Those of *Tephroseris crispa* (Waldst. ex Kit.) DC. (*Senecio rivularis)* are similar but even longer, at about 4 mm.

Life form and ecology *Jacobaea aquatica* is a biennial herb, 25–60 cm tall. It grows in fen ditches, moist grazed meadows and pastures, and on damp shaded river banks.

Distribution *Jacobaea aquatica* is endemic to Europe. It occurs across central northern Europe, avoiding the Mediterranean area and Scandinavia. It is threatened by habitat destruction and is considered extinct in eastern England.

Senecio aquaticus Hill (gbif.org)

5.2.30 ***Petasites hybridus*** (L.) G.Gaertn., B.Mey. & Scherb. Asteraceae

Morphology *Petasites hybridus* achenes are about 2.0–2.5 mm long and 0.5 mm wide. They are longitudinally ribbed and somewhat flattened, with a well-marked pappus attachment.

Life form and ecology *Petasites hybridus* is a rhizomatous perennial that can form large colonies on river and stream sides, in wet meadows, and in swampy hollows in riparian woodland. It grows on calcareous or siliceous soils with neutral pH. It is nitrophilic and can become a pest in damp cultivated areas.

Distribution *Petasites hybridus* occurs across temperate northern Europe, in the lowlands of Scandinavia south of the Arctic Circle, and around the Black Sea, becoming sparse in Iberia and European Russia. It is an introduced invasive plant in eastern North America.

Petasites hybridus (L.) G.Gaertn., B.Mey. & Scherb. (gbif.org)

5.2.31 ***Sonchus palustris*** L. Asteraceae

Morphology The pale brown achenes of *Sonchus palustris* are about 4.0 mm long and about 1.5 mm wide. They have four prominent ribs, making them squarish in cross section. There are one to two minor ribs between the major ones. The surface is wrinkly.

Life form and ecology *Sonchus palustris* is a short-lived perennial herb, 1–3 m tall. It is found in tall vegetation on wet, nitrogen-rich, open soil in grazing meadows, wet woods, shoreline swamps, and eutrophic reed swamps.

Distribution *Sonchus palustris* occurs in eastern England (Plant Atlas 2020) and in temperate northern Europe reaching Denmark and the west Swedish coast and western Baltic Sea. It occurs across Russia and Central Asia. It is naturalised in Canada.

Sonchus palustris L. (gbif.org)

5.2.32 *Impatiens glandulifera* Royle Balsaminaceae

Morphology The seeds of *Impatiens glandulifera* are broadly elliptic, obovoid to globose. The apex is obtuse or indented and the base is constricted into the short stalk. There is a longitudinal suture down the ventral side. The seeds are large, about 4 mm long and wide. The surface cells are small, forming prominently rounded tubercles, giving a characteristic surface.

Life form and ecology *Impatiens glandulifera* is an annual that grows up to 2 m tall. Explosive seed pods disperse the ripe seeds up to 7 m from the plant. It colonises river and stream banks and damp woods where it frequently out-competes the native vegetation, resulting in bank erosion when the plants die in winter. It is best controlled by pulling up or cutting before flowering.

Distribution *Impatiens glandulifera* is native to the Indian Himalayas but it has been introduced and has spread invasively throughout the Northern Hemisphere and New Zealand.

Impatiens glandulifera Royle (gbif.org)

5.2.33 *Alnus glutinosa* (L.) Gaertn. Betulaceae

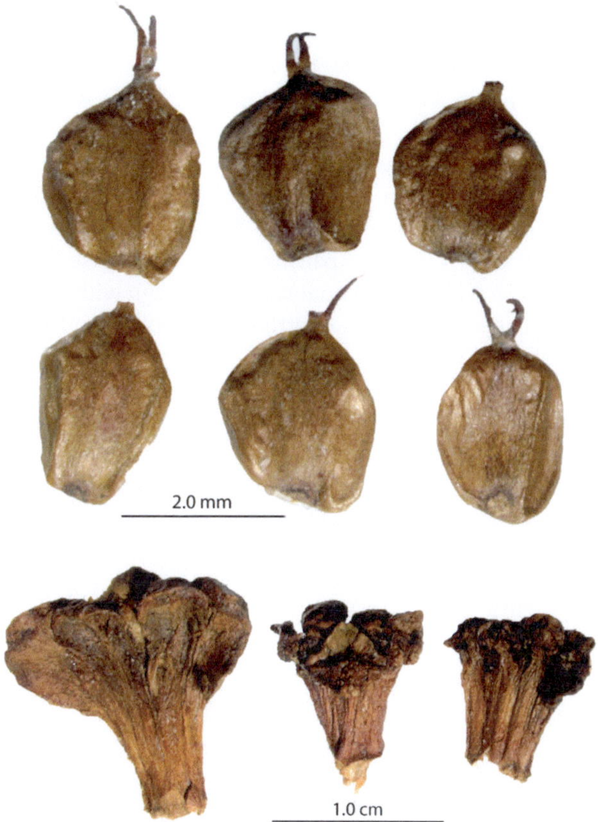

Morphology The flattened nutlets of *Alnus glutinosa* are up to about 2 mm long and wide and are variable in shape. They have distinctive corky 'wings' on each side that are adaptations to floating on water. The two stigmas are characteristic of Betulaceae, and they or their position are usually preserved at the top. The nutlets are subtended by cone scales (or bracts). These are woody and variable in shape, but always widest at the top. Sometimes intact cones are recovered. Wood remains have been identified from *Alnus* fen peat and from archaeological sites.

Life form and ecology *Alnus glutinosa* is a small relatively short-lived (about 60 years) tree often branched from the base. It bears male and female catkins, and is wind pollinated. The nutlets are released from the female cones in autumn. They are adapted for floating and are dispersed by water. *A. glutinosa* grows in wet, mildly basic habitats, usually along streams or at lake shores, but occasionally in low-lying wet hollows, growing on mesotrophic, moderately nutrient rich silt or gleyed soil, or mesotrophic peat. It can withstand drying out if the deep roots can reach water. Bacteria (*Actinomyces alni*) in nodules on the roots fix atmospheric nitrogen. McVean (1953) provides an extensive biological flora of *A. glutinosa.*

Distribution *Alnus glutinosa* is widespread throughout temperate Europe and western Asia extending to about 70°N, but rarer in the Mediterranean region. It is introduced in eastern North America, South America, southeast Australia, and New Zealand.

Quaternary history *Alnus glutinosa* and *A. incana* spread through Europe during the Holocene, probably from several glacial refugia in southern and western Europe. Scandinavia was probably colonized from a single refugium in northeast Europe (Douda et al. 2014). *Alnus* expansion was characterised by a synchronous increase of *Alnus* pollen within individual regions, though with regional differences in the times of the expansion, probably controlled by habitat availability. It expanded greatly in the British Isles in the mid Holocene (e.g. Godwin 1975; Tallantire 1992) as more suitable wet habitats became available due to increased precipitation and sea-level rise (Bennett 1990). Tallantire (1972) attempted to reconstruct the Holocene history of both *A. glutinosa* and *A. incana* in Fennoscandia. A late-glacial interstadial (Allerød) record of wood has been interpreted as glacial survival in North Yorkshire, UK (Young et al. 2021).

Alnus glutinosa (L.) Gaertn. (gbif.org)

5.2.34 *Alnus incana* (L.) Moench. Betulaceae

Morphology The flattened nutlets of *Alnus incana* are about 2 mm long and wide. They resemble those of *Alnus glutinosa* and are just as variable in shape, although they may be slightly larger. The distinctive corky 'wings' on each side are adaptations to floating. The two characteristic stigmas or their position, are usually preserved at the top. The fruits are subtended by woody bracts that form a cone, similar to *A. glutinosa*. Tallantire (1973) and others have separated fruits of *A. glutinosa* and *A. incana* but give no diagnostic characters. We are unable to differentiate the species using our own reference material.

Life form and ecology *A. incana* is a small, relatively short-lived (60–100 years) tree often multi-stemmed and reproducing at 10–20 years of age. It grows on wet nutrient-rich soils in floodplain forest, on wet mountain slopes up to 1500 m, lake and stream shores, and wet roadsides. It can also colonise screes and shallow stony slopes with subsurface water. *A. incana* is able to spread by clonal growth.

Distribution *Alnus incana* is a boreal montane tree. In northern Europe, *A. incana* ssp. *incana* covers the east Baltic region and all of Scandinavia reaching the northern treeline at greater than 70°N. It continues eastwards across European Russia to western Siberia. It is more patchy in the southern European mountains. *A. incana* ssp. *rugosa* occurs across eastern boreal North America, and *A. incana* ssp. *tenuifolia* occurs in northwest North America.

Quaternary history Tallantire (1972) attempted to reconstruct the Holocene history of both *A. glutinosa* and *A. incana* in Fennoscandia.

Alnus incana (L.) Moench (gbif.org)

5.2.35 *Myosotis scorpioides* L. Boraginaceae

Morphology The shiny, ovoid nutlets of *Myosotis scorpioides* are about 1.5 mm long and 1.0 mm wide and have an obscure border. They are obtuse at the top and have a marked attachment area at the base. The nutlets of *M. sylvatica* are similar but are a little larger, bordered, and have a keel on one face.

Life form and ecology *Myosotis scorpioides* is a rhizomatous and stoloniferous perennial that grows on fertile soils in wet meadows and in marshes and fens by streams and ponds. It can sometimes form mats in shallow water.

Distribution *Myosotis scorpioides* is native across Europe to Siberia and the western Himalaya. It is widely naturalized elsewhere, particularly in eastern North America.

Quaternary history Records of *Myosotis scorpioides* nutlets are rare (Godwin 1975).

Myosotis scorpioides L. (gbif.org)

5.2.36 *Nasturtium microphyllum* (Boenn. ex Rchb.) Rchb. Brassicaceae

Morphology *Nasturtium microphyllum* seeds are about 1 mm long, slightly smaller than those of *N. officinale.* They are irregularly round. The radicle and cotyledons are about equal in length. The surface is covered in a reticulum, which is much finer than that of *N. officinale.*

Life form and ecology *Nasturtium microphyllum* is a perennial watercress, forming patches rooting at the nodes. It is very similar to *N. officinale.* It grows at the edges of shallow, still to flowing, mildly acidic to base-rich lakes, rivers, canals, and ditches. It is cultivated as watercress.

Distribution *Nasturtium microphyllum* tends to occur further north than *N. officinale* although it is confined to Denmark in Scandinavia. It is widespread in Europe but less common in the south and extends eastwards to Poland. It is cultivated and naturalised in Africa, boreal North America, southeast Australia, and New Zealand.

Nasturtium microphyllum (Boenn. ex Rchb.) Rchb. (gbif.org)

5.2.37 ***Nasturtium officinale*** W.T.Aiton (Syn. *Rorippa nasturtium-aquaticum*) Brassicaceae

Morphology The rounded seeds of *Nasturtium officinale* are about 1 mm long. The radicle is slightly longer than the cotyledons and curls round them. The surface is covered with a coarse reticulum. Seeds of *N. microphyllum* are similar but are slightly smaller and have a much finer reticulum (Clapham et al. 1962; see photomicrographs in Stace 2019, p. 424 and Schou et al. 2023 p. 559).

Life form and ecology *Nasturtium officinale* is a rhizomatous perennial (watercress) that grows in shallow water and on wet ground associated with lowland (<300 m) marshes, ponds, lakes, ditches, and rivers (especially smaller streams). It is highly versatile in terms of substrate (although it does not favour peat) and water chemistry and can occur in both eutrophic and relatively nutrient-poor water, as well as highly calcareous and mildly acidic situations (Howard and Lyon 1952; Preston and Croft 1997). *N. officinale* seeds are abundantly produced. Fresh seeds float and can form a short-lived seed bank (Howard and Lyon 1952).

Distribution *Nasturtium officinale* has a near world-wide distribution. It is found throughout most of Europe north to Sweden (although it is absent from Norway, Iceland, and Finland) south to the entire Mediterranean region, and it is also widespread in Asia extending along the Himalaya through China to Japan. It is thought to have been introduced (largely for cultivation) to North America, South Africa, Australia, and New Zealand (Howard and Lyon 1952).

Quaternary history Seeds of N. *officinale* have been recorded in the late Holocene in Britain (Godwin 1975). Seeds of *N. microphyllum* have been tentatively recorded in the Irish Late-Glacial (Godwin 1975).

Nasturtium officinale R.Br. (gbif.org)

5.2.38 *Rorippa amphibia* (L.) Besser Brassicaceae

Morphology The seeds of *Rorippa amphibia* have the typical Brassicaceae structure, but this is almost hidden beneath a thick reticulate testa which is whitish in colour. However, this wears away with age and fossilization. The seeds are about 1.0 mm in diameter and irregularly angular.

Life form and ecology *Rorippa amphibia* is a rhizomatous perennial about 1 m tall. It grows at the edges of calcareous, typically eutrophic, freshwater habitats such as rivers, lakes, reservoirs, floodplain wetlands, ponds, and ditches. It is tolerant of water-logged situations where high winter water levels recede in spring-summer to leave damp marginal areas, although it can also live under well-drained conditions (Stift et al. 2008). *R. amphibia* produces abundant seeds.

Distribution *Rorippa amphibia* is widespread across the temperate and boreal Eurasia, decreasing to the north and towards the Mediterranean. It is introduced in eastern North America, China (Zhang et al. 2009), and New Zealand (Garnock-Jones 1978).

Rorippa amphibia (L.) Besser (gbif.org)

5.2.39 ***Rorippa islandica*** (Oeder) Borbás and ***Rorippa palustris*** (L.) Besser Brassicaceae

Rorippa islandica seeds

Rorippa palustris seeds

Morphology The rounded seeds of *Rorippa islandica* are about 0.8 mm long. The radicle and cotyledons are about the same length. The surface is distinctive, covered by regular lines of papillose cells. The seeds of *Rorippa palustris* are very similar, and not distinguishable from those of *R. islandica*. These two species are very similar and have only been differentiated since 1968 (Plant Atlas 2020).

Life form and ecology *Rorippa islandica* is an annual or biennial herb up to 60 cm tall. It occurs in open muddy wet meadows on clay soils, especially where water stands in winter, by ditches and lake and pond shores, and in damp weedy habitats. *R. palustris* is an annual, up to 60 cm tall. It is commonly found along ditches, on mud by ponds that dry up in summer, in open, nutrient-rich meadows, and in damp weedy places. It forms a long-lived seed bank (Plant Atlas 2020).

Distribution *Rorippa islandica* has a boreal-montane distribution in Europe, Asia, and North America. It is scattered in north and west Britain. *R. palustris* has a more southerly continuous circumpolar boreal distribution between 50° and 60° N (Klimešová et al. 2004). In Europe, it becomes less frequent in northern Fennoscandia and Siberia and in the Mediterranean region. *R. palustris* is naturalised in parts of Central and South America and Australia. It is regarded as native in New Zealand (Garnock-Jones 1978).

Quaternary history *Rorippa islandica* seeds have been recorded in the Late-Weichselian and Early Holocene in Britain (Godwin 1975). *R. palustris* seeds were recorded in Early Holocene peats under the North Sea (Brown et al. 2018).

Rorippa islandica (Oeder) Borbás (gbif.org)

Rorippa palustris (L.) Besser (gbif.org)

5.2.40 *Rorippa sylvestris* (L.) Besser Brassicaceae

Morphology The rounded seeds of *Rorippa sylvestris* are about 0.8 mm long. The radicle and cotyledons are about the same length. The surface bears regular lines of papillose cells. These seeds can hardly be distinguished from those of *R. islandica* and *R. palustris*.

Life form and ecology *Rorippa sylvestris* is perennial herb up to 50 cm tall. It grows on the margins and banks of rivers, streams, canals and ditches, by lakes and ponds, and in depressions in pastures on damp bare ground, often flooded in winter. It can also be a weed on drier, cultivated ground. It is a vigorous pioneer species, intolerant of competition, spreading by seed and by rhizomes (Plant Atlas 2020).

Distribution *Rorippa sylvestris* is distributed across the temperate Northern Hemisphere and is widely naturalized elsewhere.

Rorippa sylvestris (L.) Besser (gbif.org)

5.2.41 *Subularia aquatica* L. Brassicaceae

Morphology *Subularia aquatica* seeds are typical of Brassicaceae with a curved embryo covered by the testa. The seeds are 0.8–1.0 mm long and about 0.5 mm wide. The cotyledons are wider than the radicle with a rounded end that is usually darker. The radicle is narrower and extends beyond the cotyledons, and usually has a pointed tip, sometimes of a darker colour. The cells are distinct and isodiametric and are particularly visible in fossilised material where the embryo has decayed. The seeds are produced in a typical Brassicaceae pod (Stace 2019). The valves of Brassicaceae pods are sometimes preserved in sediments, but they cannot be identified to species.

Life form and ecology *Subularia aquatica* is a small isoetid most frequently found in shallow water (usually less than 1 m) at the edges of acid oligotrophic lakes. It is an annual (rarely biennial) and reproduces by seed. The rosettes have thin linear leaves that taper to a point, about 10 cm long. *Subularia* does not form a turf but occurs scattered in other isoetid vegetation or on open silt, gravel, or stones.

Distribution *Subularia aquatica* has a circumboreal distribution, although it is rare in Asia. It is common in Fennoscandia, Scotland, and Iceland. It is less frequent in northwest England, Wales, western Ireland, Denmark, and northwest Europe. It is absent from The Alps, but is widespread in the Pyrenees, and it also occurs in central Spain. It is recorded from southern Greenland and through Canada from the maritime provinces across to British Columbia and then westwards into Alaska.

Quaternary history *Subularia aquatica* has been recorded in Late Glacial and Early Holocene sites in Britain (Godwin 1975) and in Early Holocene sediments in western and northern Norway (e.g. Birks and van Dinter 2010), and particularly at Kråkenes, western Norway, where it was a prominent component of an isoetid succession (Birks 2000). *S. aquatica* was an early colonist in newly formed lakes in Greenland (Fredskild 1992).

Subularia aquatica L. (gbif.org)

5.2.42 *Butomus umbellatus* L. Butomaceae

Morphology The seeds of *Butomus umbellatus* are 1.6–2.0 mm long, cylindrical or flattish, and strongly ribbed longitudinally. The wavy ribs can become wing-like. The isodiametic cells are prominent and rounded on their outer faces.

Life form and ecology *Butomus umbellatus* is a rhizomatous thin-leaved perennial rush up to 1.5 m tall. It grows as a usually emergent plant in shallow, nutrient-rich, stagnant water up to 2 m deep, or on banks of still or flowing water where it favours fluctuating water levels (Hroudová et al. 1996). It is typically associated with highly calcareous and eutrophic conditions although it tends to decline under severe eutrophication and shading. The buoyant seeds are dispersed by water. Its tubers are edible.

Distribution *Butomus umbellatus* is widespread through temperate and boreal Eurasia and in North Africa. It is introduced in temperate-boreal North America where it can be highly invasive (Gaskin et al. 2021).

Quaternary history In spite of their characteristic appearance, *B. umbellatus* seeds are seldom recorded from sediments. It was recorded in glacial-period sediments with other reedswamp plants at Lake Zeribar, Iran (Wasylikowa 1967).

Butomus umbellatus L. (gbif.org)

5.2.43 *Lobelia dortmanna* L. Campanulaceae

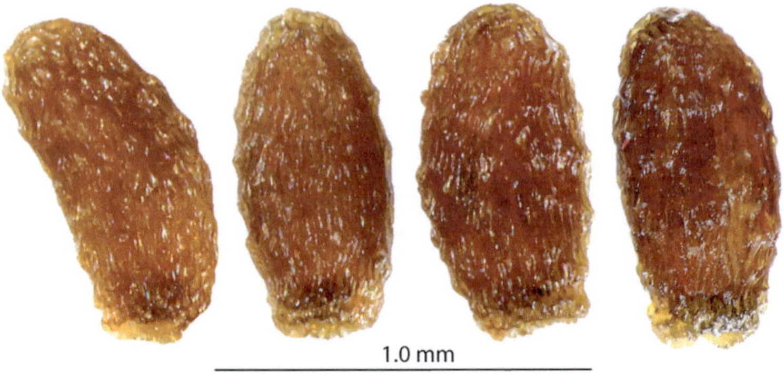

Morphology *Lobelia dortmanna* seeds are very small, less than 1 mm long and about 0.4 mm wide, but very characteristic. The basal attachment is flat and well marked. The cells of the testa are elongated and parallel. Their ends protrude slightly, giving a wavy appearance to the surface. They curve around the rounded top of the seed. (See Lamentowicz and Milecka (2004) for SEM photographs).

Life form and ecology *Lobelia dortmanna* is a tufted submerged isoetid with flat linear leaves up to 10 cm long containing two air canals and recurved at the tips. These bring CO_2 from the sediments and deliver oxygen to them that supports mycorrhizal fungi (Schou et al. 2023). *L dortmanna* is intolerant of shade and has low competitive ability. The flowering spikes emerge above water. Each capsule contains many seeds which sink when released to form a seed bank for up to 30 years (Schou et al. 2023). It grows in shallow water <2 m deep on coarse sand, silt, or organic lake mud where it forms an underwater turf, either alone or with other isoetids such as *Isoetes* spp., *Subularia aquatica*, and *Littorella uniflora,* and other macrophytes. It grows in soft-water lakes with low nutrient concentrations. (See Preston and Croft (1997) for more details). Oligotrophic lakes with this vegetation can be called 'Lobelia lakes' (Lamentowicz and Milecka 2004). *Lobelia* and other isoetids decline as a result of eutrophication (e.g. Salgado et al. 2010). The main reason for the dramatic loss of soft-water Lobelia lakes in Denmark and elsewhere in northwest Europe is overgrowth by larger macrophytes during eutrophication (Nielsen et al. 2023).

Distribution *Lobelia dortmanna* has an amphiatlantic distribution. It is widespread in lowland Fennoscandia south of 70°N. In the British Isles it has a northern and western distribution related to acid-rock geology. It is present throughout Denmark, the Low Countries and northern mainland Europe, but is absent further south. It occurs in temperate eastern Canada and USA north of Massachusetts, and in British Columbia.

Quaternary history *L. dortmanna* seeds are frequently recorded in Holocene sediments from Lobelia lakes in northeast Poland (Lamentowicz and Milecka 2004). They were recorded from Loch Leven in Scotland before eutrophication occurred (Salgado et al. 2010).

Lobelia dortmanna L. (gbif.org)

5.2.44 ***Valeriana officinalis*** L., ***Valeriana excelsa*** ssp. ***sambucifolia*** (J.C.Mikan ex Pohl) Holub., ***Valeriana dioica*** L. Caprifoliaceae (formerly Valerianaceae)

Valeriana excelsa ssp. *sambucifolia* achenes

Valeriana officinalis achenes

Valeriana dioica achenes

Morphology The characteristic achenes of *Valeriana officinalis* and *Valeriana excelsa* ssp. *sambucifolia* are indistinguishable from each other. The former are 2.5–4.0 mm long and the latter are 3.0–3.5 mm long (Mossberg and Stenberg 2018) and thus the sizes overlap. Those of *V. dioica* are similar, but slightly smaller, about 2.5 mm long. *Valeriana* achenes are flattened and obovoid, widest at the base tapering to a narrow apex and surrounded by a marginal rib. The dorsal side has three prominent ribs, and the ventral side has one prominent medial rib. The pappus at the top for wind dispersal of the fruits is not preserved.

Life form and ecology *Valeriana officinalis* is a perennial herb up to 2 m tall. It grows on damp or wet nutrient-rich soil in fens, ditches, lakesides, waste places, and roadsides, *V. excelsa* ssp. *sambucifolia* is similar (indeed is often not distinguished from *V. officinalis*) and grows in similar places but usually on wetter soil (Mossberg and Stenberg 2018). *Valeriana dioica* is a smaller plant of calcareous mires, marshy grassland, water-meadows, base-rich flushes, willow fen-carr, and alder woods (Plant Atlas 2020).

Distribution *Valeriana officinalis* occurs widely across the Northern Hemisphere in boreal and temperate regions. *V. excelsa* ssp. *sambucifolia* has a similar distribution but in Scandinavia is more widespread in Norway and Finland and grows further north than *V. officinalis* (Mossberg and Stenberg 2018). It is uncommon in Sweden. *V. dioica* is widespread in Europe, reaching central Russia, but with its northern limit in southern Sweden and northern England. It occurs widely across North America.

Valeriana officinalis L. (gbif.org)

Valeriana excelsa subsp. sambucifolia (J.C.Mikan ex Pohl) Holub

Valeriana dioica L.

5.2.45 ***Silene flos-cuculi*** L. (Syn. *Lychnis flos-cuculi*) Caryophyllaceae

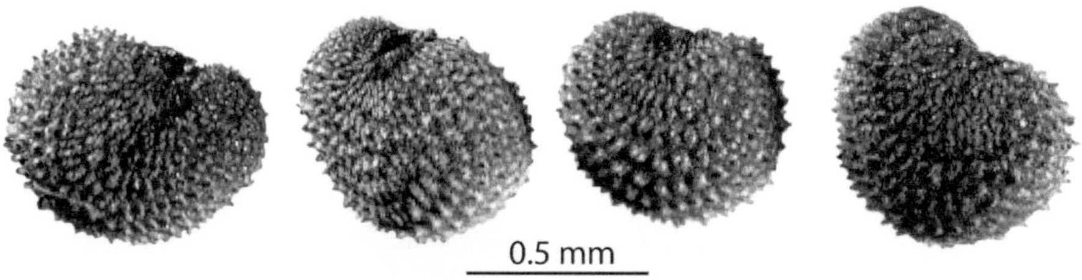

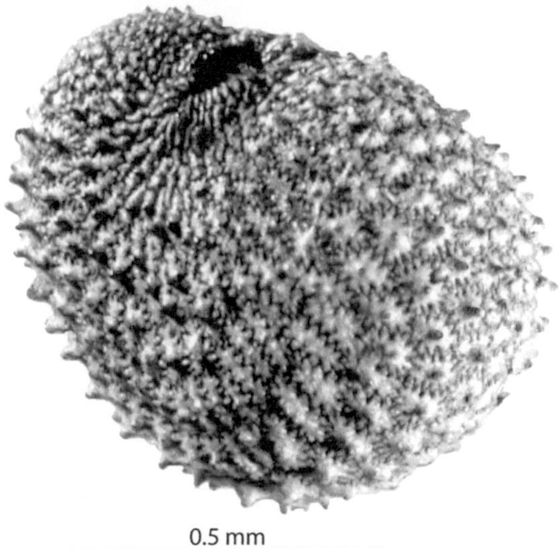

Morphology A capsule of *Silene flos-cuculi* contains many seeds that are dispersed when the capsule is shaken. The seeds are typical of the Caryophyllaceae, with a curved embryo, distinct micropyle, and a characteristic cell pattern. They are about 0.7–1.0 mm long, grey, and bear prominent tubercles, each arising from a cell with a wavy wall, interlocking with adjacent cells. The cell walls are distinct, and the interlocking connections are pointed.

Life form and ecology *Silene flos-cuculi* is a perennial herb up to 0.75 m high. It grows in damp and wet places such as wet meadows, marshes, and wet woodlands. Plants are usually interspersed in vegetation. Seeds can be secondarily spread by water (Hauser and Loeschcke 1994).

Distribution *Silene flos-cuculi* occurs across Europe from Iceland, northern Norway, western Russia, and south to northern Spain, and also in eastern Canada and northeast USA.

Quaternary history *Silene flos-cuculi* seeds have been recorded in Quaternary sediments (e.g. Hulme and Durno 1980; Havinga et al. 1992). They can be distinguished with care from other similar Caryophyllaceae seeds.

Lychnis flos-cuculi L. (gbif.org)

5.2.46 ***Ceratophyllum demersum*** L. and ***Ceratophyllum submersum*** L. Ceratophyllaceae

Ceratophyllum demersum fruit

Ceratophyllum submersum fruits

Ceratophyllum demersum leaf

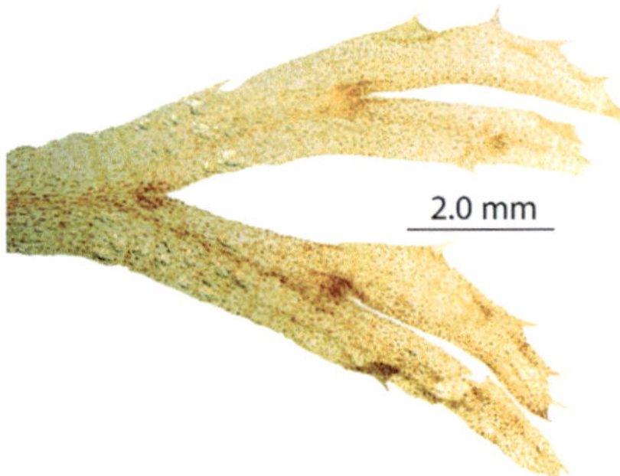

Ceratophyllum submersum leaf

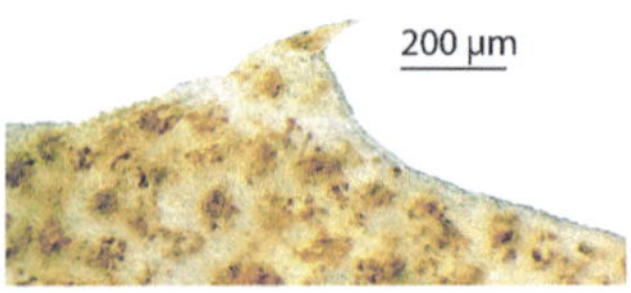

Ceratophyllum demersum spine

Ceratophyllum submersum spine

Morphology *Ceratophyllum* fruits are elliptical, 4–5 mm long excluding the spines, and about 3 mm broad exluding the spines. *C. demersum* fruits usually have three long spines and spine morphology is variable. Spines are short or lacking in *C. submersum.* Fruits are sparsely produced and are most often found as fossils in lake sediments in warm climates.

Leaf fragments are often found in lake sediments and are recognised by their branched tubular shape with clusters of dark cells and marginal spines. The fossil leaves of both species are similar. The individual spines can be found as microfossils on pollen slides.

Life form and ecology The submerged stems with numerous side shoots can reach lengths of 1–3 m, making a single specimen appear as a large, bushy mass, or, in deeper water, like an underwater forest. Leaves are in whorls of six to twelve. Each leaf is 8–40 mm long, usually branched into two to eight thread-like segments bearing spiny teeth; they are stiff and brittle and render the plant rough to the feel. The plants are wholly submerged elodeids, lacking roots, and with submerged flowers and fruits. *C. demersum* grows in shallow (10–30 cm) to deep (3 m) eutrophic base-rich lowland lakes or slow-flowing streams and rivers, with water of pH 6–7, usually with high conductivity and nutrient content. *C. submersum* tends to grow in shallower water and can tolerate slightly brackish conditions. *C. demersum* is often a pioneer in new water bodies and both species can form extensive vegetation. In lakes impacted by human eutrophication, such as by the addition of sewage or agricultural fertilizers, *C. demersum* can become an aggressive weed, reducing the biodiversity of the other aquatic flora (e.g. Clarke et al. 2014) and impeding water flow.

Distribution *C. demersum* has a cosmopolitan distribution, but is frequently introduced, becoming a pest in New Zealand. *C. submersum* is predominantly Eurasian but occurs worldwide except for North America. However, the probably synonymous *C. muricatum* is native in southeast USA and the Caribbean. *C. demersum* is widespread in North America and grows further north in Europe (to 70°N in Norway) than *C. submersum,* which prefers summer-warm habitats and extends to nearly 60°N in Sweden.

Quaternary history *Ceratophyllum* has been recorded in deposits of Eocene and Palaeocene age and in all interglacial deposits in the British Isles since the Cromerian (Godwin 1975). Its fruits are infrequently produced so it is underrepresented. It seems that *Ceratophyllum* fruits readily when it is a pioneer in new lakes rich in nutrients, either after glacial retreat in the Early Holocene, or in new man-made lakes. It frequently plays a role in Early Holocene successions in eutrophic lakes throughout Europe, the Middle East, and North America (e.g. Watts 1978; Gałka et al. 2024). Fruits occur in the earliest Holocene deposits but with time *Ceratophyllum* ceases to fruit and it is only represented by its leaf remains, demonstrating its vegetative persistence (Watts 1978; Birks 1980). In the mid-Holocene *Ceratophyllum* expanded north of its present range to 66°N, especially up the Gulf of Bothnia (Lang 1994). Its presence is often evidence of eutrophication in recent lake sediments,

Ceratophyllum demersum L.

Ceratophyllum submersum L.

5.2.47 **Characeae** S.F.Gray

Characeae, often called Charophytes, are fresh- or brackish-water macroalgae. They are elodeids and can form underwater 'meadows'. These are important habitats for aquatic organisms such as diatoms and invertebrates (e.g. Rodrigo et al. 2021). The female reproductive organs are oogonia, which produce oospores. The oospores are readily preserved in sediments and can be very abundant. They are less than 1 mm long and have a characteristic shape, mostly longer than broad, and consist of five fused spiral membrane rows that form ridges (Haas 1994).

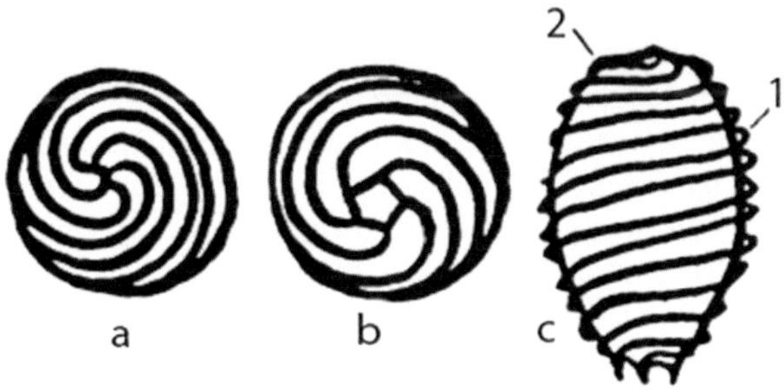

Generalised diagram of a charophyte oospore free from lime encrustation: (a) apical view (distal end); (b) basal view (proximal end); (c) lateral view. 1, ridge; 2, shoulder. Modified from Haas (1994)

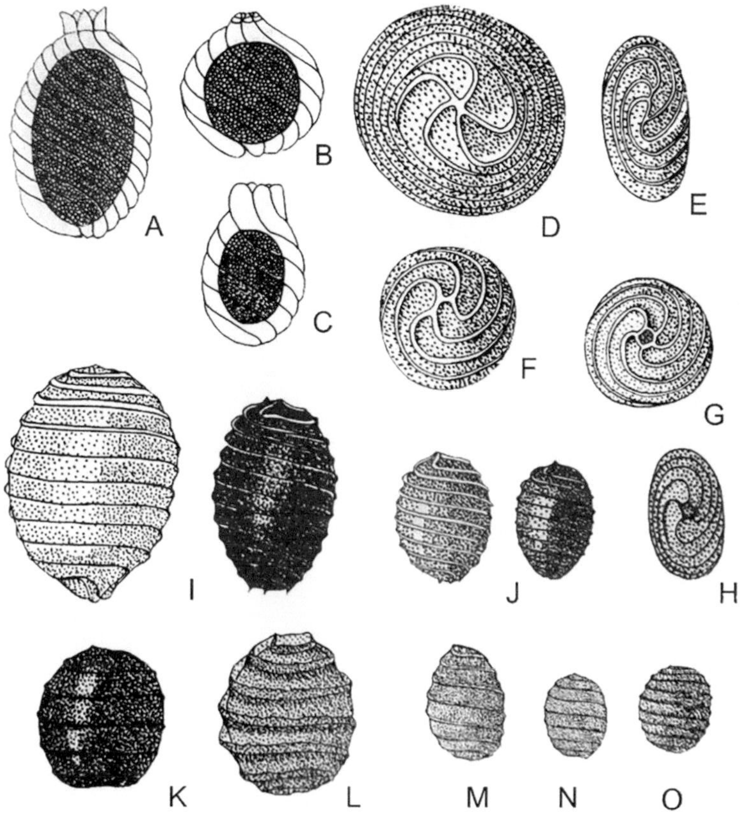

Characeae oospores. A *Chara*, B *Nitella*, crown remaining, C *Nitella*, crown fallen off, D *Chara*, E *Nitella*, F *Tolypella*. The pattern of the ridges is not constant. Oospore base: G *Chara*, H *Nitella*. Oospore size and shape: I *Chara hispida*, with and without calcareous encrustation, J The same thing for *Chara vulgaris,* L. Uncalcified oospores: K *Nitella syncarpa* (Thuillier) Chevalier, L *N. capillaris* (Krocker) J. Groves et Bullock-Webster, the broad ribs are winged, M *N. mucronata* (A. Braun) Miquel. N *N. tenuissima* (Desvaux) Kützing, O *Tolypella glomerata* (Desvaux in Loiseleur-Deslongehchamps) Leonhardi. A-C, I-O x 30, D-H x 40. Modified from Krause (1997)

The Characeae consist of 2 subfamilies, Nitelleae, containing *Nitella*, and Chareae, containing *Chara*, *Tolypella*, *Nitellopsis*, and *Lycnothamnus* in northwest Europe. The majority of fossil records are of *Nitella* and *Chara*. The subfamilies have distinct oospores. Those of *Nitella* are almost square in front view, flattened rather than round in side view, and with fewer than ten spirals and prominent ridges. Those of Chareae are longer than broad, 'egg' or 'cigar' shaped sometimes with distinct shoulders at the top. There are many spiral ridges in lateral view, and the ridges are not so prominent as in *Nitella* and can be almost lacking. They are difficult to identify to species, as there is a lot of variation. A key and descriptions of the oospores are presented by Haas (1994), and illustrations and descriptions by Krause (1986, 1997) and Blindow and Werner (1990). Depending on the research question, an identification to *Chara* or *Nitella* is usually sufficient. As shown in the table below, *Chara* spp. are characteristic of calcareous base-rich water and often have carbonate encrustation. *Nitella* spp. are characteristic of softer, more acid waters.

pH-values	5.6	5.8	6.0	6.2	6.4	6.6	6.8	7.0	7.2	7.4	7.6	7.8	8.0	8.2	8.4	8.6
Species																
Chara aspera								+	+	+	+	+	+	+	+	
Chara braunii							+	+	+	+	+					
Chara contraria								+	+	+	+	+	+	+	+	
Chara globularis group							+	+	+	+	+	+	+	+		
Chara hispida group								+	+	+	+	+	+	+	+	
Chara strigosa									+	+	+	+	+	+		
Chara tenuispina									+	+	+	+	+			
Chara tomentosa									+	+	+	+	+	+		
Chara vulgaris								+	+	+	+	+	+	+	+	+
Lychnothamnus barbatus					+	+	+	+	+	+	+	+	+	+	+	+
Nitella capillaris			+	+	+	+	+	+	+							
Nitella confervacea						+	+	+	+	+	+	+				
Nitella flexilis			+	+	+	+	+	+	+	+	+	+	+			
Nitella gracilis			+	+	+	+	+	+								
Nitella hyalina						+	+	+	+	+	+	+	+			
Nitella rnucronata						+	+	+	+	+	+					
Nitella opaca			+	+	+	+	+	+	+	+	+					
Nitella syncarpa		+	+	+	+	+	+	+	+	+	+	+	+			
Nitella tenuissirna					+	+	+	+	+	+	+	+	+			
Nitella translucens			+	+	+	+	+	+	+	+						
Nitellopsis obtusa								+	+	+	+	+	+	+	+	
Tolypella glomerata									+	+	+	+	+	+	+	
Tolypella intricata										+	+	+				
Tolypella prolifera								+	+	+	+	+	+	+		

pH ranges of central European charophyte species. Optimal pH values (if known) are marked by shading (from Haas 1994).

Keys to living species of Charophytes in Europe are given by Krause (1997) and in Sweden by Blindow and Werner (1990) with excellent illustrations, including oospores. Moore (1986) for the British species gives keys to plants but little information on oospores. Langangen (2007) gives especially good descriptions of habitats and distributions for the Fennoscandian species and a comprehensive list of literature references. Langangen (1974) gives detailed ecological notes (lake type, preferred water depth, bottom sediments, specific conductivity, pH, water colour, chloride ions, calcium ions) about Norwegian species and their distribution in Norway. A comprehensive volume 'Charophytes of Europe' (Schubert et al. 2024) contains detailed information on Charophyte taxonomy, ecology, and distributions, with excellent SEM illustrations of the oospores and a key to the whole plants.

Here we shall take the common species *Chara virgata* in the *C. globularis* group as an example of typical *Chara* oospores. As an example of *Nitella* oospores, we illustrate some subfossil material.

5.2.47.1 *Chara virgata* Kütz. (Syn. *Chara globularis* var. *virgata* (Kütz.) R.D.Wood) Characeae

Morphology The oospores are ovoid to elliptical, with 11 or more ridges, and no "shoulder-like" structure below the apical pole (Haas 1994). Because they are similar to *Chara globularis* and other species (Haas 1994), the oospores of *Chara virgata* are usually grouped together as *Chara globularis*-type. They are black or brown, approximately 0.62–0.77 mm long and 0.46–0.55 mm wide.

Life form and ecology *Chara virgata* is a perennial or annual elodeid, usually about 30 cm long. It grows in brackish or fresh shallow water in ponds, lakes, and streams, on substrates including peat, marl, and sand. It grows in soft or calcareous water, often rich in electrolytes. Although they occur most frequently in shallow water up to 5 m deep at the margins of lakes, *Chara* spp. can grow at depths of 10 m or more in clear water (Langangen 1974). The oospores are dispersed by water currents and waterbirds and can be found in abundance at least 150 m away from living plants (Birks 1973). They can be abundant in sediments under water up to 7.5 m deep (Dieffenbacher-Krall and Halteman 2000).

Distribution *Chara virgata* is found in throughout Europe and has a cosmopolitan distribution.

Quaternary history *Chara* species or Characeae oospores are frequently recorded and can be abundant in Quaternary sediments (e.g. Wasylikowa et al. 2006; Birks 2001). They are seldom identified to species (e.g. Zalat 1996). This is difficult and the ecological information gained can be slight, as most *Chara* spp. have similar ecological requirements.

Chara virgata Kütz (gbif.org)

5.2.47.2 *Nitella* C.Agardh. Characeae

Nitella oospores 'large', probably *N. flexilis* from Kråkenes Lake, western Norway

Nitella oospores 'small', probably *N. opaca* group, from Kråkenes Lake, western Norway

Morphology The oospores of *Nitella* are about as broad as long and are laterally flattened. They have 10 or fewer spirals and the ridges are prominent. At Kråkenes, western Norway, two sizes of *Nitella* oospores were recovered (Birks 2000), shown in the photographs above. The large ones are probably *N. flexilis*. The small ones are in the *N. opaca* group (see Fig. 4 in Krause 1986).

Life form and ecology *Nitella* spp. are elodeids, about 30 cm long. They typically grow in low productivity 'Lobelia lakes' characterised by isoetid vegetation (Langangen 2007). The water is often brown due to dissolved organic carbon, with a pH of about 7 or less and low in electrolytes. Although they usually grow in shallow water, they can grow at depths of up to 20 m in clear-water lakes (Haas 1994; Langangen 1974). The oospores are dispersed by water currents and waterbirds. *Nitella* oospores can be abundant up to 150 m away from living plants (Birks 1973) and in sediments under water up to 7.5 m deep (Dieffenbacher-Krall and Halteman 2000). *Nitella* spp. do not occur above the treeline in Scandinavia.

Distribution Both *Nitella flexilis* and *N. opaca* have cosmopolitan distributions and are common throughout Europe.

Quaternary history There are many records of Characeae and *Nitella* oospores from Quaternary sediments but very few instances of named species. Birks et al. (2001) recorded *N. opaca* from a Tunisian lake.

Nitella C.Agardh, 1824 (gbif.org)

5.2.48 *Crassula aquatica* (L.) Schönl. Crassulaceae

Morphology The tiny seeds of *Crassula aquatica* are about 0.4 mm long and 0.2 mm wide. The cells of the testa are arranged in a longitudinally elongated reticulum with raised, slightly wavy margins and sunken centres. Each centre has a faint ladder pattern.

Life form and ecology *Crassula aquatica* is a small, 2–6 cm tall, procumbent, slightly succulent annual, with creeping stems forming inconspicuous lax to dense patches. It photosynthesises using Crassulacean Acid Metabolism. It grows on bare sandy soil or wet mud, sometimes submerged up to 25 cm deep, at the edges of mesotrophic freshwater to brackish or saline lakes, ponds, seashores and brackish pools (Preston and Croft 1997; Mossberg and Stenberg 2018).

Distribution *Crassula aquatica* is widespread in southern Finland, southern Sweden and all along the Swedish coast and less commonly at the Norwegian coast up to the Arctic Circle. It has a circumboreal disjunct distribution. There is one extant site in Scotland, and it is rare in continental northwest mainland Europe and Iceland. It is rare in eastern Asia. It occurs across temperate and boreal North America.

Quaternary history *Crassula aquatic* seeds are recorded in an Early Weichselian interstadial in northwest Finland where, with other taxa, they provide evidence of a warmer climate there than today (Väliranta et al. 2009).

Crassula aquatica (L.) Schönl. (gbif.org)

5.2.49 ***Bolboschoenus maritimus*** (L.) Palla (Syn. *Scirpus maritimus* L.) Cyperaceae

Morphology The shiny nutlets of *Bolboschoenus maritimus* are 2.5–4.0 mm long and 1.5–2.4 mm wide. They are broadly obovate, tending to be trigonous below and biconvex above with rounded edges and a rounded apex with a persistent style base. They are smooth, with a distinct polygonal network of cells with depressed margins.

Life form and ecology *Bolboschoenus maritimus* is a rhizomatous sedge-like perennial that grows up to 1.2 m tall. It can form dense stands in coastal habitats on wet silty substrates in both saline and brackish conditions. It can also grow inland in the upper reaches of tidal rivers where silt is deposited, and in brackish or saline habitats, such as wet marshy flats, seasonal and permanent wetlands, and lake and pond margins (Hroudová et al. 2007).

Distribution *Bolboschoenus maritimus* is cosmopolitan in coastal and inland regions but avoids the arctic and boreal zones, reaching southern Norway, Sweden, and Finland. It is absent from Iceland and Greenland.

Scirpus maritimus L. (gbif.org)

5.2.50 *Carex* L. Cyperaceae

Carex L. is one of the largest genera of flowering plants. Although they occur in most habitats, *Carex* spp. are particularly prominent in wetlands. As a consequence, their remains are frequently found in lake and mire sediments. We only consider here selected wetland species of northwest Europe as the remains of these are most likely to be found in lake and mire sediments here.

Morphology *Carex* nutlets (achenes) are enclosed in a utricle (perigynium), which can be tightly adherent or loose (inflated). This allows the fruits to float until the utricle decays. Fruits of many wetland *Carex* species can float for several months (Praeger 1913), allowing for wide dispersal. The utricle usually decays during fossilisation, although it can be preserved in favourable conditions. Utricles can provide useful characters, particularly the number of veins visible or surface hairs, but they are not usually diagnostic in fossil condition. The inflated utricles of *Carex rostrata* are sometimes preserved separately from the fruits and can be identified by the 12–20 narrow nerves (Dickson 1970).

Carex fruits (nutlets or achenes) are biconvex (lenticular) or three-angled (trigonous) depending on how many styles there are in the flower. The endosperm is usually lost during fossilisation, and the fruit walls can become flattened and sometimes translucent and ghost-like. The surface of many fossil achenes bears a distinctive cell pattern of square or hexagonal cells that can best be seen with overhead light under a stereomicroscope when the nutlets are almost dry. Sometimes a long beak or style is preserved (e.g. *C. rostrata, C. vesicaria, C. pseudocyperus*) but most often only the style base is present, if at all.

Carex nutlets can be identified to species or groups of species on their size and shape. Nilsson and Hjelmqvist (1967) provide a comprehensive key, descriptions, and drawings of the south Scandinavian species of *Carex*. Berggren (1964) also provides keys to the northwest-European (Fennoscandia and Iceland) Sections within the Genus and to the species within the Sections, and an index to nutlet shapes. The species are well illustrated by photographs. The study of subgenus *Phacocystis* (biconvex achenes) by Jiménez-Majías and Martinetto (2013) provides diagnostic tables and photographs and concludes that wall thickness, cell size, outline, style, and base characteristics are sufficient to assign a fossil to a morphological type. The photographic guide to Sedges of the Northern Forest (Jenkins 2019) contains images of North American *Carex* utricles and some nutlets, and the naked nutlets of other genera in the Cyperaceae.

The literature below has been used to provide information on the morphology of the *Carex* spp. considered here.

Berggren G (1964) Atlas of seeds and small fruits of Northwest-European plant species with morphological descriptions. Part 2 Cyperaceae. The Swedish Natural Science Research Council, Stockholm.

Dickson CA (1970) The study of plant macrofossils in British Quaternary deposits. In: D Wlaker and RG West (eds) Studies in the Vegetational History of the British Isles. Cambridge University Press, Cambridge, pp. 233–254.

Jenkins J (2019) Sedges of the Northern Forest. A photographic guide. Cornell University Press, Ithaca.

Jiménez-Majías P, Martinetto E (2013) Toward an accurate taxonomic interpretation of *Carex* fossil fruits (Cyperaceae): a case study in Section *Phacocystis* in the western palaeartic. American Journal of Botany 100: 1580–1603.

Nilsson Ö, Hjelmqvist H (1967) Studies on the nutlet structure of south Scandinavia species of *Carex*. Botaniska Notiser 120: 460–485.

Quaternary history *Carex* achenes are found frequently in lake and mire sediments. Usually they are reported as 'lenticular' or 'trigonous' or comparable terms. Godwin (1975) gives an account of many species that have been identified through the Quaternary, including interglacials, the last glacial period, (Weichselian), and particularly in the Holocene in the British Isles. Many records have been made throughout northwest Europe since that time, but there is no comprehensive compilation available.

5.2.51 *Carex acuta* L. Cyperaceae

Carex acuta achenes inside their utricles

Achenes of *Carex acuta* showing the shape variability

Single achene showing distinct cells

Morphology The utricles are loose-fitting and are lost on fossilisation. The biconvex nutlets are about 2 mm long are and are variable in shape being elliptic or broadly obovate to almost round. They have a broad base with a callus. The surface cells are distinct but with unclear cell walls and concave lumina.

Life form and ecology *Carex acuta*, is a tufted rhizomatous perennial, fine-leaved sedge 60–90 cm tall. It grows at the margins of rivers and lakes in water up to 15 cm deep and in unimproved water meadows and marshes, usually on calcareous or neutral mineral soil. It is intolerant of prolonged soil desiccation.

Distribution *Carex acuta* has a temperate-Eurasian distribution, especially in central Europe and southern Scandinavia. It becomes less common in the north of Britain and Scandinavia. It is absent from Iceland and Greenland.

Carex acuta L. (gbif.org)

5.2.52 ***Carex appropinquata*** Schumach. Cyperaceae

Morphology The lenticular nutlets are turbinate, broadest in the middle and rather variable in shape. They are about 1.0–1.6 mm long. The utricles are tight-fitting. They are strongly ribbed and have toothed apices surrounding the stigmas.

Life form and ecology *Carex appropinquata* is a tussock-forming perennial sedge about 70 cm tall. Its main habitat is open lowland fenland. It also occurs in *Salix*-carr where however, it is vulnerable to shading and drying out as it requires moist soil. In western Europe it prefers calcareous fens but has greater acid-tolerance in more continental areas.

Distribution *C. appropinquata* is widespread in central and eastern Europe, extending north to northern Scandinavia and eastward to Siberia. It is rare in Britain.

Carex appropinquata Schumach. (gbif.org)

5.2.53 *Carex aquatilis* Wahlenb. Cyperaceae

Morphology *Carex aquatilis* nutlets are lenticular. The sides slope gently up from the base so the nutlets are parallel sided or slightly obovate. They about 1.3–1.7 mm long and 1.3 mm wide. They have a distinct hexagonal cell pattern.

Life form and ecology *C. aquatilis* is a perennial rhizomatous sedge, often forming dense turf, 15–100 cm tall. Aerenchyma in the roots allows it to survive in low-oxygen substrates like heavy mud and to withstand 1 to 2 months of submersion. It occurs in arctic tundra in unshaded wet and boggy meadows, stream banks, and at pond and lake margins as monocultures, in mixed poor-fen vegetation, and in open low *Salix* scrub. It colonises open disturbed areas. On northern mountains, it grows on poorly drained moist areas and seepage slopes and often grows on deep, wet, gently sloping peat.

Distribution *C. aquatilis* is a circumpolar boreo-arctic montane species. It occurs in northern Britain and in Ireland, and northern Scandinavia, Siberia, northern North America (Alaska, Canadian arctic, Rocky Mountains), and on Greenland. It is absent from Iceland and the mountains of central Europe.

Carex aquatilis Wahlenb. (gbif.org)

5.2.54 *Carex chordorrhiza* L.f. Cyperaceae

Morphology The utricles are smooth with prominent ribs. The nutlets are lenticular. They are elliptic with rather straight sides, narrowing sharply to the stigma. They are about 2 mm long and 1.0–1.5 mm wide.

Life form and ecology *Carex chordorrhiza* is a rhizomatous perennial sedge. It can spread by up to 70 cm per year by prostrate stems rooting from the nodes. It grows in base-poor fens and mires, usually in standing water. It can form extensive stands and be a dominant species in some boreal wetlands.

Distribution *Carex chordorrhiza* is circumpolar in boreal and subarctic areas. It occurs in Iceland, Scandinavia, northern Germany, Poland, The Alps, the Pyrenees, northern Russia, and in North America from Alaska to Greenland, and more patchily as far south as Indiana. It is rare in Scotland.

Carex chordorrhiza L.f. (gbif.org)

5.2.55 ***Carex demissa*** Hornem. (Syn. *Carex viridula* subsp. *oedocarpa*) Cyperaceae

Morphology *Carex demissa* nutlets are trigonous. They are obovate, about 1.5 mm long and up to 1 mm wide with marked angle lines. The epidermal cells are distinct. Other species within the section Ceratocystis (*C. flava* agg.) have similar nutlets.

Life form and ecology *Carex demissa* is a small perennial tufted sedge. It prefers wet, acid to neutral soils and is found at water body margins such as stony lakeshores, and in other damp open habitats, such as wet tracks, fields, and heathland, and in damp waste places in the lowlands. It also grows in base-rich submontane flushes and seepage areas, and on wet rocky hillsides.

Distribution *Carex demissia* has a sub-oceanic boreo-temperate distribution in Europe, occurring both in the northern European lowlands and in the subalpine zone of mountains. In Britain, it is frequent in Scotland and Wales, and in Scandinavia in southern Sweden and in southern and western Norway. It has a few localities in Finland. It also occurs in southeast Canada. It is introduced in New Zealand.

Carex demissa Hornem. (gbif.org)

5.2.56 *Carex diandra* Schrank Cyperaceae

Morphology The smooth dark brown utricles are 1.4–1.7 mm long and about 1 mm wide, biconvex, and tapering sharply to a stalk-like base. The beaks are as long as the bodies, with fine teeth on the margins. The nutlets are lenticular, but rounded and wide from front to back, appearing sub-trigonous. They are 1.5 mm long, obovoid, and very much narrowed towards the base. The angle line is well marked.

Life form and ecology *Carex diandra* is a rhizomatous, loosely tufted, perennial sedge. It is usually about 30 cm tall, but can reach 100 cm. It grows in calcareous to neutral flushes, springs, and fens within forests, in wetlands developed on flood plains, and in floating mats bordering alkaline lakes.

Distribution *Carex diandra* is circumboreal. It is widespread in northern Europe becoming rare at the most northern latitudes. In the British Isles, it is widespread in Scotland, northern England, Wales, and central Ireland on calcareous rocks, and in calcareous fens in East Anglia. It is widespread in the boreal zone of Scandinavia, but it avoids the acidic rocks of southern Norway and north Sweden. It is present in southern Iceland but not in Greenland. It is scattered across the boreal zone of North America. It is native in Tasmania and in New Zealand, being most widespread on the South Island.

Carex diandra Schrank (gbif.org)

5.2.57 *Carex dioica* L. Cyperaceae

Morphology The utricles of *Carex dioica* adhere tightly to the nutlets. They are 2.5–3.0 mm long, strongly veined, with a long, minutely toothed beak. The nutlets are lenticular with almost round faces, a broad base and a broad stigma base. They are 1.3–1.7 mm long.

Life form and ecology *Carex dioica* is a shortly rhizomatous perennial sedge, characteristic of very wet, mesotrophic to base-rich mires low in nitrogen. It flourishes on silty, calcareous mud by lakes and at the edges of open lime-rich flushes and springs. It occurs from sea level to low-alpine elevations.

Distribution *Carex dioica* has a circumpolar boreo-arctic montane distribution. In the British Isles, it is most common in Scotland and North Wales. It is widespread in Fennoscandia north to Svalbard, in Iceland, throughout northern Europe, and The Alps. It occurs in western North America, particularly Alaska and northwest Canada.

Carex dioica L. (gbif.org)

5.2.58 *Carex elata* All. Cyperaceae

Morphology *Carex elata* has lenticular nutlets. They are elliptical and widest across the middle, about 2 mm long and 1.0 mm wide. The surface has fine papillae. The nutlet fills about two thirds of the utricle.

Life form and ecology *Carex elata* is a tussock-forming perennial sedge up to 95 cm tall. It is a lowland plant. It favours oligotrophic or mesotrophic (sometimes eutrophic) marshy habitats, often calcareous, which include fens, the margins of lakes, ponds, rivers, and canals, ditches prone to seasonal flooding, and wet *Alnus* or *Salix* woodland.

Distribution *Carex elata* has a Eurasian temperate distribution. In the British Isles, it occurs mainly in central Ireland, England, and Wales. It is widespread in temperate Europe. It is absent from the Arctic and much of the boreal zone and is very rare in the Mediterranean region. It is rare in Norway and reaches its northern limit in south Sweden and south Finland. It is scattered across Asia to Manchuria. The very similar, perhaps synonymous species *Carex stricta* Gooden. is widespread in eastern North America.

Carex elata All. (gbif.org)

5.2.59 *Carex hispida* Willd. ex Schkuhr (Syn. *Carex acutiformis* Ehrh.) Cyperaceae

Morphology The nutlets are trigonous and obovate, about 2 mm long and 1 mm wide. The oval faces are concave and the angle lines are raised and distinct. The base is relatively broad and the stigma base may be preserved.

Life form and ecology *Carex hispida* is a rhizomatous tufted perennial sedge up to 80 cm tall. It grows in shallow base-rich, mesotrophic and eutrophic water, often occurring in extensive stands. It grows in a range of wetland habitats including river banks, lake and pond margins, marshy areas near streams, and in fen-meadows and tall-herb fens. It is shade-tolerant and grows in fen carr. It is intolerant of marked fluctuations in water level.

Distribution *C. hispida* is widespread in temperate Europe and western Asia. It is introduced elsewhere.

Carex acutiformis Ehrh. (gbif.org)

5.2.60 *Carex lasiocarpa* Ehrh. Cyperaceae

Morphology The nutlets of *Carex lasiocarpa* are trigonous, about 2 mm long, broadly oblong-ovate with more-or-less parallel margins and indistinct angle lines. The hexagonal cell pattern is distinct. The utricles are unusual in being densely grey-hairy.

Life form and ecology This species is a clonal perennial that can form extensive nearly monospecific stands on lake shores and slow-flowing rivers, usually in standing water. It generally occurs in nutrient-poor water, which may be base-rich or base-poor. Its well-developed rhizome system can form floating mats bordering acid lakes that often support other plants of wet infertile conditions, such as *Sphagnum* moss and Ericaceous shrubs. It can also colonise old peat workings and drainage channels.

Distribution *Carex lasiocarpa* has a circumboreal-montane distribution. It is absent from Iceland. In the British Isles, its occurrences are concentrated in Scotland and Wales. It is rare in Ireland.

Carex lasiocarpa Ehrh.(gbif.org)

5.2.61 *Carex nigra* (L.) Reichard Cyperaceae

Morphology *Carex nigra* nutlets are lenticular. They are usually as broad as long, about 1.4–1.8 mm, with a broad base. The surface cell pattern becomes more distinct on fossilisation. The utricle is loose-fitting with strong veins and a dark coloration. The nutlets can be variable in shape, from round to sometimes elliptical (illustrated in Jiménez-Majías and Martinetto 2013). This may reflect the variability of the plant, with several subspecies being delimited. Similar nutlets are found in *C. acuta, C. aquatilis,* and *C. elata,* and in the terrestrial *C. bigelowii*. They have been termed '*Carex nigra*-type.

Life form and ecology *Carex nigra* is a rhizomatous perennial sedge 20–70 cm tall. It occurs in a wide range of wet habitats avoiding extremely acid or basic sites. It is commonly found in shallow water at lake shores, in fens, fen-meadows, bogs, stream-sides and flushes, wet grassland and dune-slacks. It is morphologically very variable: tussock forms (var. *juncea*) may occur in stagnant, acidic sites, and rhizomatous tufted variants are known from calcareous mires.

Distribution *Carex nigra* has a boreo-temperate distribution, occurring throughout Europe, including Iceland, and western Asia reaching central Siberia, Turkey, and the Caucasus. It is also native in eastern North America but is introduced in the west.

Quaternary history *Carex nigra*-type nutlets have occasionally been identified in Holocene sediments (e.g. Birks 2000), or sometimes *C. nigra* itself has been specified (see Godwin 1975).

Carex nigra (L.) Reichard (gbif.org)

5.2.62 *Carex paniculata* L. Cyperaceae

Morphology *Carex paniculata* nutlets are lenticular, ovoid, or somewhat broadest below the middle, 1.5–1.7 mm long and about 1.0 mm wide. Their shape is variable, and three forms are illustrated. The surface cell pattern is distinct. The utricles are about 2.5–3.0 mm long and about 1.5 mm wide. They are sharply constricted at the base and taper upwards to a short beak. The walls are thick and spongy.

Life form and ecology *Carex paniculata* is a tussock-forming perennial sedge, growing about 1.5 m tall. It occurs in a wide range of mesotrophic habitats, including swamps and fens, the edges of lakes, ponds, canals and ditches, open fen-carr, and swampy woodland, usually in the lowlands. It tolerates moderate shade, but flowering and fruiting are reduced.

Distribution *Carex paniculata* has a temperate European sub-montane distribution. It occurs through most of Europe (including Britain and Ireland), south of about 62°N in Scandinavia, and southwest Finland. It reaches northwest Asia.

Carex paniculata L. (gbif.org)

5.2.63 *Carex pseudocyperus* L. Cyperaceae

Morphology *Carex pseudocyperus* nutlets are trigonous. The faces are ovate, about 1.6 mm long and about 1.0 mm broad. The angle lines are distinct. The long slender persistent style is often bent at the base. The cells are distinct and appear papillose. The utricles are lanceolate-elliptic, 4–6 mm long, the body grading into a long beak forked at the top. The 12–20 nerves are very distinct.

Life form and ecology *Carex pseudocyperus* is a lowland loosely tufted perennial sedge about 90 cm tall. It grows in eutrophic or mesotrophic open-water swamps, at the margins of ponds and slow-flowing streams, or in ditches. It can be a component of *Phragmites* reed-swamps. It can tolerate some shade and can grow in swampy woods and by wooded pools.

Distribution *Carex pseudocyperus* is widely distributed in temperate northwest Europe and southern Scandinavia and eastwards in western Asia. It occurs in eastern North America in the boreal-temperate zone. It is a lowland plant in England, Wales, and Ireland.

Carex pseudocyperus L. (gbif.org)

5.2.64 *Carex riparia* Curt. Cyperaceae

Morphology *Carex riparia* nutlets are trigonous, about 2 mm long and about 1.5 mm wide. They are slightly obovate or straight-sided. The angles are rounded. The cells are fairly distinct. The utricle is shiny and inflated, with marked veins.

Life form and ecology *Carex riparia* is a large, tufted sedge up to 130 cm tall. It has far-creeping rhizomes and forms large stands. It grows beside slow-flowing rivers, in ditches, around ponds, and in wet woodland. It may be dominant in swamps, especially with standing water in spring. It also grows in tall-herb fens, with other tall sedges.

Distribution *Carex riparia* has a temperate European - western Asia distribution. It is widespread in lowland northwest Europe, England, southern Sweden, and southern Finland. It is rare in Ireland, Scotland, southern Norway, and North Africa.

Carex riparia Curtis (gbif.org)

5.2.65 *Carex rostrata* Stokes Cyperaceae

Morphology The elliptic or obovate nutlets of *Carex rostrata* are trigonous, about 2 mm long and 1.0–1.5 mm wide. The angles are rounded, being sharper near the base. The polygonal cells are distinct and obvious in fossil specimens when viewed with overhead lighting. The curved style is often persistent. The utricle is inflated and shiny, with distinct veins and a bidentate beak. The utricle can be preserved empty or encasing the nutlets (Dickson 1970).

Life form and ecology *Carex rostrata* is a rhizomatous sedge up to 100 cm tall, with glaucous leaves. It grows in shallow water beside oligotrophic or mesotrophic lakes with pH 4.5–6.5. It also grows in acidic fens and occasionally in swampy woodland.

Distribution *C. rostrata* is widespread throughout boreal and temperate regions of the Northern Hemisphere.

Quaternary history *Carex rostrata* is frequently recorded as fossils in Quaternary lake sediments.

Carex rostrata Stokes (gbif.org)

5.2.66 *Carex vesicaria* L. Cyperaceae

Morphology The elliptic nutlets of *Carex vesicaria* are trigonous, about 2.0–2.5 mm long and 1.5–2.0 mm wide. The angles are rounded, and often protruding. The curved stigma can be persistent. The utricle is inflated and shiny with a short divergent beak.

Life form and ecology *C. vesicaria* is a rhizomatous sedge up to 120 cm tall. It grows in lowland mesotrophic mires, often at the edges of lakes, streams, and canals that have standing water in winter.

Distribution *C. vesicaria* has a Holarctic distribution throughout the Northern Hemisphere. It extends from the Mediterranean area to northernmost Scandinavia.

Quaternary history *Carex vesicaria* nutlets have been recorded in late-Holocene sediments in the Drava valley, Croatia, formed during the overgrowth of an oxbow lake (Gałka et al. 2024).

Carex vesicaria L. (gbif.org)

5.2.67 ***Cladium mariscus*** (L.) Pohl. Cyperaceae

Morphology The nutlets of *Cladium mariscus* are enclosed in loose shiny utricles composed of air cells with thin walls that can float for 15 months or more (Praeger 1913; Ridley 1930). The nutlets inside are ovate, urn-shaped, about 1 mm long and 1 mm wide, with a wide base, an obtusely pointed apex, and a rough surface texture. They sink when the utricle is removed. Both utricles and nutlets can be found as fossils and the nutlets are instantly recognisable.

Life form and ecology *Cladium mariscus* is a tall (up to 2.5 m) reed. It is a rhizomatous perennial, growing in unshaded fens or beside lakes, ponds, and rivers with mesotrophic base-rich water 15–45 cm deep. It can also grow in more acidic habitats overlying base-rich substrates. It can develop large pure beds that can extend as floating mats, or be mixed with other fen plants. *Cladium* grows best where mean summer temperatures are 14–16 °C. Its underground growing points are killed by temperatures of –10 °C (von Post 1925; Conway 1942). During active growth, thc growing points cannot withstand any degree of frost.

Distribution *Cladium mariscus* has a nearly worldwide distribution in tropical and temperate regions south of 60° N. It is intolerant of low winter temperatures and needs a long frost-free period. Hence it thrives in oceanic areas. Other factors that limit its distribution are historical and local factors, such as habitat availability and long-term persistence of suitable habitats.

Quaternary history The distinctive subfossil nutlets of *Cladium mariscus* have been widely recorded in Quaternary sediments. The rhizome fragments are also distinctive, having a black cortex containing a parallel bundle system inside a bright brick-red matrix. *Cladium* peat can often form thick layers, frequently as a late stage in the hydroseral infilling of lakes or lagoons (Godwin 1975). *Cladium* pollen is also distinctive and has been widely differentiated from other Cyperaceae pollen (e.g. Hájková et al. 2013; Salmina 2004). Von Post (1925) demonstrated that fossil records of *Cladium* in Sweden were most abundant and occurred beyond its current climatic limit in the early and mid Holocene. This was one of the first demonstrations of an early to mid-Holocene temperature maximum.

Cladium mariscus (L.) Pohl (gbif.org)

5.2.68 *Cyperus fuscus* L. Cyperaceae

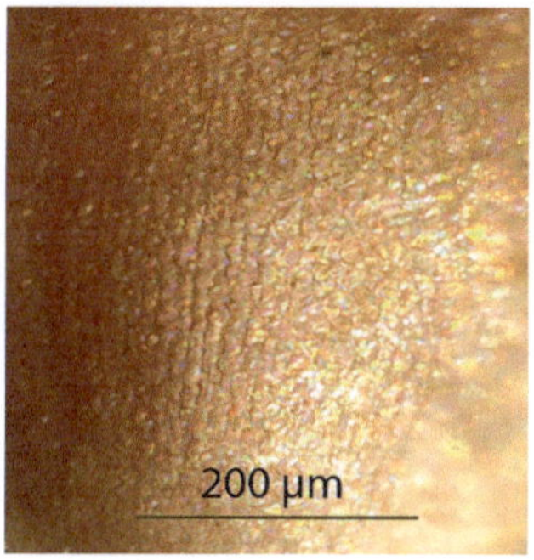

Morphology The nutlets of *Cyperus fuscus* are small, 0.7–0.8 mm long and about 0.4 mm wide. They are trigonous and obovoid, with paler ridges of spongy tissue on the angles. The style base is prominent and the base of the nutlet is narrowed to a short stalk. The minutely papillose cells are aligned longitudinally.

Life form and ecology *Cyperus fuscus* is a small tufted annual sedge, up to 25 cm tall. It grows in nutrient-rich, moist, open, muddy or sandy, humus-rich soil around ponds and ditches. It requires a disturbed habitat, usually by grazing animals, that is flooded in winter, but that dries out in summer. This ephemeral draw-down habitat is widely threatened. The fruits can form a long-lived seed bank. (See Böchelmann et al. 2017).

Distribution *Cyperus fuscus* is widely distributed through temperate Europe, where it reaches its northern limit in England, Denmark, and south Sweden and its southern limit in North Africa. It occurs throughout Asia to China and Thailand. It is introduced in North America where it is naturalized across the United States and in eastern Canada.

Cyperus fuscus L. (gbif.org)s

5.2.69 *Cyperus longus* L. Cyperaceae

Morphology The nutlets of *Cyperus longus* are trigonous with rounded edges. They are about 1 mm long and narrowly elongate obovoid. The cells are isodiametric and papillose. The stigma base is persistent, and the nutlets narrow at the base to a broadened attachment point.

Life form and ecology *Cyperus longus* is a perennial sedge 50–120 cm tall, with creeping rhizomes forming vigorous patches. It grows in shallow mesotrophic to eutrophic water or on damp soil in marshy pond-sides, by ditches, and in base-rich flushes, often near the sea or on sea cliffs.

Distribution *Cyperus longus* is widespread through central Europe and the Mediterranean region, and through Asia as far as eastern Kazakhstan, Pakistan, and northwest India. It is widespread in Africa, usually as ssp. *tenuiflorus*.

Cyperus longus L. (gbif.org)

5.2.70 *Eleocharis acicularis* (L.) Roem. & Schult. Cyperaceae

Morphology *Eleocharis acicularis* achenes are small (0.9 mm long, 0.5 mm wide) and obovate. The conical-pyramidal style base is spongy and articulated to the achene. The 3–4 perianth bristles are not usually preserved in fossil material. The achenes are shiny and yellowish with about 12 conspicuous longitudinal ribs with transverse striations in between. Similar fruits/seeds are found in some *Juncus* spp. and *Isolepis setacea*. The ribs and striations are more conspicuous than in *Juncus* spp. *Isolepis* has a rounder shape and more numerous ribs and lacks the spongy style base.

Life form and ecology *E. acicularis* has tufts of stems up to about 15 cm tall, each topped by an inflorescence, arising from shortly creeping rhizomes. It grows in wet muddy and sandy fine-grained soils in non-shaded habitats at the margins of still water that are exposed when water levels become lowered, or sometimes on disturbed ground such as wet tracks. It can form an underwater turf in slowly flowing shallow water, where its stems are longer and it does not flower or fruit. It requires acidic to slightly alkaline soft water (pH 6.0–7.2).

Distribution *Eleocharis acicularis* is widespread throughout Europe, in central and southeast Asia, and in North America. It ranges from Arctic Scandinavia and Russia to Spain and from Arctic Canada and Alaska to northeast South America as far south as Ecuador. It also occurs in Australia, where it is probably introduced.

Eleocharis acicularis (L.) Roem. & Schult. (gbif.org)

5.2.71 *Eleocharis mamillata* (H.Lindb.) H.Lindb. Cyperaceae

Morphology The yellow achenes of *Eleocharis mamillata* are broadly obovate with fairly distinct cells. They are about 1.5 mm long and about 1 mm wide. They are unequally trigonous with rounded angles. The style base is broad and spongy, rapidly narrowing to a short spike. The perianth bristles are longer than the achene and densely barbed. They are not usually preserved in fossil material.

Life form and ecology *Eleocharis mamillata* has a far-creeping rhizome from which arise tufts of leafless stems with terminal inflorescences. It grows in shallow oligotrophic, usually calcareous water at the margins of upland rivers in gravel and silt in sheltered embayments on the shores of freshwater lakes and shallow ponds, and in oligotrophic fens and ditches.

Distribution *Eleocharis mamillata* is widespread in the temperate zone of Eurasia and North America. It is common in southern Norway, Sweden, and Finland and eastern Europe, although extending to the Pyrenees. It is more scattered in Denmark, northern Britain, and western Europe.

Eleocharis mamillata (H.Lindb.) H.Lindb. (gbif.org)

5.2.72 *Eleocharis multicaulis* (Smith) Desvaux Cyperaceae

Morphology The achenes of *Eleocharis multicaulis* are rounded-triangular, obovate in side view, 1.2–1.5 mm long and 0.8 mm wide, and narrowing towards the base. The cells are fairly distinct. The style base is spongy and broadly triangular in side view and is narrowly constricted at the junction with the achene. The perianth bristles are of different lengths, shorter than the achene, and sparsely barbellate. They are usually lost in fossil material.

Life form and ecology *Eleocharis multicaulis* has shortly creeping rhizomes from which arise dense clusters of stems, bearing terminal inflorescences. It grows in wet acid habitats in oligotrophic soligenous mires, around small pools in *Sphagnum*-dominated mires, at margins of oligotrophic lakes, and in seasonally inundated wet heaths and dune slacks.

Distribution *Eleocharis multicaulis* has a sub-oceanic temperate distribution in Europe. It is widespread in the northern and western British Isles. It occurs mainly in northwest Europe including southern Scandinavia and reaches southern Spain and northern Morocco. It is common in the Azores.

Eleocharis multicaulis (Sm.) Desv. (gbif.org)

5.2.73 *Eleocharis palustris* (L.) Roem. & Schult. Cyperaceae

Morphology The orange to brown achenes of *Eleocharis palustris* are broadly obovate and biconvex, 1.2–1.5 mm long and 1.0–1.4 mm wide. The cells are small but rather distinct. The style base is conical, constricted at the junction of the achene but forming a wide rim at the base. It is persistent in fossil material. The four perianth bristles are shorter than the achene and barbellate. They are lost during fossilisation.

Life form and ecology *Eleocharis palustris* has far-creeping rhizomes giving rise to sparse to dense tufts of stems up to 90 cm high but usually less, terminating in inflorescences. It is indifferent to soil type. It grows beside or in shallow water in ponds, ditches, and river margins, even in swiftly flowing water, often forming large populations in upland pools and lakes and in soligenous soaks.

Distribution *Eleocharis palustris* is widely distributed in Europe and across the Northern Hemisphere in the boreal-temperate zone. It avoids the Arctic. It extends to northern and southern Africa including the Azores and the Canary Islands, and to northwest South America.

Quaternary history *Eleocharis palustris* has been recorded frequently in Quaternary lake sediments.

Eleocharis palustris (L.) Roem. & Schult. (gbif.org)

5.2.74 ***Eleocharis quinqueflora*** (Hartmann) O.Schwarz Cyperaceae

Morphology The greyish achenes of *Eleocharis quinqueflora* are obovate and three-sided with rounded angles. They are about 1.5–1.7 mm long and 1 mm wide. The epidermal cell pattern is very clear with isodiametric cells arranged in more-or-less vertical rows above the black interior cells. The style base is confluent with the achene and not wider at the base but parallel-sided and black above. The 6 perianth bristles are densely finely barbellate. They are lost in fossil material.

Life form and ecology *Eleocharis quinqueflora* forms loose to dense tufts of stems about 40 cm tall, each topped by an inflorescence, arising from short, creeping rhizomes. It grows in wet peaty mires in base-rich habitats such as base-rich marshes and fens, stony and muddy areas with seeping water such as spring-heads, calcareous flushes on peaty soils, and in a variety of moist meadows in calcareous areas. It also grows in sedge communities in dune slacks and seepage areas in upper saltmarshes. *E. quinqueflora* requires open sites and is often dependent on grazing, cutting, or other minor disturbance such as along wet paths.

Distribution *Eleocharis quinqueflora* has a boreal-temperate distribution in the Northern Hemisphere. It is widespread across Europe including Iceland and Greenland, northern Asia (Siberia, China, Kazakhstan, Himalayas, etc.), and North America (Canada, northern and western USA). There are isolated populations in Chile.

Eleocharis quinqueflora (Hartmann) O.Schwarz (gbif.org)

5.2.75 *Eleocharis uniglumis* Schult. & Zinserl. Cyperaceae

Morphology The yellowish obovate achenes of *Eleocharis uniglumis* are roundly biconvex, about 1.4–1.7 mm long and 1.0–1.5 mm wide. The cells are distinct. The style base is relatively large and spongy, triangular in side view with a dark tip. It is constricted at the base, and not confluent with the achene.

Life form and ecology Small clusters of stems up to 70 cm tall bearing terminal inflorescences arise from a shortly creeping rhizome. *Eleocharis uniglumis* is primarily a coastal species in upper tidal flats and marshes, but it can also grow inland in fen meadows and in upland calcareous marshes where water conductivity is high.

Distribution *Eleocharis uniglumis* has a temperate and boreal distribution throughout the Northern Hemisphere, including Iceland. It is largely coastal in Scandinavia, but it is widespread throughout northwest Europe from the Low Countries to Spain.

Eleocharis uniglumis (Link) Schult. (gbif.org)

5.2.76 ***Isolepis cernua*** (Vahl) Roem. & Schult. Cyperaceae

Morphology The achenes of *Isolepis cernua* are round to slightly obovate with a small apiculus. They are small, 0.5–1.2 mm long and about 1 mm wide. The surface is minutely tuberculate.

Life form and ecology *Isolepis cernua* forms graceful tufts of stems up to 30 cm tall, each topped by an inflorescence. It is found mostly near the sea in wet peaty or sandy soil in seepage areas in the upper salt marsh, around lagoons and slow flowing brackish water, on coastal rocks and boulder beaches, and in dune slacks. More rarely inland, it grows beside ditches, in grazed wet meadows, and around lake margins.

Distribution *Isolepis cernua* is mainly coastal in western and northern Britain and Ireland, reaching southern Scotland. In Europe it is mainly western and around the Mediterranean. It is native to many regions of the world, including southern Australasia, southern Africa, the Azores, Canary Islands, western North America, and South America.

Isolepis cernua (Vahl) Roem. & Schult. (gbif.org)

5.2.77 ***Isolepis fluitans*** (L.) R.Br. (Synonyms: *Eleogiton fluitans*; *Scirpus fluitans*) Cyperaceae

Morphology The obovate achenes of *Isolepis fluitans* are two-sided and biconvex with sharp angles. They are obovate and narrow basally into a stipe-like end. They are 0.7–1.5 mm long and about 0.7 mm wide. The surface is smooth and the cells are indistinct, giving a granular appearance.

Life form and ecology *Isolepis fluitans* is a short-lived perennial elodeid aquatic plant, usually growing as floating or submerged mats, rarely terrestrial. Branching stems arise from shortly creeping rhizomes and bear flaccid leaves about 10 cm long and terminal inflorescences. It grows in acidic and oligotrophic waters in peaty streams and ditches and in shallow water at lake shores.

Distribution In the British Isles, *Isolepis fluitans* is most frequent in the north and west. It is widespread in northwest Europe, including western Denmark. Its northern limit is in southwest Sweden and southeast Norway. It extends through western France to southwest Spain. It is frequent on the Azores and occurs in central and southern Africa and Madagascar and eastwards to Australasia. It is absent in Finland, Asia, North and South America, Greenland, and Iceland.

Isolepis fluitans (L.) R.Br. (gbif.org)

5.2.78 *Isolepis setacea* (L.) R.Br. Cyperaceae

Morphology The rounded triangular achenes of *Isolepis setacea* are obovate and very small, about 0.8 mm long and 0.6 mm wide. The surface has numerous longitudinal ridges with transverse striations, the cell centres being sunken. The ridges merge at the top into a short apiculum. The achenes resemble those of *Eleocharis acicularis*, but they are wider and not spindle shaped and the pattern is more robust.

Life form and ecology *Isolepis setacea* forms small perennial tufts or short turf about 10–25 cm tall. An involucral bract that exceeds the spikelet subtends the terminal inflorescences. It grows in open, damp, generally acidic sites often subject to winter flooding. It is a pioneer species of disturbed ground, such as in damp cattle grazed water meadows and on the margins of ditches and ponds. It also occurs in dune slacks and in upper saltmarshes, and occasionally in calcareous mires.

Distribution *Isolepis setacea* is widespread in the British Isles and throughout Europe. It reaches its northern limit in southern Sweden and Norway. It extends eastwards through the Caucasus and the Himalaya into western China. It is also native to central and southern Africa, and possibly Australasia. It is introduced in western North and South America.

Isolepis setacea (L.) R.Br. (gbif.org)

5.2.79 ***Scirpoides holochoenus*** (L.) Soják Cyperaceae

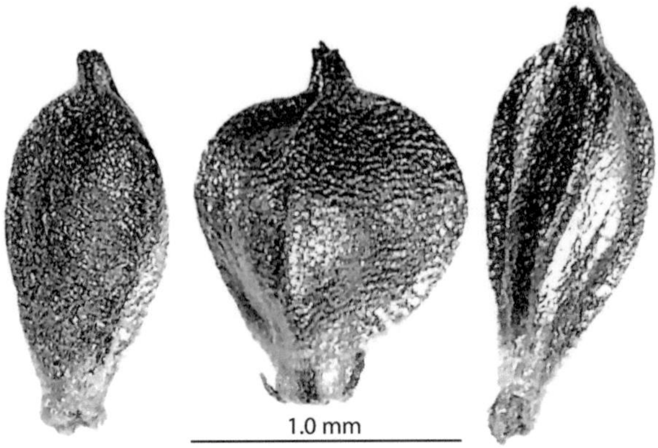

Morphology The nutlets of *Scirpoides holochoenus* are obovoid and are flattened trigonous, with one flat side. They are 1.0–1.8 mm long and 0.5–0.75 mm broad. The surface appears minutely reticulate. The style base is persistent.

Life form and ecology *Scirpoides holochoenus* is a rhizomatous perennial rush up to 90 cm tall, with distinctive branched inflorescences bearing globose flower heads. It grows in damp sandy places usually by the sea, in dune systems, or in river valleys, where its habitat is often threatened.

Distribution *Scirpoides holochoenus* is widespread throughout the Mediterranean basin and along the coasts of the Black Sea. It extends eastwards to Pakistan, southwest Siberia, and southern Mongolia. It is particularly widespread in western and southern France, northern Italy, and the Iberian Peninsula. It reaches its northern limit in southwest England, The Netherlands, and northern Germany.

Scirpoides holoschoenus (L.) Soják (gbif.org)

5.2.80 *Scirpus sylvaticus* L. Cyperaceae

Morphology The relatively small nutlets of *Scirpus sylvaticus* are oval and compressed-trigonous, with one flat side. They are 0.7–1.4 mm long and 0.5–0.8 mm wide with a prominent style base. They are smooth and usually pale brown.

Life form and ecology *Scirpus sylvaticus* is a robust rhizomatous rush up to 120 cm tall. It is characteristic of wet alder-dominated woodlands, but it also occurs in open areas, such as along streams and river margins, and by lakes whose shores are periodically flooded. It prefers eutrophic silty soils, often rich in iron and with high nitrogen content.

Distribution *Scirpus sylvaticus* has a temperate Euro-Siberian distribution. It occurs throughout northwest Europe including England and southern Scotland, lowland Scandinavia south of the Arctic Circle, and eastwards across temperate Asia to Japan. It avoids the Mediterranean area.

Quaternary history Seeds have been recorded in late-Holocene sediments in the Drava valley southern Hungary, formed when an oxbow lake was becoming overgrown (Gałka et al. 2024).

Scirpus sylvaticus L. (gbif.org)

5.2.81 *Schoenoplectus* (L.) Palla (Syn. *Scirpus lacustris*) Cyperaceae

Morphology *Schoenoplectus lacustris* nutlets are about 2.5 mm long and 1.8–2.5 mm wide. They are broadly ovoid and compressed trigonous, with one flat side. The style base is persistent. They are dark grey to black, and retain this colour on fossilisation. The surface is smooth and the small cells are rather distinct.

Life form and ecology *Schoenoplectus lacustris* is a tall (up to 350 cm) rhizomatous perennial club rush. It grows on silt, clay, peat or gravel in standing or slow-flowing fresh water, such as ponds, lakes, canals, dykes and slow-moving rivers. It tolerates a wide range of conditions from eutrophic and base-rich to oligotrophic and base-poor. It usually occurs in water 0.3–1.5 m deep but can also be found in deeper water.

Distribution *Schoenoplectus lacustris* belongs to the Euro-Siberian temperate element. It occurs throughout temperate Europe and into North Africa, and across Siberia and central Asia to China and Japan. It occurs as far north as the Arctic Circle in lowland Scandinavia. It introduced in North America.

Quaternary history *Schoenoplectus lacustris* nutlets have been recorded frequently in Quaternary sediments (Godwin 1975).

Schoenoplectus lacustris (L.) Palla (gbif.org)

5.2.82 ***Schoenoplectus tabernaemontani*** (C.C.Gmel.) Palla Cyperaceae

Morphology The obovoid nutlets of *Schoenoplectus tabernaemontani* are usually lenticular or plano-convex with one flat side. They are similar to those of *S. lacustris* but the latter is usually trigonous. They are about 2.0–2.2 mm long and 1.2–1.7 mm wide. The style base is persistent. The surface is smooth, and the small cells are rather indistinct.

Life form and habitat *Schoenoplectus tabernaemontani* is a rhizomatous creeping rush up to 150 cm tall. It usually grows on clay or peat near the coast in shallow brackish water and marshes. It can also occur in eutrophic water in shallow ponds, ditches, and slow rivers, often with *Typha* spp.

Distribution *Schoenoplectus tabernaemontani* is cosmopolitan in the temperate zone, not extending north of the Arctic Circle. It is known as *Scirpus validus* in USA. Its distribution is often coastal, due to its tolerance of brackish water.

Schoenoplectus tabernaemontani (C.C.Gmel.) Palla (gbif.org)

5.2.83 ***Schoenoplectus triqueter*** (L.) Palla Cyperaceae

Morphology *Schoenoplectus triqueter* has ellipsoid to obovoid nutlets that are 2–2.5 mm mm long and 1.5–2.5 mm broad. They are shiny and flattened with no sharp angles. The style base is persistent.

Life form and ecology *Schoenoplectus triqueter* is a rhizomatous perennial club-rush often forming extensive clonal stands. Its triquetrous stem can be up to 150 cm tall. In Britain and Ireland where it is rare and declining, it grows mainly in soft nutrient-rich mud in estuaries and tidal rivers, and it may sometimes be submerged by brackish water. Elsewhere in Europe it can grow in freshwater marshes.

Distribution *Schoenoplectus triqueter* occurs throughout central Europe, from Iberia to Southern England and Ireland. It extends across Asia to China and Japan, and southwards to Africa. It is introduced in Oregon and Washington and in South Africa.

Schoenoplectus triqueter (L.) Palla (gbif.org)

5.2.84 *Aldrovanda vesiculosa* L. Droseraceae

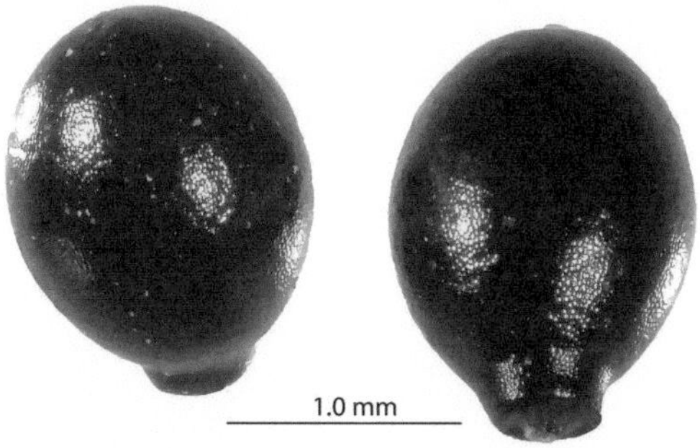

Morphology The seeds of *Aldrovandra vesiculosa* are round, black, and shiny, about 1.4–1.6 mm long. There is a thick basal foot and an operculum at the base. The surface cells are small and distinct. Seed-set is rare and requires high temperatures. The seeds probably form a seed bank (Adamec 2018a, b).

Life form and ecology *Aldrovandra vesiculosa* is a submerged, rootless aquatic plant. The leaves are in whorls, and each terminates in a toothed snapping insect trap, giving it the name 'waterwheel plant'. It grows in shallow dystrophic still water such as lakes, peaty fishponds, pools in bogs and fens etc. (Cross 2012; Adamec 2018a). It increases through vegetative fragmentation and overwinters by turions. Plant fragments and dormant seeds are probably dispersed by waterbirds. It needs a narrow combination of factors for successful growth: high CO_2 concentration in shallow water, freedom from competition, a thick layer of slowly decomposing plant litter on the bottom, and abundant zooplankton as prey (Cross 2012; Adamec 2018a and references therein). It is critically endangered in Europe, by eutrophication, decreasing water levels, and habitat loss.

Distribution *Aldrovandra vesiculosa* has a wide distribution across Eurasia, from Europe across to Bangladesh, China, and Japan It also occurs in Australia and Africa south of the equator. Its distribution seems to correlate with migration routes of waterbirds (Cross 2012). However, it is extremely rare and there may be as few as 50 actual sites (Cross 2012). It was introduced to artificial habitats in North America.

Quaternary history It is probable that *Aldrovanda* is a relict genus from the Tertiary, and there are several Tertiary and interglacial records in Europe and beyond (Štorchová et al. 2023). A seed of *Aldrovandra vesiculosa* has been recovered from a late-Holocene lake deposit in southeast Poland where it grows nearby today, and one seed from late-Holocene lake sediments in Tanzania (Gałka et al. 2015). Three were recorded in late-Holocene sediments in Bulgaria (Beug and Tonkov 2017).

Aldrovanda vesiculosa L. (gbif.org)

5.2.85 *Elatine* L. Elatinaceae

Elatine alsinastrum L.

Elatine hexandra (Lapierre) DC

Elatine hydropiper L.

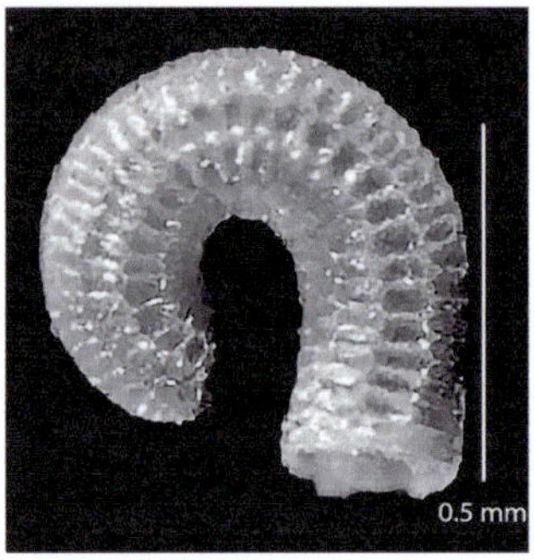

Elatine hydropiper L. Scanning electron micro-cope photo

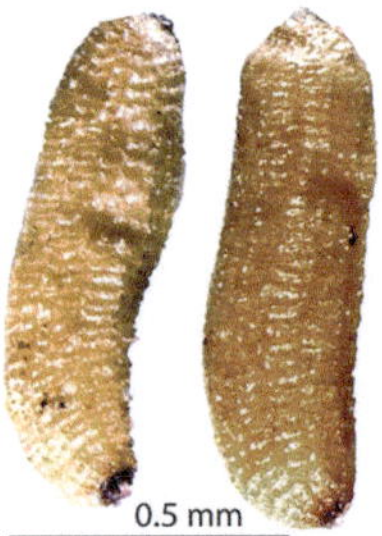

Elatine orthosperma Düben

Elatine triandra Schkuhr

Elatine hexandra (Lapierre) DC, *Elatine hydropiper* L., *Elatine alsinastrum* L., *Elatine triandra* Schkuhr., and *E. orthosperma* Düben are the five most common European *Elatine* species. Due to close similarities in their seed morphology, the species are dealt with together for comparison. *E. orthosperma* is closely related to *E. hydropiper* but has straight seeds.

Morphology All *Elatine* species are abundant seed producers (Salisbury 1967) and have bean to fish-hook shaped seeds that are very small, less than 1.0 mm long. They have a highly characteristic pitted surface ornamentation. *E. hydropiper*, without exception, produces hook-shaped seeds (Brinkkemper et al. 2008), in contrast to the weakly curved seeds of *E. alsinastum, E. hexandra, E. orthosperma,* and *E. triandra*. The morphological study by Popiela et al. (2017) of seeds from all 10 European *Elatine* species showed that the angle of curvature and the number of pits in the middle row were the key characters that differentiate them (see Table below). Brinkkemper et al. (2008) studied Dutch seed material. They proposed that it was possible to distinguish seeds of *E. hexandra* with a scalariform, ladder-like pattern of surface pits (clearly wider than high) from *E. triandra* with a honey-comb pattern (pits slightly wider than high). Further, according to Brinkkemper et al. (2008), *E. hexandra* seeds are generally wider than 0.2 mm, whereas those of *E. triandra* are maximally 0.2 mm wide (see Table 1), although the authors note that care needs to be taken not to increase measured widths by flattening seeds beneath coverslips on microscope slides. The shape of the pits on the seed surface of *E. alsinastrum* resembles that of *E. hexandra. E. alsinastrum* differs from all species by its larger size (0.7–0.9 mm length).

Differentiating morphological characters of *E. hexandra, E. triandra*, and *E. alsinastrum* (from Brinkkemper et al. 2008 and Popiela et al. 2017).

	Seed Length (mm) [a]	Seed Length (µm) [b]	Seed Width (mm) [a]	Width of pits (µm) [b]	No. of pits in middle row [b]	Shape of pits [a]	Shape of pits [b]
Elatine hexandra	0.5–0.6	(593) 656–697 (760)	0.2–0.25	(223) 283–322 (381)	(16) 19–21 (24)	Clearly wider than high (scalariform or "ladder-like")	Hexagonal and rectangular pits
Elatine triandra	0.45–0.65	(328) 467–560 (700)	0.15–0.2	(158) 201–231 (274)	(16) 20–23 (28)	Slightly wider than high (honey-comb shaped)	Net-work shaped pits
Elatine alsinastrum	–	(708) 799–859 (950)	–	(230) 290–330 (391)	(18) 21–23 (26)	–	Rectangular-shaped pits

[a] Data from Brinkkemper et al. (2008)

[b] Data from Popiela et al. (2017)

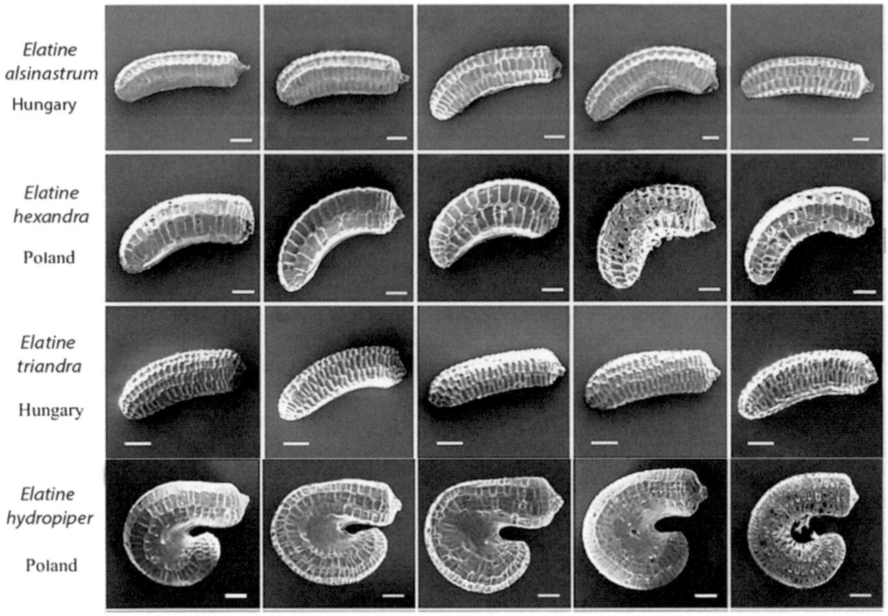

Scanning Electron Microscope photographs of seeds of *E. alsinastrum, E. hexandra, E. triandra* and *E. hydropiper*. Scale bar 200 μm. Modified from Popiela et al. (2017)

Life form and ecology *Elatine* species are small, weakly competitive, usually annual, isoetid plants (<10 cm high). They usually grow in open shallow water on shores of lakes, rivers, and ponds. The substrate is gyttja usually mixed with clay, or sand or silt with fine organic detritus. *Limosella aquatica, Subularia aquatica, Lobelia dortmanna, Eleocharis acicularis,* and other isoetids are common associates. Their abundant seeds are usually dispersed by water currents or on the feet of birds (Odland 2001) where they readily become attached by means of surface tension aided by their seed-coat pattern. Overgrowth, water-level management, eutrophication, and other habitat destruction universally threaten their habitats.

E. hexandra, and *E. hydropiper* favour periodically exposed habitats, with exposure triggering high seed production resulting in substantial seed banks. However, underwater seed production can occur through cleistogamous fertilisation, but fewer seeds are produced (Odland 2001). Although also found in lakes and ponds, *E. alsinastrum* and *E. triandra* are more strongly associated with disturbed ephemeral habitats. The former is often found in flooded meadows, dried out puddles, and wet roadsides on nutrient-rich substates (Popiela et al. 2013). *E. triandra* can also occur in dried out fish ponds, temporary ponds in continental climates, and rice paddy fields (Popiela et al. 2015) and also in water 10–30 cm deep. This association with highly modified habitats could suggest that both *E. alsinastrum* and *E. triandra* were associated in the wild with naturally fluctuating wetlands in the past.

E. hexandra tends to favour oligotrophic to mildly eutrophic sites and is negatively affected by eutrophication (Salgado et al. 2010), whereas *E. hydropiper*, *E. orthosperma*, *E. triandra*. and *E. alsinastrum* are more typically associated with mesotrophic to eutrophic water (Preston and Croft 1997; Odland 2001; Popiela et al. 2015). *E. hydropiper, E orthosperma,* and *E. triandra* are tolerant of brackish water, and palaeoecological and archaeobotanical data for northwest Europe suggest strong associations with former freshwater/tidal marsh habitats (Brinkkemper et al. 2008).

Distribution *E. alsinastrum* is a Euro-Siberian sub-Mediterranean species that is absent from Britain. It occurs east to the Ural Mountains, north to southern Finland and south to southern Spain and Crete. It is rare and threatened throughout its range (Popiela et al. 2013). Elatine alsinastrum L. (gbif.org)

E. hexandra is confined to Europe but is widespread from southern Scandinavia to Portugal and southern Spain, and from western Ireland and the Azores to the northern part of the Balkan Peninsula. Elatine hexandra (Lapierre) DC. (gbif.org)

E. hydropiper is widespread in northern and central Europe but less common in north and western Scandinavia and Britain. Elatine hydropiper L. (gbif.org)

E. orthosperma is widespread in Finland and the Baltic coast of Sweden, and in south Sweden and southeast Norway. It has scattered records in Norway as far north as 69°N. It can occur in mixed populations with *E. hydropiper* but has a more northerly distribution (Odland 2001). Elatine orthosperma Düben (gbif.org)

E. triandra is a widespread species occurring on all continents except Antarctica, but it is most frequent in southern Scandinavia, particularly Finland and the Baltic coast of Sweden, and in central Europe. It is absent from Britain and has a scattered distribution in southern and western Europe, including Iberia (Popiela et al. 2015). Elatine triandra Schkuhr (gbif.org)

5.2.86 *Equisetum* L. Equisetaceae

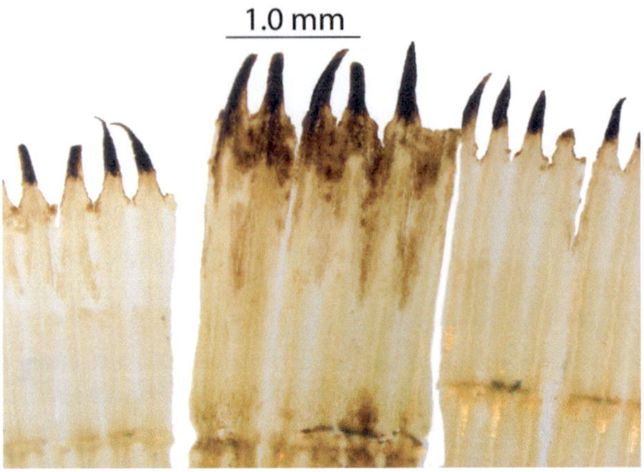

Morphology The nodal sheaths of *Equisetum* stems are sometimes preserved in sediments. They are very characteristic, with parallel bands of vascular tissue corresponding to the number of grooves on the stem, each ending in a dark point at the node. *Equisetum* rhizomes can sometimes be preserved (Godwin 1975; Walker 1966; see Mauquoy and van Geel 2007). Such rhizomes penetrate the substrate deeply and are probably much younger than the surrounding sediment.

Life form and ecology Two *Equisetum* species grow in aquatic habitats, *E. fluviatile (= E. limosum)* and *E. palustre.* They are rhizomatous perennials with unbranched or sparsely branching stems up to 60 cm tall. They grow as reeds in shallow water at the margins of ponds and lakes, often forming dominant stands. *E. palustre* also grows on damp soil in bogs, fens, marshes, wet heaths, woods, and meadows.

Distribution Both *E. fluviatile* and *E. palustre* are widespread throughout the Northern Hemisphere but rare or absent in the Southern Hemisphere.

Equisetum fluviatile L. (gbif.org)

Equisetum palustre L. (gbif.org)

5.2.87 *Eriocaulon aquaticum* Druce Eriocaulaceae

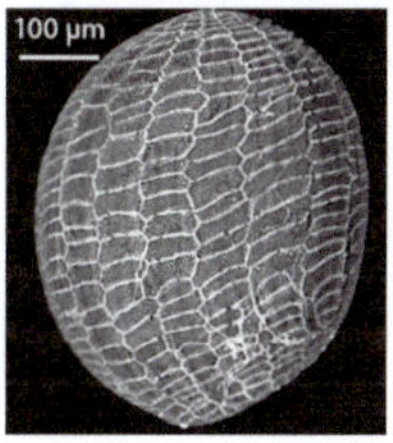

Scanning electron photomicrograph of an *E. aquaticum* seed modified from Zona et al. (2012)

Morphology The seeds of *Eriocaulon aquaticum* are sphaerical to ellipsoidal, 0.6–0.7 mm long and 0.4–0.5 mm wide. The cells of the testa are elongated quadrangular to hexagonal, with a length:width ratio of approximately 1:3 to 1:4. They are aligned in vertical rows. The margins of adjacent cells form raised ridges (Zona et al. 2012). These appear faintly on the seed surface seen under a stereomicroscope.

Life form and ecology The perennial rosettes of *E. aquaticum* have thin, often translucent narrow leaves up to 10 cm long, that taper to an acute point. They have prominent veins and conspicuous cross hatched veinlets, especially near the base. The rosette produces a single stalk with a dense round whitish inflorescence on top. *E. aquaticum* is an isoetid that grows in shallow oligotrophic water rich in dissolved organic carbon on peat or inorganic substrates at the edges of lakes and pools usually surrounded by acid, often *Sphagnum*-rich bog. It may grow with other isoetids and emergent sedges and rushes.

Distribution *Eriocaulon aquaticum* is restricted in Europe to northwest Scotland and western Ireland but it is widespread north of the Great Lakes in eastern North America.

Eriocaulon aquaticum Druce (gbif.org)

5.2.88 *Lathyrus palustris* L. Fabaceae

Morphology The globose seeds of *Lathyrus palustris* are 3–5 mm in diameter. The well-marked hilum extends halfway or more around the seeds in our material, but other accounts say it extends a quarter to a third of the circumference (e.g. Bojňanský and Fargašová 2007; Cappers et al. 2012). Fresh seeds have a dark mottled pattern on the smooth surface.

Life form and ecology *Lathyrus palustris* is a scrambling perennial herb. It grows in base-rich fens that can be flooded in winter, reedbeds, and fen-meadows, and also in bushy fen carr. More rarely it occurs on marshy riverbanks and on seashores.

Distribution *Lathyrus palustris* has a boreo-temperate distribution across the Northern Hemisphere. It reaches the Arctic Ocean and the northern Mediterranean coast. In the British Isles, its main areas are in central Ireland and East Anglia.

Lathyrus palustris L. (gbif.org)

5.2.89 *Fontinalis antipyretica* Hedw. Fontinalaceae

Morphology *Fontinalis antipyretica* is an aquatic moss. It has long stems bearing three rows of leaves that are folded inwards forming a sharp keel. The illustrated shoot bears capsules among the leaves which are supporting epiphytic micro-algae. The drawing of a leaf is modified from E-Flora BC.

Life form and ecology *Fontinalis antipyretica* grows submerged in a wide range of shallow, cold, still, slowly, or rapidly flowing water, including lakes and ponds, canals, streams and rivers. It is usually attached to stones, rocks, or waterside trees, but sometimes occurs in loose masses on the bottom of shallow lakes. It prefers neutral to acid water, a pH of around 8.4 in limestone lakes being the maximum tolerated.

Distribution *Fontinalis antipyretica* occurs across Europe from the Arctic to southern Iberia, in parts of Asia, Greenland, and the seaboards of North America.

Quaternary history Dickson (1973) lists fossil records in the British Isles from the Late Glacial and through the Holocene. He states that fossils seldom bear leaves but that the long straight black stems are distinctive.

Fontinalis antipyretica (Renauld & Cardot) Cardot, 1891 (gbif.org)

5.2.90 *Myriophyllum alterniflorum* DC. Haloragaceae

Modern nutlets

Fossil nutlets from Måkeskittjern, Utsira, western Norway

Morphology The fruit of *Myriophyllum alterniflorum* is a schizocarp that splits to release four individual nutlets. The nutlets (1.5 × 0.5 mm) are subcylindrical with flatter sides where the four nutlets adjoined in the fruit. There is a round plug at the top of the nutlet which opens on germination, leaving a round hole. This is characteristic of fossils. Normally, the membranous outer layers are lost during fossilisation, leaving the hard smooth endocarp. The surface is smooth except for the basal part, which can be covered in wart-like protrusions that can be observed with good preservation. The nutlets are smaller and more slender than those of *M. spicatum* and *M. verticillatum*.

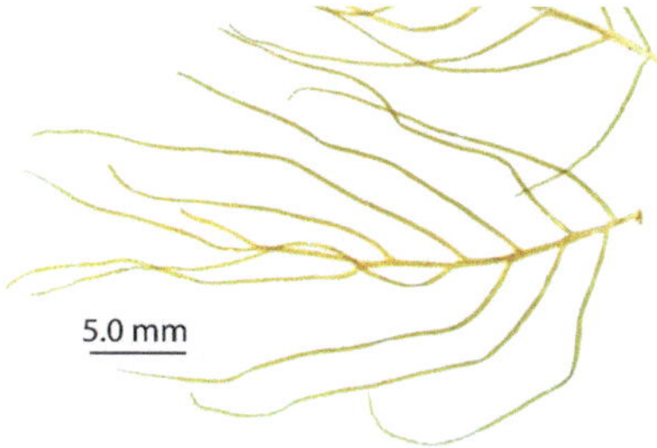

M. alterniflorum leaves are finely divided with thin flexible segments. Fossil leaf fragments are readily recognisable but cannot be identified to species. Their abundance has to be estimated on a frequency scale, as they cannot be counted.

Life form and ecology *Myriophyllum alterniflorum* is a perennial elodeid of lakes, ponds, and both slow and fast flowing rivers. It grows on a range of nutrient-poor substrates including silt, sand, stones, peat, and clay at water depths of typically 0.3–3 m (Preston and Croft 1997). Despite being able to use bicarbonate in photosynthesis (Maberly and Madsen 2002), *M. alterniflorum* prefers more acidic waters than *M. verticillatum* and *M. spicatum*. It can also grow in relatively basic oligotrophic waters, but is unable to survive in eutrophic conditions, which excludes it from many lowland aquatic habitats today. *M. alterniflorum* has also been lost from many base-poor upland European lakes due to acidification where it is replaced by species such as *Juncus bulbosus* and *Sphagnum* spp. *M. alterniflorum* rather infrequently produces seeds, so their importance for dispersal of the plant is poorly understood. It lacks specialised turions but is able to regenerate from short lengths of shoot. Colonisation of new sites is mainly through stem fragments transported by birds.

Distribution *Myriophyllum alterniflorum* is widely distributed in the boreal and temperate Northern Hemisphere, in Europe, North Africa, western Russia, eastern North America and Greenland. It has recently been recorded in China (Yu et al. 2002). In Europe, it is most frequent in the north and west, although it occurs as far south as Sicily.

Quaternary history *Myriophyllum alterniflorum* fruits are frequently recorded in both glacial and Holocene sediments (e.g. Godwin 1975). It is an early Holocene colonist in northwest Europe (e.g. Birks 1980; Birks and Mathewes 1978; Watts 1978; Fredskild 1992). At Kråkenes Lake in western Norway, the fruits (and pollen) are recorded from about 11,250 to 10,800 cal yr BP, when their record stops. However, *Myriophyllum* leaf fragments are recorded both during that time and subsequently to the top of that record attesting to its persistence in a vegetative state. Fruiting was frequent while the lake water was moderately rich in nutrients and base-cations but ceased on acidification and nutrient depletion (Birks 2000).

Myriophyllum alterniflorum DC. (gbif.org)

5.2.91 *Myriophyllum spicatum* L. Haloragaceae

Morphology The fruit of *Myriophyllum spicatum* is a smooth-surfaced schizocarp that splits into four individual nutlets. The outer fleshy pericarp decays, revealing the hard endocarp that is preserved in sediments. The fruits are about 2 mm long and 1.2–1.5 mm wide. They are distinctively trigonous with two flat sides where they originally joined with the other three nutlets, and an outer convex side ornamented with tooth-like processes or with short stumps when less well preserved. At the top is a circular plug that opens on germination, leaving the round hole seen in fossil nutlets.

Vegetative remains may be preserved in sediments. Aerial leaves on the emergent inflorescence are up to 6 mm long and 0.5 mm wide, robust, and densely pinnately divided. There is a swollen gland at the apex of each pinnule. Submerged leaves are filamentous and finely pinnately divided, giving the plant a bushy appearance. Leaf fragments may be recovered in sediments, but they cannot be identified to species. They cannot be counted, and their abundance has to be estimated on a frequency scale.

Life form and ecology *Myriophyllum spicatum* is a perennial elodeid species of shallow (usually <4 m), lowland, alkaline (pH 7 or more) and mesotrophic lakes, ponds and slow-flowing rivers. The vegetative plant is submerged but the flowering spike is aerial. It grows in oligotrophic through to eutrophic waters, preferring high conductivity, but it declines if eutrophication becomes severe (Davidson et al. 2005). *M. spicatum* can also tolerate brackish conditions such as in brackish lakes and freshwater-tidal marsh habitats. Seeds can float for a few hours to a day and detached inflorescences for longer. Ridley (1930) proposed that they may be dispersed on the feathers or feet of water fowl. Guppy (1897) suggested that germination may be increased by passage through the alimentary systems of wildfowl. However, reproduction from seed is rare and colonisation and establishment of new sites is mainly through stem fragments and axillary buds transported by birds. It rarely overwinters as turions but dies down to the roots in winter.

Distribution *Myriophyllum spicatum* is widespread in boreal and temperate regions of the Northern Hemisphere. Its native status in South Africa was confirmed by DNA analysis (Weyl et al. 2016). *M. spicatum* is an aggressive, much hated invasive species in North America, where it has also hybridised with the indigenous *Myriophyllum sibiricum* (Moody and Les 2007).

Quaternary history The characteristic fruits (often together with pollen) of *M. spicatum* are frequently recorded in Late-Weichselian and Early-Holocene sediments (e.g. Godwin 1975). The fruits can be confused with *M. verticillatum*, but the latter are smaller and never spiny. *M. spicatum* is also recorded as a pioneer in new or disturbed water bodies.

Note. *Myriophyllum aquaticum* is widely introduced but spreads vegetatively, only female plants being known outside South America. The leaves are more robust than other *Myriophyllum* species.

Myriophyllum spicatum L. (gbif.org)

5.2.92 ***Myriophyllum verticillatum*** L. Haloragaceae

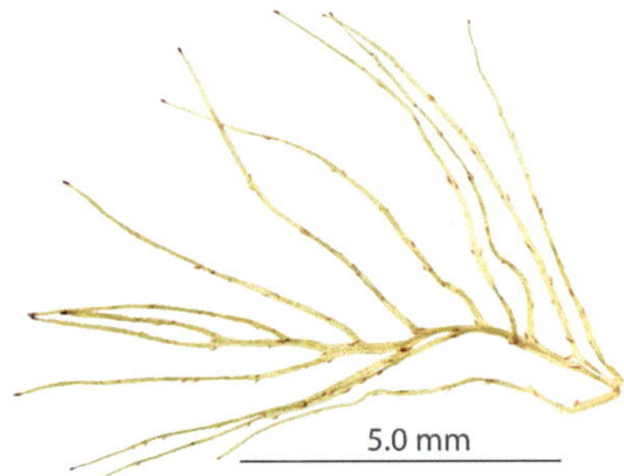

Morphology The fruit of *Myriophyllum verticillatum* is a schizocarp that splits into four nutlets. These are trigonal with two flat ventral sides and a slightly convex dorsal side. Unlike *M. spicatum* the margins do not possess tooth-like protrusions. They are 1.8–2.2 mm long and about 1.0–1.5 mm wide, wider and rounder than *M. alterniflorum*. The surface is smooth to slightly rough. Fossil seeds recovered from lake sediments show a hole where the plug used to be. The leaves are finely pinnate and minutely toothed.

Life form and ecology *Myriophyllum verticillatum* is a perennial elodeid of lowland still or slow flowing calcareous and mesotrophic-eutrophic water with a sandy or loamy substrate. It declines with excessive nutrient-enrichment (Ayres et al. 2008). *M. verticillatum* is unable to use bicarbonate as a carbon source in photosynthesis and so it is often found at a lower pH than *M. spicatum*, but at a higher calcium concentration than *M. alterniflorum* (Hutchinson 1970). It often grows over fen peat and is highly characteristic of fenlands, but it can also grow on inorganic substrates. *M. verticillatum* produces seeds but overwinters as turions that develop in late summer (Weber and Noodén 2005) and these are the main means of dispersal, usually by birds.

Distribution *Myriophyllum verticillatum* is widespread in temperate climates of the Northern Hemisphere including Europe, North America, North Africa, China, and India (Yu et al. 2002; Moody and Les 2010). In Europe *M. verticillatum* extends from northern Scandinavia to the Mediterranean and in Asia east to Kamchatka and Japan and south to Afghanistan and Kashmir (Yu et al. 2002).

Myriophyllum verticillatum L. (gbif.org)

5.2.93 *Elodea canadensis* Michx. and *Elodea nuttallii* (Planch) H.St.John Hydrocharitaceae

Elodea canadensis leaf

Elodea nuttallii leaf

Morphology *Elodea* seeds are unknown in Europe. It is possible to recover leaves from very recent sediments (e.g. Zhao et al. 2006), but they quickly decay. The leaves of *Elodea canadensis* are on average about 2 mm wide with obtuse to sub-acute tips and small marginal spines. The leaves of *E. nuttallii* are narrower, less than 1 mm wide with acute apices and larger marginal spines.

Life form and ecology *Elodea* spp. are eponymous elodeids. The long, branched stems are anchored by roots to the sediment in water up to 3 m deep and often form dominant stands. Plant fragments vigorously regenerate forming dense colonies and the plants are spread by birds, boats, or anglers to other waterbodies. They overwinter as prostrate shoots on the mud. *Elodea canadensis* prefers still or slow-flowing water that is mesotrophic to eutrophic, and often calcareous. *E. nuttallii* prefers eutrophic water in otherwise similar habitats.

Distribution *Elodea canadensis* and *E. nutttallii* are both native to temperate North America. They are introduced and have become widespread and often invasive elsewhere. *E. canadensis* is common in central Eurasia, southeast Australia, and New Zealand. *E. nuttallii* is widespread in western Europe and in Japan. They both have scattered occurrences throughout the world.

Elodea canadensis Michx. (gbif.org)

Elodea nuttallii (Planch.) H.St.John

5.2.94 *Hydrocharis morsus-ranae* L. Hydrocharitaceae

Morphology The oval seeds of *Hydrocharis morsus-ranae* are about 1.5 mm long often with remains of the stigma at the top. When dry, they are covered by a reticulum of comparatively large lumina with high walls. The centres of the lumina are tubercles that collapse when dried. They have spirally thickened walls (Scribailo and Posluszny 1985). In fossil seeds, the walls break down, and the spiral thickenings give the seed a tangled hairy appearance (Kelly 1964). Seed set is infrequent and needs a warm climate (Scribailo and Posluszny 1985; Preston and Croft 1997).

Life form and ecology The floating rosettes of *Hydrocharis morsus-ranae* are found in shallow, mesotrophic, calcareous water in ponds, canals, and drainage ditches, often in open reedswamp (Preston and Croft 1997) where competition from large nymphaeids is not strong and pollution is minimal (Cook and Lüönd 1982). It reproduces mainly by stolons from the rosettes and less frequently by seeds and overwintering turions (Cook and Lüönd 1982).

Distribution The main distribution of *Hydrocharis morsus-ranae* is across Europe and central Asia east to Lake Baikal. It occurs as far north as central Sweden and Finland but is almost absent from Norway. It occurs in Britain as far north as southern Scotland and into northwest Ireland. It is rare south of The Alps and in Iberia. It is introduced in northeast USA and Canada.

Quaternary history *Hydrocharis morsus-ranae* seeds have been recorded in interglacials in the British Isles (Kelly 1964; Godwin 1975).

Hydrocharis morsus-ranae L. (gbif.org)

5.2.95 *Lagarosiphon major* (Ridl.) Moss Hydrocharitaceae

Morphology All *Lagarosiphon* plants in western Europe are female and therefore only reproduce vegetatively. The leaves may be preserved in recent lake sediments. Leaves are narrowly acute at the apex, minutely denticulate, recurved and with a strongly visible central vein (Sell and Murrell 1996–2018). Individual cells on either side of the central vein are also visible.

Life form and ecology *Lagarosiphon major* is a perennial elodeid, rooted on the bottom. The stems can extend up to 3 m. It grows in ponds, canals, lakes, and slow-moving rivers. The plant can form dense masses that exclude other species and can be an invasive pest. Vegetative reproduction is by plant fragments dispersed by water (Centre for Ecology and Hydrology 2004).

Distribution *Lagarosiphon major* is native to southern Africa but was introduced as an aquarium escape to northwest Europe where it is considered naturalised but invasive. Only female plants are known beyond its native range.

Lagarosiphon major (Ridl.) Moss (gbif.org)

5.2.96 *Najas flexilis* (Willd.) Rostk. & W.L.E.Schmidt
Hydrocharitaceae (Najadaceae)

Fossil *Najas flexilis* fruits

Morphology The fruit of *Najas flexilis* is a one-seeded achene which is distinctively spindle-shaped. Fruits vary in colour from dark red through to brown and yellow and are 2.0–3.5 mm long and 0.7–1.0 mm wide (Triest 1988). They have a distinctive, reticulate surface pattern composed of shallow, square to hexagonal pits (areoles) which run in longitudinal rows (generally 30–40 in each row—Triest 1988). When in good condition, the fruits are characteristically smooth and shiny. They often split longitudinally, and even small fragments can be identified.

Najas flexilis s.l. occurs as diploids and tetraploids, which have different seed shapes (Les et al. 2015).

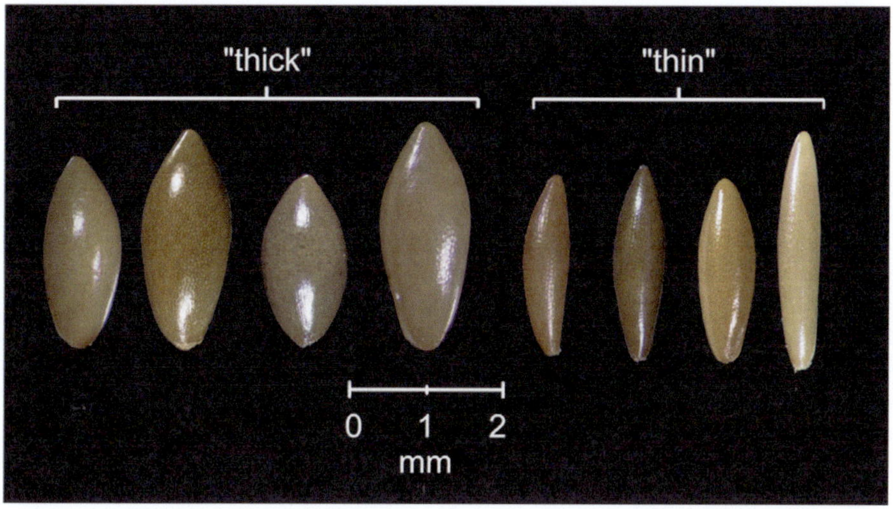

Seed size and colour variants in *Najas flexilis s.l.* The 'thick' phenotypes correspond to the "diploid series" and the 'thin' to the "tetraploid series". Modified from Les et al. (2015).

The leaves possess minute sharp spines that can preserve well in sediments (Gałka et al. 2012; Bishop et al. 2019).

Leaf tip of *Najas flexilis* with spines. Enlargement of two spines on the leaf margin. Drawing of a spine, modified from Nilsson (1961).

Life form and ecology *Najas flexilis* is a monoecious annual elodeid and produces abundant seeds. Seeds can be numerous in lake sediments even if the plant is relatively rare (Birks 1973; Bishop et al. 2018). *N. flexilis* grows in oligotrophic to mesotrophic clear water in lowland lakes where it favours soft silty substrates. Although it can grow in shallow water, it prefers water deeper than 1 m. It is an obligate CO_2 user and is unable to metabolise bicarbonate for photosynthesis. Therefore, it is typically found in water of low pH. As daytime pH in eutrophic waters is often >8, free CO_2 would be largely unavailable to *N. flexilis* for photosynthesis (Bishop 2019). This means that it is highly susceptible to eutrophication, which can be a major cause of its decline (Preston and Croft 1997; Bishop et al. 2019). The occurrence of *N. flexilis* is also restricted by acidification and increased humic acids that can be produced, for example, by the invasion of spruce into the catchment, and by successional lake infilling by fen and bog vegetation (Gałka et al. 2012). Lake-drainage and pollution also cause loss of habitat.

Distribution *Najas flexilis* has a circumboreal distribution across northern Europe, North America (where it is most common), and northern Asia. In Europe it occurs in central Europe, southern Finland, Sweden, and Norway, Latvia, Russia, and the northern and western British Isles, extending south to Switzerland. It is rare and declining in many countries. It is probably extinct in Germany and Poland (Gałka et al. 2012).

Quaternary history *Najas flexilis* was widespread in Britain and northern Europe during the Early and mid-Holocene (e.g. Backman 1948; Godwin 1975; Hultén and Fries 1986 species 127; Lang 1994; Bennike et al. 2001), reaching beyond its present range in areas with warm summers, such as northern Finland. Its seeds are unaffected by cold winters in continental climates. It retreated southwards in Europe and to the west in the British Isles after the Holocene thermal maximum, about 6000 years ago. It is also commonly recorded in Interglacial deposits (e.g. Godwin 1975).

Najas flexilis (Willd.) Rostk. & W.L.E.Schmidt (gbif.org)

5.2.97 *Najas marina* L. Hydrocharitaceae (Najadaceae)

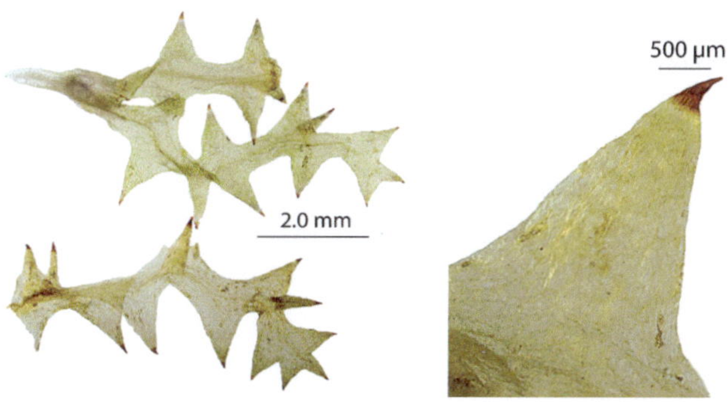

Morphology The fruit of *Najas marina* is a one-seeded achene which is distinctively tooth-shaped, often splitting longitudinally into two halves. Smaller fragments are recognizable. The seeds vary in colour from brown to yellow and are generally 4–6 mm long and 1.2–2.2 mm wide. They have a characteristic reticulate surface formed by pits arranged either irregularly or sometimes more regularly.

Leaves possess teeth that end in a sharp yellow to brown spine that can be preserved in sediments (e.g. Ayres et al. 2008).

Life form and ecology *Najas marina sensu lato* has been split into several subspecies and their fruits have been distinguished (see discussion in Lang 1994; Schou et al. 2023). It is an annual elodeid that lives in lakes, ponds, canals, and slow flowing rivers in generally alkaline fresh to brackish water. It is a thermophilous species associated with mesotrophic to eutrophic conditions with relatively high conductivity. It grows in water up to 1.5 m deep and can develop an extensive root system allowing plants to anchor themselves in soft sediments (Handley and Davy 2002; Ayres et al. 2008). Plants are dioecious with male and female flowers occurring on different individuals. In the absence of male plants, seeds can be produced by apomixis (Handley and Davy 2000). As an annual, *N. marina* produces large numbers of seeds, which can be super-abundant in lake sediments and form a seed bank. Due to a tough seed coat, Agami and Waisel (1984) suggested that fruits could be carried long distances by fish or water birds, hence explaining the wide distribution of *N. marina.*

Distribution *Najas marina* has a global distribution, occurring in Europe, Asia, northern and central Africa, North and Central America, Australia, and several oceanic islands, spanning temperate, sub-tropical, and tropical climates (Stuckey 1985). In Europe *N. marina* is a southern species reaching its northern limit in eastern England and the Baltic coast of Finland and Sweden (Hultén and Fries 1986).

Quaternary history *Najas marina* seed coats are thick and resistant and hence they preserve well in sediments. It is frequently recorded in Holocene and in interglacial and Pliocene sediments in Europe, but rarely from colder periods. It reached its maximum extent at the mid-Holocene thermal maximum, beyond its modern range, especially in Britain (Lang 1994), but then its range contracted with summer cooling. It also became predominantly a species of brackish rather than fresh water near its northern limit in the late Holocene probably as former lakes became acidified or overgrown (e.g. Backman 1941; Godwin 1975; Bennike et al. 2001; Gałka et al. 2024).

Najas marina L. (gbif.org)

5.2.98 *Najas minor* All. Hydrocharitaceae (Najadaceae)

Two *N. minor* seeds from Italian Bronze Age sediments

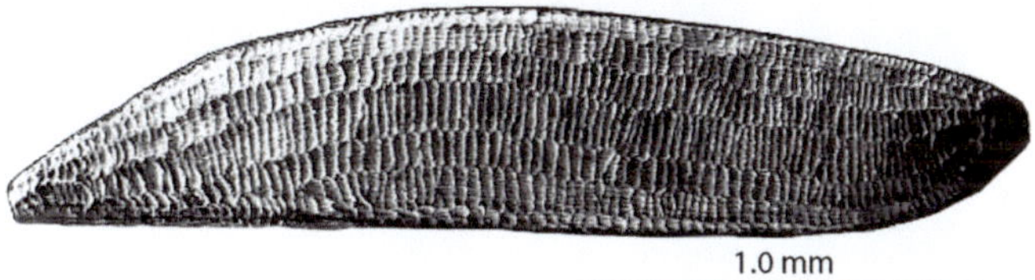

Scanning electron micrograph of a ca. 9250 year old *N. minor* seed from Fakse Bugt, Denmark, modified from Bennike et al. (2001)

Morphology *Najas minor* has elongated spindle-shaped achenes, about 2.5–3.0 mm long and 0.5 mm wide. They are often slightly bent. The characteristic longitudinal ribs are composed of scalariform cells, much broader than long.

Life form and ecology *Najas minor* is an annual elodeid. It grows in stagnant or slowly flowing water in ponds, lakes, ditches, reservoirs, and streams in water up to 4 m deep. It prefers a soft, muddy substrate and mesotrophic to eutrophic and sometimes mildly brackish water.

Distribution *Najas minor* is native to Europe, western Asia, and northern Africa, particularly central and eastern Europe. It is thermophilous, not growing north of 45° N in western Europe and 58° N in Russia and is absent in Britain and Scandinavia. It is introduced in North America by multiple introductions (Les et al. 2015).

Quaternary history *Najas minor* is characteristic of the floras of interglacials in southeast England (Godwin 1975) and Europe (Backman 1951). It is absent from Britain in the Holocene (Godwin 1975). It has been recorded in the Early Holocene north of its present range in Denmark, Finland, and Russia (Rasmussen et al. 2020; Backman 1951) reflecting the warm and long summers then. It was widespread in the southwest Baltic Sea before the marine transgression about 8000 BP (Bennike et al. 2001). Within its present range, it has a record from Poland in the Early Holocene and was reduced by lake dystrophication (Fajer et al. 2012). It was a pioneer in late-Holocene oxbow lakes in the Drava valley on the Hungarian-Croatian border (Gałka et al. 2024).

Najas minor All. (gbif.org)

5.2.99 *Najas tenuissima* (A.Braun ex Magnus) Magnus Hydrocharitaceae (Najadaceae)

Morphology The spindle-shaped achenes of *Najas tenuissima* are about 2–3 mm long. The surface cells are diagnostic, elongated lengthwise (2–7 times longer than broad) and arranged in longitudinal rows (see drawings in Backman 1950; Katz et al. 1965; and photographs in Tralau 1962).

Life form and ecology *Najas tenuissima* is an annual elodeid. It grows in shallow clearwater lakes with a low pH, often one to three meters deep, on sand or gravel. The main threat to its existence is competition from vigorous macrophytes responding to eutrophication, which leads to infilling of shallow water bodies and destruction of its habitat (Vargot et al. 2016).

Distribution *Najas tenuissima* is one of the rarest Tertiary relic aquatic plants with a disjunct range today. 50% of its localities are in southern Finland. Single locations are recorded in European Russia, eastern Siberia, and in the Far East to Japan (Vargot et al. 2016). It is relatively thermophilous.

Quaternary history *Najas tenuissima* was distributed beyond its current range in the warm Early Holocene and in interglacials (Tralau 1962). Recently, it was found far north of its present range in an early Weichselian (last glacial) interstadial in Finland (Väliranta et al. 2009). Fossil seeds were found in Early Holocene sediments in the Moscow region (Vargot et al. 2016) and in 31 lakes in western Finland north of its present range (Backman 1950).

Najas tenuissima (A.Braun ex Magnus) Magnus (gbif.org)

5.2.100 *Stratiotes aloides* L. Hydrocharitaceae

The same seed from both sides

Stratiotes aloides leaf tip

Stratiotes aloides leaf spines. Modified from Gałka (2010)

Fossil seeds of *Stratiotes aloides*, modified from Gałka (2010)

Morphology The seeds of *Stratiotes aloides* are semi-cylindric and usually curved at the micropylar end. The dorsal side is raised, and the ventral side is flat. The seeds can have longitudinal ridges or be smooth. They are 6.5–9.8 mm long (mean 8.5 mm) and 2.0–3.1 mm wide (mean 2.6 mm). Fresh seeds are covered by fugacious mucilaginous hairs and are released by decay of the fruit wall as a gelatinous mass that sinks. Because *Stratiotes* is dioecious, seed set is uncommon. Seeds are ripe in October. Dispersal is probably by zoophagy by ducks and fish (Smolders et al. 1995). More details can be found in Arbor (1920), Cook and Urmi-König (1983), and Gałka (2010).

The leaf spines of *Stratiotes aloides* are characteristic, about 0.5–1.0 mm long. They can be retained on small-mesh sieves (100 μm) during macrofossil analysis (Bennike and Hoek 1999). They are the main source of sedimentary records, especially in northern Europe, as seeds are rarely found (Gałka 2010).

Life form and ecology *Stratiotes aloides* has floating rosettes with long hanging roots or is loosely rooted in mud. It produces new rosettes from stolons and characteristically forms extensive carpets in ponds, ditches, sluggish canals, and oxbow lakes, or less frequently in sheltered bays or inlets of lakes. It does not tolerate water-level fluctuations. It overwinters at the bottom and floats up as new leaves develop containing gas bubbles. After flowering, the fruits descend below water, where the mucilage round the seeds swells and causes dehiscence. In autumn, the plants sink to the bottom again, due to the weight of waterlogged old leaves (Cook and Urmi-König 1983).

Stratiotes grows over nutrient-rich, soft, muddy aerobic or anaerobic substrates. It prefers water of high alkalinity and moderate pH. It accumulates phosphorus and is used as a green fertilizer. It does not tolerate eutrophication and has died out in many places due to habitat destruction.

Distribution *Stratiotes aloides* is widespread in central and eastern Europe through to Siberia. It is native in England, Sweden, Finland, Denmark, the Low Countries, north Germany, Poland, the Po plain in Italy, and the countries along the Danube. It reaches 68°N in Finland, Male plants are rare or absent north of a line from Cherbourg to St Petersburg and mainly occur further south (Cook and Urmi-König 1983).

Quaternary history The genus *Stratiotes* has a long evolutionary history back to the Eocene (Chandler 1922: Sculthorpe 1967). Both seeds and leaf spines have been reported from every British interglacial (Godwin 1975). Holocene seed records of *S. aloides* appear to be restricted to Poland, where they have been recorded from five localities (Gałka 2010), except for a record in western Siberia (Lapateva et al. 2023) and at the Hungarian-Croatian border (Gałka et al. 2024). Records of leaf spines have extended the Late-Glacial and Holocene range to northern Finland, southern Sweden, and the Baltic Sea (Bennike and Hoek 1999). They speculate that *Stratiotes* survived the last glacial maximum in southern Europe or Asia. It probably spread to Poland and The Netherlands by the Late Glacial and arrived in Fennoscandia in the Early Holocene.

Stratiotes aloides L. (gbif.org)

5.2.101 *Hypericum elodes* L. Hypericaceae

Morphology The seeds of *Hypericum elodes* are very small, about 0.8 mm long. They are cylindrical to ellipsoidal, longer than they are wide, with longitudinal ribs with narrow scalariform cells between them (Robson 1981; Sell and Murrell 1996–2018).

Life form and ecology *Hypericum elodes* is a perennial stoloniferous herb up to 40 cm tall. It typically grows in acidic, nutrient-poor water in bogs, fens, and pond and river shores. It is usually found in shallow water that is prone to drawdown (Murphy 2002) but can also be terrestrial or form floating mats in deeper water (Lansdown, 2011b). The micro-structure of the seeds traps air, thus enabling dispersal by floating on water.

Distribution *Hypericum elodes* has a western European distribution. At its northern limit it occurs in the Outer Hebrides northwest Scotland, and extends south to Iberia, the Azores, and North Africa. It is absent in Scandinavia.

Hypericum elodes L. (gbif.org)

5.2.102 *Iris pseudacorus* L. Iridaceae

Morphology *Iris pseudacorus* seeds come in a variety of shapes including flattened, D-shaped, or circular depending on their position in the capsule. They have a slightly rough reddish-brown testa and are large, about 6.5–8.5 mm in diameter.

Life form and ecology *Iris pseudacorus* is a rhizomatous perennial up to 1.6 m high. It grows in marshes, swampy woodland, and at the edges of ponds, rivers, and ditches. It often grows in shallow water but can also be found on saturated but unsubmerged soils. It can grow in a range of soil types and can tolerate a range of pH conditions. Plants normally occur in large clonal patches. Seeds are primarily dispersed by water, where they remain buoyant for several months and are normally deposited at the high-water mark (Coops and van der Velde 1995; Sutherland 1990).

Distribution *Iris pseudacorus* is widespread throughout Europe reaching northernmost Scandinavia, North Africa, and temperate western Asia. It is absent from Iceland and Greenland. It is introduced throughout North America, areas of South America, South Africa, India, the Korean peninsula, Japan, and Australia. It can become invasive where it is introduced.

Quaternary history Fossil seeds are fairly easy to identify due to their size and texture and are occasionally recorded in Holocene sediments (e.g. Godwin 1975).

Iris pseudacorus L. (gbif.org)

5.2.103 *Isoetes lacustris* L. and *Isoetes echinospora* Durieu Isoetaceae

Isoetes lacustris megaspores

Isoetes echinospora megaspores

Fossil *Isoetes* megaspores probably *I. lacustris*

Morphology The spores of *Isoetes* form in sporangia at the bases of the leaves, either megasporangia with female megaspores, or microsporangia with male microspores. The megaspores are very characteristic and can sometimes occur in great abundance. They are small, about 300 µm diameter, and white. They have a prominent triradiate mark where they joined in the tetrad. There is usually a raised ridge around the equator joining the ends of the trilete mark (cf. *Selaginella*). The outer white coat is brittle and can be lost during preservation. The chitinous inner wall often survives. It is thin and usually yellowish, with the trilete mark still visible. The megaspores often split. *I. echinospora* megaspores have prominent spikes whereas in *I. lacustris* these are irregular bumps and protuberances. However, the megaspores of the two species are not always easy to tell apart.

Life form and ecology *Isoetes* spp. are isoetids, namely they are short, tufted perennials living in shallow water in lakes. They form colonies, often together with other isoetids like *Subularia aquatica*, *Littorella uniflora*, and *Lobelia dortmanna*, forming an underwater turf on sandy or stony substrates. *I. echinospora* also grows on soft silt substrates.

Both species grow in cool, clear water in shallow lakes and are early colonisers. *I. lacustris* tends to grow in deeper (>3–8 m) water than *I. echinospora* (2–3 m), as it is sensitive to ice-scour (Rørslett and Brettum 1989), although it can also be found in very shallow water at the lake shore. The slow-growing rosettes of *Isoetes* spp. have high root biomass and efficient air-conducting tissue which enables them to live in oligotrophic, nutrient-poor water. They utilise dissolved CO_2 using CAM (Crassulacean Acid Metabolism) (Arts 2002; Rørslett and Brettum 1989). They discourage competition and shading from more vigorous macrophytes by exuding oxygen from their roots into the sediment, thus reducing nutrient availability (Smolders et al. 2002). Both species have declined in recent decades due to anthropogenic water acidification, eutrophication, and also to reduced water transparency due to humification and dystrophication induced by the spread of conifers (*Picea, Abies*) or *Fagus* in the catchment (e.g. Moravcová et al. 2021).

Distribution *I. lacustris* is distributed throughout Fennoscandia and western and northern Britain. It is restricted in central Europe, the Pyrenees, and central Spain to mountain lakes (Moravcová et al. 2021). There are records from south Greenland and western Iceland. It is widespread in eastern Canada and the Great Lakes region.

I. echinospora has a similar distribution to *I. lacustris* in Europe but is absent from central Spain and more widespread in Iceland. In addition to a similar distribution in eastern North America, it also occurs in the west, especially in British Columbia and central and southern Alaska.

Quaternary history *I. echinospora* was present in the Late Glacial in Europe (Lang 1994) and spread to lakes in northwest Europe and Scandinavia in the earliest Holocene. At this time, it colonised lakes in Greenland (Fredskild 1992) and western Norway (Birks 2000, unpublished data). *I. lacustris* expanded later in the Holocene in central Europe than in northern Europe (Lang 1994; Moravcová et al. 2021). At Kråkenes Lake (western Norway) *I. echinospora* expanded first and was later replaced by *I. lacustris* (Birks 2000). In other locations in western Norway, both species occurred together. *Isoetes* spp. show an overall decline after the mid-Holocene, caused by sediment accumulation, dystrophication, and human disturbance leading to eutrophication and turbid water (e.g. Arts 2002).

Isoetes lacustris L. (gbif.org)

Isoetes echinospora Durieu (gbif.org)

5.2.104 *Juncus* L. Juncaceae

Juncus seeds are characteristic for the genus. They are small, 0.5–1.0 mm, and ovate to ellipsoidal often yellowish with a darker area at each end (Dickson 1970). Körber-Grohne (1964) provides an important key and illustrated descriptions for the identification of northwest European *Juncus* seeds. Fresh seeds have a transparent coat that is lost on fossilisation. It can be gently removed after soaking in water or dilute NaOH. The testa has an outer layer composed of vertical rows of narrow cells. The pattern and structure of these cells are diagnostic and are best observed by placing seeds on filter paper and observing them with overhead lighting during drying. Alternatively, the seeds may be mounted and examined under a high-power microscope. However, the outer cell layer can be lost during fossilisation, leaving an inner layer of plain more-or-less isodiametric cells which are not diagnostic. Therefore, the ability to identify *Juncus* seeds to species is dependent on good preservation.

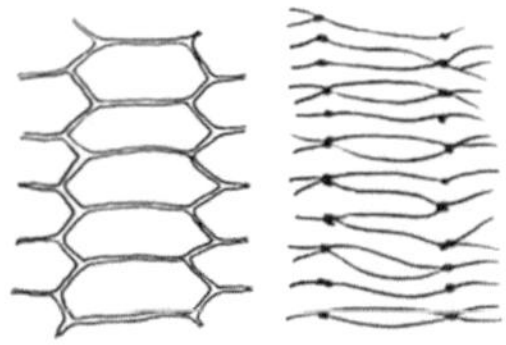

Surface cell patterns of *Juncus effusus*-type and *Juncus articulatus*-type. Modified from Körber-Grohne (1964)

Juncus spp. seeds can occur in great abundance in sediments. They can be recognised by their small size and characteristic yellow colour, often with darker areas at each end. They can resemble *Typha* fruits which are also small, elongated, and yellow. Besides the few that are aquatic, *Juncus* spp. typically grow in wetland habitats (Rose 1989). We have included them here because they produce myriads of seeds that are well dispersed due to their small size and that are well represented in lake and fen sediments.

5.2.105 *Juncus acutiflorus* Ehrh. ex Hoffm. Juncaceae

Drawing of the outer cells of the seed wall in surface view at high magnification, and toothed projections in profile. Modified from Körber-Grohne (1964)

Morphology The small seeds of *Juncus acutiflorus* are about 0.5 mm long and 0.3 mm wide (Körber-Grohne 1964). The cells are arranged in vertical rows and have punctate thickenings at the corners, which form short projections with bifurcate or toothed tips. In subfossil seeds, the cells are more prominent with thickened walls, but the projections may be lost (Körber-Grohne 1964).

Life-form and ecology *Juncus acutiflorus* is a long-lived rhizomatous rush. It grows on acid soil in fens, valley bogs, moorland, damp acid grassland, wet woods, marshes, dune-slacks and the margins of rivers and ponds.

Distribution *Juncus acutiflorus* is widespread in the British Isles and northwest Europe north of The Alps and Pyrenees. It is scattered in Spain, Ireland, Denmark, and rare in southern Sweden, Norway, and Finland. It is absent from Iceland.

Juncus acutiflorus Ehrh. ex Hoffm. (gbif.org)

5.2.106 *Juncus articulatus* L. Juncaceae

Drawings of cells of outer seed wall and toothed projections in profile at high magnification. Modified from Körber-Grohne (1964)

Morphology The seeds of *Juncus articulatus* are about 0.5 mm long and 0.25 mm wide. They are yellow with a distinct darker projection at the top. The yellow colour is usually retained in fossil material. The underlying cells of the testa are quadrate. The surface cells are arranged in vertical rows. They are very narrow and have distinct toothed projections at the cell junctions, only visible in well preserved specimens. See Körber-Grohne (1964).

Life form and ecology *Juncus articulatus* is a rhizomatous perennial rush growing in tufts or in tight mats. It usually grows on acid soil in damp grassland, marshes, fens, dune slacks, wet tracks, and in shallow water in ponds or slow-flowing rivers.

Distribution *Juncus articulatus* has a circumboreal distribution. It occurs throughout Europe, Iceland, the Azores, reaching 71°N in Norway. It is introduced and widely naturalised in the Southern Hemisphere.

Quaternary history *Juncus articulatus* seeds were recorded in abundance in the early Holocene at Kråkenes Lake, western Norway (Birks 2000). There are scattered Holocene records in England (Godwin 1975). *Juncus articulatus* seeds are hard to distinguish from seeds of *Juncus acutiflorus* even when well preserved (Dickson 1970) resulting in records of *J. articulatus/acutiflorus.* In common with *J. articulatus, J. bufonius*, and others, the outer layer of diagnostic cells is easily lost, leaving a membrane of large isodiametric cells (Dickson 1970).

Juncus articulatus L. (gbif.org)

5.2.107 *Juncus bufonius* L. Juncaceae

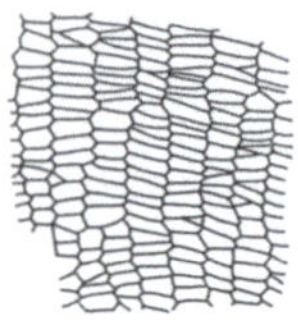

Drawing of the cells of the seed surface at high magnification. Modified from Körber-Grohne (1964)

Morphology The seeds of *Juncus bufonius* are oval, about 0.5 mm long and about 0.3 mm wide. The cells are in longitudinal rows. They are small (10–25 µm broad and 4–10 µm long) and usually not thickened (Körber-Grohne 1964).

Life form and ecology *Juncus bufonius* is an erect or prostrate diffusely branching perennial up to 30 cm tall. It grows on damp soils on wet tracks, arable land, and waste ground, pond edges, etc.

Distribution *Juncus bufonius* is widespread in the Northern Hemisphere, in Europe from the Mediterranean to northernmost Norway. It is not on Svalbard or Greenland.

Juncus bufonius L. (gbif.org)

5.2.108 *Juncus bulbosus* L. Juncaceae

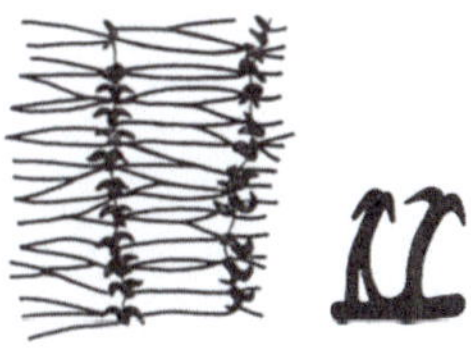

Drawings of cells of the outer seed wall, and recurved spikes in side view, Modified fom Körber-Grohne (1964)

Morphology The seeds of *Juncus bulbosus* are about 0.5 mm long and 0.3 mm wide. Observed using a high-powered microscope, the narrow scalariform cells have bowed walls that are often branched but are not thickened. At their junctions, they bear bifurcated projections with reflexed points (Körber-Grohne 1964).

Life form and ecology *Juncus bulbosus* is a small (30 cm) tufted perennial. It grows in damp habitats on acid to neutral soils, such as beside ponds and in dune slacks. It can also grow underwater (var. *fluitans*) in acid streams and lakes. In southern Norway this variety is blocking waterways since the 1980s and nuisance growth has also been documented in Europe by Schneider et al. (2013).

Distribution *Juncus bulbosus* is widespread in Europe south of the Arctic Circle, including Iceland and the Azores. It also occurs in North America. It is naturalised in Australia and New Zealand.

Quaternary history Godwin (1975) lists records of *Juncus bulbosus* seeds from late-Weichselian and late-Holocene sites in the British Isles.

Juncus bulbosus L. (gbif.org)

5.2.109 ***Juncus conglomeratus*** L. and ***Juncus effusus*** L. Juncacaeae

Juncus conglomeratus

Enlargement showing the cell pattern of *J. conglomeratus*

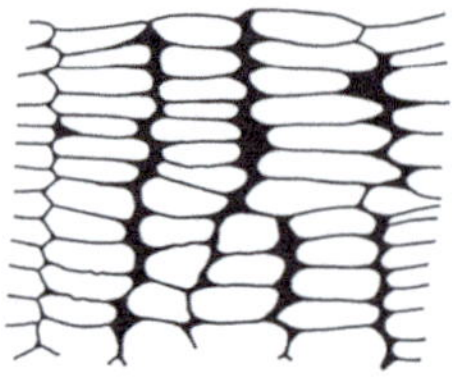

Drawing of the cells modfied from Körber-Grohne (1964)

Juncus effusus

Enlargement showing the cell pattern of *Juncus effusus*

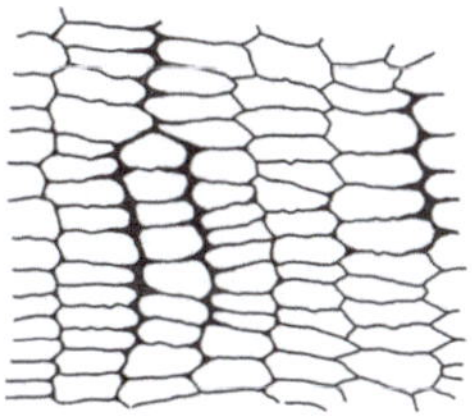

Drawing of the cells modified from Körber-Grohne (1964)

Morphology It is almost impossible to distinguish the seeds of these two species. They are similar in size and shape, about 0.5 long and 2.5 mm wide. The cells are in narrow longitudinal rows, 25–50 µm wide. They are narrow, 5–20 µm, with largely unthickened walls. Those of *J. effusus* may be slightly wider (Körber-Grohne 1964).

Life form and ecology Both species form large tough clumps, up to 150 cm tall. They both commonly occur on ill-drained acid soils in damp grassland, marshes, bogs, and margins of ponds and lakes. *J. effusus* forms a community with *Molinia caerulea* and *Sphagnum* spp. that is widespread in upland Britain.

Distribution *J. conglomeratus* is widespread in Europe south of the Arctic Circle. *J. effusus* has a cosmopolitan distribution with its main abundance in Europe, southeast Asia, and North America. It is naturalised in Australia and New Zealand.

Quaternary history *Juncus effusus*-type (includes *J. conglomeratus*) was recorded in the Early Holocene at Kråkenes Lake, western Norway (Birks 2000). The type has also been recorded at a number of sites in the British Isles (Godwin 1975).

Juncus conglomeratus L.

Juncus effusus L.

5.2.110 *Juncus filiformis* L. Juncaceae

Juncus filiformis seeds still encased in the transparent coat which is not preserved in sediments. The large plain cells are visible

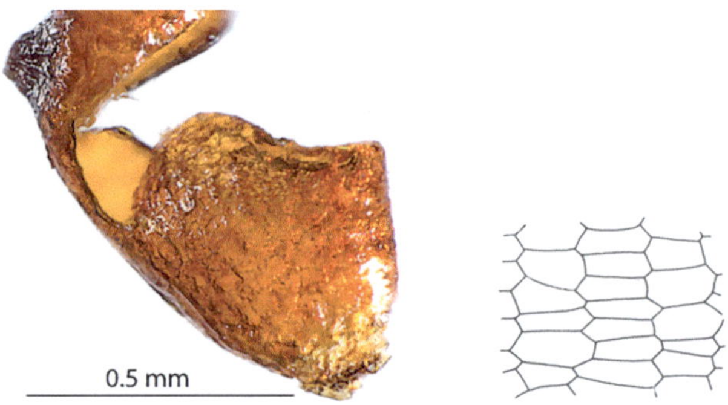

Left: Seed treated by soaking in dilute alkali. The large cells are clearly visible. A high-powered microscope is needed to observe them in detail. Right: Drawing of cells (15–35 μ × 50–70 μ) modified from Körber-Grohne (1964)

Morphology The seeds are very small, about 0.5 mm long. They are relatively broad, with an oval shape. The cells are arranged in longitudinal rows and are large and rectangular or square (15–35 μ × 50–70 μ), with thin walls. They resemble the underlying cells of other *Juncus* species. They are best observed under a high-power microscope.

Life form and ecology *Juncus filiformis* forms perennial rhizomatous mats on acid soil in damp meadows and along damp sandy tracks, in swampy woods, and on stony and silty lake shores. It occurs from sea level to 1400 m in Norway.

Distribution *J. filiformis* has a circumboreal distribution. It is common throughout Fennoscandia, northern Iceland, and Central Europe, The Alps, and the Pyrenees. It is rare in the British Isles and absent from Ireland.

Juncus filiformis L. (gbif.org)

5.2.111 *Juncus subnodulosus* Schrank Juncaceae

Drawing of the outer cells of the seed wall at high magnification, and a toothed projection in profile. Modified from Körber-Grohne (1964)

Morphology The seeds of *Juncus subnodulosus* are rather round, about 0.5 mm long and 0.3 mm wide. Pronounced punctate thickenings occur on both latitudinal and transverse cell walls, composed of small spikes with two equal teeth when seen under a high-powered microscope (Körber-Grohne 1964).

Life form and ecology *Juncus subnodulosus* is a strong, rhizomatous perennial growing in dense stands usually on oligotrophic, base-rich soils. It is locally common in calcareous fens with e.g. *Cirsium palustre*, especially when intermittently mown, in dune-slacks, in ditches, and by water, often on base-rich peat. It may also grow fully submerged.

Distribution *Juncus subnodulosus* occurs through northwest Europe south of about 55°N. It is rare in southern Norway and Sweden except for Öland, and rare in Scotland, Ireland, and western Iberia. It is introduced elsewhere.

Juncus subnodulosus Schrank (gbif.org)

5.2.112 *Lycopus europaeus* L. Lamiaceae

Morphology The nutlets of *Lycopus europaeus* are pyramidal in shape with two flat sides where they were next to the other three on the receptacle (right fruit) and a rounded outer face. There are small transparent tubercles on the upper part of the two inner faces that are lost on fossilisation. A characteristic spongy collar surrounds each nutlet, which allows them to float almost indefinitely until the tissue decays (Praeger 1913). They are approximately 1.5–1.8 mm long and 1.1–1.4 mm wide (see Moon and Hong 2006).

Life form and ecology *Lycopus europaeus* is a rhizomatous herb up to 1 m high that can form large clonal patches. It grows on wet soil at the edges of rivers, ditches, and ponds, usually just above the waterline, and in fens and marshes. It can grow on a range of soil types and can tolerate a range of pH conditions. The fruits are dispersed by water or by wildfowl (Soons et al. 2008).

Distribution *Lycopus europaeus* is widely distributed throughout Europe and western Asia in the temperature and boreal zones, reaching about 62°N, although fruits have been recorded north of this in Finland (Godwin 1975). It is naturalised in North America, where it has the potential to become invasive.

Quaternary history *Lycopus europaeus* has been recorded in Quaternary sediments, including glacial and interglacial deposits (e.g. Godwin 1975). Fossil *Lycopus europaeus* nutlets are easy to identify due to their unique shape.

Lycopus europaeus L. (gbif.org)

5.2.113 *Mentha aquatica* L. Lamiaceae

Morphology *Mentha aquatica* seeds are ovoid to sub-trigonous with a finely verrucose surface. They have a characteristic indented attachment scar at the base where 4 seeds meet (typical of Lamiaceae). They are approximately 0.8–1.0 mm long and 0.6–0.7 mm wide.

Life form and ecology *Mentha aquatica* is a rhizomatous perennial up to 1.5 m tall that can form large clonal patches. It grows on the margins of rivers, ponds, and lakes, and in wetlands and marshes. It usually grows above the waterline but it is tolerant of intermittent flooding. It can also grow on drier waste-ground such as roadsides or field edges. It thrives best on clay soils over a range of pH conditions. The seeds are dispersed by animals or water.

Distribution *Mentha aquatica* is widespread throughout temperate lowland Europe, North Africa, and west Asia. It is introduced throughout much of the rest of the world.

Quaternary history *Mentha aquatica* is reported from Quaternary sediments (e.g. Cappers 1993b; Stachowicz-Rybka 2015; Lodwick 2017; Brown et al. 2018; Godwin 1975).

Mentha aquatica L. (gbif.org)

5.2.114 *Stachys palustris* L. Lamiaceae

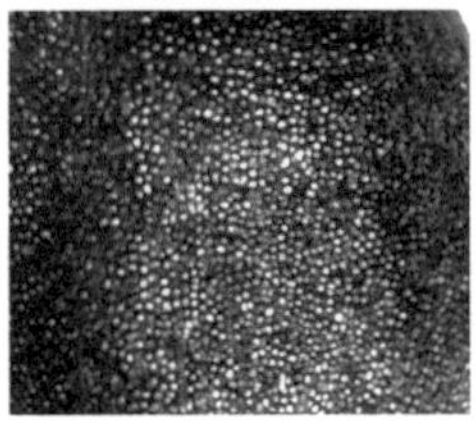

Morphology The nutlets of *Stachys palustris* are obovate to spherical and slightly flattened where they were adjacent to the other 3 nutlets. There is a shallow groove on the ventral side up from the base. The cells of the testa are small and distinct. The size of the nutlets is approximately 1.5 × 2.0 mm.

Life form and ecology *Stachys palustris* is a rhizomatous perennial 30–120 cm tall. It grows on damp soils near lakes, ponds, rivers, and ditches and in damp meadows, as well as in drier sites such as arable land (Taylor and Rowland 2011). The seeds are dispersed mainly by water where they can float for a long period (Praeger 1913; Geertsema 2005).

Distribution *Stachys palustris* is native to much of boreo-temperate Eurasia but is less frequent in Mediterranean areas. It is introduced in North America.

Quaternary history The characteristic nutlets of *Stachys palustris* have been recorded in Quaternary sediments (e.g. Godwin 1975; Bajzáth 1996; Tonkov et al. 2008; Gałka et al. 2024).

https://www.gbif.org/species/2927245

5.2.115 *Utricularia intermedia* Hayne Lentibulariaceae

Utricularia intermedia leaf, showing the spines

Morphology *Utricularia intermedia* seeds are tiny, about 0.5 mm in diameter. In contrast to those of *U. vulgaris* and *U. minor,* they are round or only slightly angular. The surface is covered in round pits and the attachment scar is prominent.

Utricularia intermedia leaves are finely divided and bear tiny simple spines similar to those of *U. vulgaris.* The presence of spines in sediments testifies to the presence of *Utricularia.*

Life form and ecology *Utricularia intermedia* is a perennial carnivorous elodeid that uses stolons to anchor itself to substrates. It usually grows in shallow (0–30 cm), nutrient-poor, acid and dystrophic peat-stained water with optimal pH values from 5.5 to 7.0 but always with high free-CO_2 concentrations of 0.8–1.5 mM (Adamec 2020). It can be found in bog and fen pools, flushes, slow flowing streams, and the edges of lakes. It can also grow in more calcareous water over a peat substrate (Preston and Croft 1997). It favours upland environments, and it can be found up to 1450 m elevation (Adamec 2020). Although *U. intermedia* does set seed, as for *U. vulgaris* and *U. minor*, turion production is thought to be its main means of propagation. *U. intermedia* is often found growing with *U. minor* (Thor 1988). Other European species of *Utricularia* (*U. ochroleuca*, *U. stygia*, *U. bremii* (Adamec 2020), and *U. australis* (Kamayama et al. 2005)) do not set seed.

Distribution *U. intermedia* has a circumboreal distribution, concentrated in the Nordic Countries and northwest Scotland, although it is absent from Iceland (Thor 1988). It is rare elsewhere in Europe, extending southwards into southern France and northern Italy and eastwards into northwest Russia. It occurs across Asia into the Kamtchatka peninsula, eastern Siberia, and northeast China. In North America, *U. intermedia* is found throughout Alaska and most of Canada extending east into Labrador and Newfoundland and south into northern California and Pennsylvania (Adamec 2020).

Utricularia intermedia Hayne (gbif.org)

5.2.116 *Utricularia minor* L. Lentibulariaceae

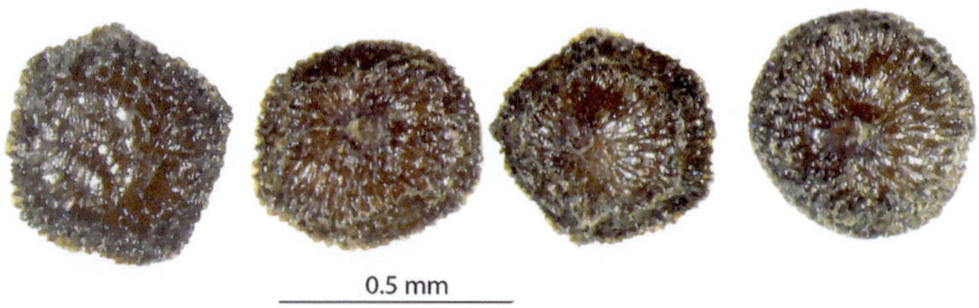

Morphology The seeds of *Utricularia minor* are very similar to those of *U. vulgaris*. They are about 0.5 mm diameter and prismatic, 5–6 sided, but have less-pronounced ridges than *U. vulgaris,* appearing more rounded. *U. minor* leaf segments lack teeth or bristles.

Life form and ecology *Utricularia minor* is a perennial rootless carnivorous elodeid most often growing in acid, generally oligotrophic, humic-rich waters, although it can also grow in nutrient-poor, base-rich environments (Thor 1988; Adamec 2020). It is associated with the stagnant conditions offered by bog pools, flushes, ditches, and the quiescent edges of small lakes where it grows in shallow water (Preston and Croft 1997). This species is very sensitive to human-induced degradation of freshwater wetland environments by hydrological impacts and invasive species, exacerbated by climate change, and it has declined in many areas. *U. minor* is not a large seed producer and reproduces vegetatively by turions (Neid 2006).

Distribution Taxonomic challenges may confuse the distribution of *U. minor* with the extremely similar *Utricularia bremii* (Adamec 2020), although the latter is not thought to set seed (Bobrov et al. 2022). *U. minor* is a widespread circumboreal species found throughout the Northern Hemisphere although it also occurs in Papua New Guinea (Neid 2006). In Europe *U. minor* extends into northern Scandinavia and Iceland and south into the Mediterranean region where it becomes less common (Preston and Croft 1997). It occurs across Asia, occurring in the Himalaya of India, Nepal, and Bhutan, and in Myanmar and Japan. In North America it extends south into California and the Intermountain Region (Neid 2006).

Utricularia minor L. (gbif.org)

5.2.117 *Utricularia vulgaris* L. Lentibulariaceae

Utricularia vulgaris leaf with bristles on the teeth

Utricularia vulgaris turions

Morphology The tiny seeds, about 0.5 mm in diameter, of *Utricularia vulgaris* are highly characteristic. They are prismatic with flat sides bordered by raised cells and a slightly convex top with the attachment scar. See Prausová et al. (2022) for more photographs.

The leaves of *Utricularia vulgaris* are pinnately branched and the narrow segments are toothed, each tooth bearing a tiny characteristic bristle. It is possible that these can be recovered from sediments and analysts would do well to keep an eye out for them. Further, it is also possible that parts of the turions, which are composed of densely appressed leaves, could be recognised in sediments.

Life form and ecology *U. vulgaris* is a perennial, free-floating, carnivorous elodeid. It favours base-rich, often humic conditions in lowland lakes, ponds, river backwaters, and ditches. It prefers oligotrophic-mesotrophic water and declines with increased eutrophication. This is the major cause of its decline throughout much of its range (Preston and Croft 1997; Prausová et al. 2022). *U. vulgaris* produces fruits only occasionally, and turions are thought to be its main means of reproduction (Prausová et al. 2022). The closely related *U. australis* appears not to set seed.

Distribution The species is widespread throughout lowland Europe being less common in the Mediterranean region. It is absent from Iceland (Thor 1988). It extends across Siberia and the Tibetan plateau (Bobrov et al. 2022). It also occurs widely across North America.

Utricularia vulgaris L. (gbif.org)

5.2.118 ***Lythrum hyssopifolia*** L. Lythraceae

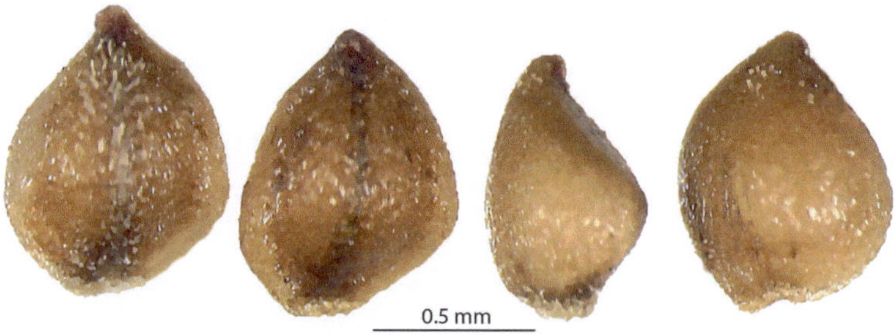

Morphology The small seeds of *Lythrum hyssopifolia* are broadly obovoid and slightly flattened. The long hilum is marked by a dark line. The surface cells are slightly raised, but the whole seed is smooth. The seeds are about 0.9 mm long and 0.7 mm wide.

Life form and ecology *Lythrum hyssopifolia* is a small erect or prostrate annual, 5–25 cm tall. It grows on bare mud or among open vegetation in hollows where water collects in winter. Its seeds may survive many years (Preston 1999) resulting in its erratic appearance when conditions become favourable, usually after soil disturbance. For example, it was found at Heydon Pond in Norfolk, England after pond restoration, having previously been thought to be extinct there (Sayer and Parmenter 2021).

Distribution *Lythrum hyssopifolia* has a Euro-Siberian southern-temperate distribution. It reaches its northern limit in Britain and southern Scandinavia. It is widely naturalized outside this range.

Lythrum hyssopifolia L. (gbif.org)

5.2.119 *Lythrum portula* (L.) D.A.Webb Lythraceae

Morphology The seeds of *Lythrum portula* are small (about 0.5 mm). They are ovate, with the outer side convex and the inner side concave. The micropyle is about 0.1 mm up from the base. The surface cells are distinct and somewhat swollen, appearing as a coarse reticulum (Jonsell and Karlsson 2010).

Life form and ecology *Lythrum portula* is a creeping annual herb up to 0.3 m high. It grows at muddy margins of pools or puddles in open communities or on bare wet ground where it is favoured by disturbance. It avoids calcareous soils. Dispersal is by wind, water, or animals.

Distribution *Lythrum portula* is widespread in Europe and southern Scandinavia and extends to western Asia and North Africa. It is introduced in the western Americas and New Zealand.

Quaternary history *Lythrum portula* seeds have been reported from Medieval-age sediments in England (Hall et al. 1995).

Lythrum portula (L.) D.A.Webb (gbif.org)

5.2.120 *Lythrum salicaria* L. Lythraceae

Morphology The seeds of *Lythrum salicaria* are elongated, about 1 mm long, and are often angular due to their position within the capsule. The surface cells are distinct and elongated. Those near the base are rectangular and somewhat swollen. The seeds are readily identifiable from these characters.

Life form and ecology *Lythrum salicaria* is a rhizomatous perennial up to 2 m tall. It can form large colonies. It grows in wetland habitats including marshes and swamps bordering ponds, lakes, and streams, often as part of the early successional vegetation. It prefers moist, disturbed, neutral to weakly acidic soils. Seeds are primarily dispersed by wind and further by water and waterfowl.

Distribution *Lythrum salicaria* occurs throughout Europe and temperate Asia. It is introduced and is a problematic invasive fen plant in much of North America and in South Australia.

Quaternary history *Lythrum salicaria* has been recorded in Quaternary sediments (e.g. Wasylikowa 1967; Brayshay and Dinnin 1999; Brown et al. 2018; Godwin 1975; Gałka et al. 2024).

Lythrum salicaria L. (gbif.org)

5.2.121 ***Trapa natans*** L. Lythraceae

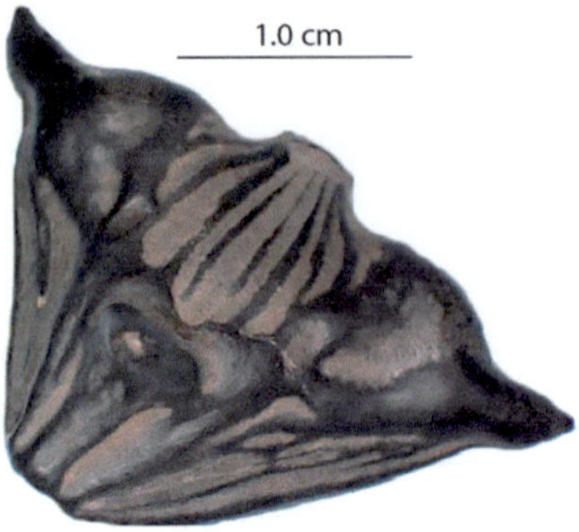

Nut

Fossil ‘harpoons’

Chinese decoration made of *Trapa* nuts

Morphology The unmistakeable nuts of *Trapa natans* are about 3 cm wide. They usually have four projecting spikes, sometimes only two, formed from the sepals, and a characteristic ribbed pattern. The much smaller 'harpoons' (sometimes called spines or barbs) are also unmistakeable. They are projections of the fruit spikes, forming spines with backward-pointing branchlets. They are usually much more numerous in sediments than the large nuts (Miotk-Szpiganowicz and Gałka 2009).

Life form and ecology *Trapa natans* is an annual plant, forming neat floating rosettes of angular leaves on the surface that are rooted in organic mud by a long stem, up to 3–5 m long, bearing finely divided assimilatory roots and linear submerged leaves. Each rosette petiole has a swelling containing aerenchyma. The flowers are produced above the water, but when fertilised, bend down so the fruit matures under water. Dispersal can occur by water currents, animal transport, and human agency. A ripe fruit sinks to the bottom, where it can remain viable for up to 12 years. Seeds can overwinter in frozen lakes in northern climates if covered by sediment. *Trapa* requires warm summers with water temperature above 20 °C to flower and set seed. It primarily occurs in non-polluted nutrient-rich, high pH lakes with a constant water level. It does not tolerate salinity. A detailed account of *Trapa* biology and ecology is given by Palm et al. (2024).

The seeds are edible (water chestnuts) and *Trapa* has been cultivated in China and the Indian subcontinent for at least 3000 years. The plant has many other uses. However, where it is introduced, it can become a significant invasive weed, reducing faunal and floral biodiversity (e.g. US Fish and Wildlife Service 2019; CABI.org n.d.).

Distribution *Trapa natans* is widespread in central and southern Europe and western Russia, especially in more continental areas south of the 18 °C summer isotherm. It extends south to Africa and east to the Indian subcontinent and southeast Asia (China, Japan, etc.) where it is widely cultivated for its nuts. It is naturalized and invasive in northeast USA (US Fish and Wildlife Service 2019; CABI.org n.d.) and it has also been introduced but strictly controlled in Australia. It is hard to eradicate because of the longevity of the seeds in the lake mud.

Quaternary history Holocene fossil records are common in central and eastern Europe. Mid-Holocene fossils are frequently recorded in Denmark, southern Sweden north to Uppsala (60°N), and southern Finland as far as 64°N, and also in the Baltic states (Lang 1994). There is one record in eastern England (Flenley et al. 1975). These records are well beyond its current northern limit at about 53°N. Samuelsson noted these records in 1934 and inferred that summer temperatures were 1.5 °C warmer in Sweden and Finland during the mid Holocene, providing early evidence for a mid-Holocene thermal maximum. The now numerous fossil occurrences, especially in Finland, confirm this. The increased distribution and abundance may have been enhanced by transport and cultivation of its nuts by prehistoric humans. However, it would not have grown beyond its thermal limit at the time. Its decline since the mid Holocene was most likely due to decreased temperatures in the late Holocene associated with natural acidification. *Trapa* fruits and harpoons occurred in oxbow lake sediments in the Drava valley in southern Hungary in the Medieval Warm Period (Gałka et al. 2024).

Trapa natans L. (gbif.org)

5.2.122 *Menyanthes trifoliata* L. Menyanthaceae

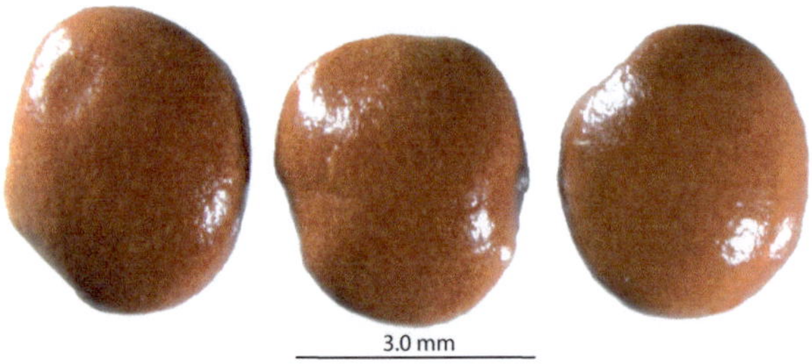

Morphology The 'bog beans' of *Menyanthes trifoliata* are easily seen in fresh sediment. They are large (3.0 × 2.5 mm), usually orange-coloured, and shiny, although they can turn black in sediments. They are rounded in face view with an indentation at the position of the micropyle. They are more flattened in side view. They readily split into halves, revealing the thick wall.

Life form and ecology *Menyanthes trifoliata* is an emergent rhizomatous perennial up to 50 cm tall with large trifoliate leaves. It grows in shallow water at lake shores, pools, or in slow-flowing rivers. It also occurs in swamps, flushes, and dune-slacks, and in bog pools. It tolerates a wide range of water chemistry but is intolerant of shade. It prefers peaty substrates and forms transitional mire communities, often with *Carex rostrata, Potentilla palustris,* and *Sphagnum* spp.

Distribution *Menyanthes trifoliata* occurs widely across the boreal and temperate zones of the Northern Hemisphere.

Quaternary history "The polished, kidney-shaped seeds of *Menyanthes trifoliata* are perhaps the most abundant and most easily recognizable of all Quaternary plant fossils" Godwin (1975, p. 313).

Menyanthes trifoliata L. (gbif.org)

5.2.123 *Nymphoides peltata* (S.G.Gmel.) Kuntze Menyanthaceae

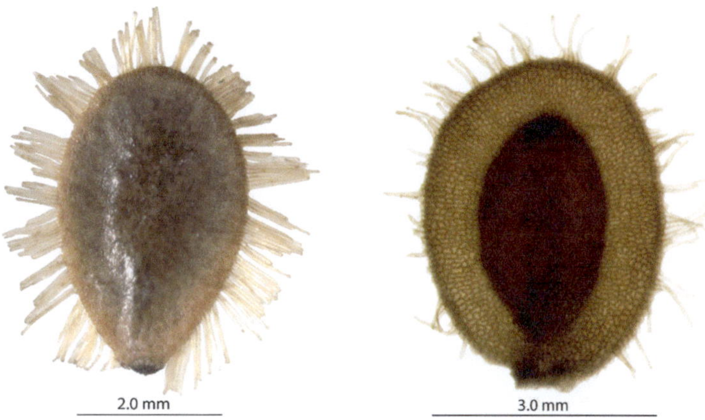

Nymphoides peltata seed photographed with back lighting

Morphology The fruit of *Nymphoides peltata* is a 1.2–2.5 cm beaked capsule that contains the many flat, smooth seeds (Cook 1990). The seeds are ovate and flat, about 3 mm wide and 4 mm long and about 1 mm thick. The margin bears a fringe of thin bristles. These are hollow, and together with the hydrophobic seed surface allow the seeds to float. The bristles, and the minute projections on them can cause the seeds to become entangled in floating mats and also become attached to waterfowl which disperse them to other water bodies. The seeds do not survive ingestion. They are resistant to desiccation (Smits et al. 1989; Cook 1990).

Life form and habitat *Nymphoides peltata* is a perennial nymphaeid with rounded floating leaves spreading to 3 m, and long branching stolons with adventitious roots. The stolons build large patches, and broken fragments can develop into new plants. *N. peltata* typically grows in lakes, ponds, reservoirs, and slow-moving rivers but it can also grow on damp mud in swamps, and wetlands, especially those that are flooded in winter. It grows best on mineral substrates such as clay. It occurs primarily in eutrophic, alkaline water less than 3 m deep. It requires basic water to provide calcium for the production of floating leaves (Naul and Mikulyuk 2009 and references therein). The ecology and invasiveness of *N. peltata* are discussed in detail by Naul and Mikulyuk (2009).

Distribution *Nymphoides peltata* is native to Eurasia including China and India. It is the only temperate species in the otherwise tropical or subtropical genus. It is introduced and invasive in North America, and also in New Zealand and southern Scandinavia (Naul and Mikulyuk 2009).

Quaternary history *Nymphoides peltata* seeds have been recorded in Quaternary sediments (e.g. Bhandari et al. 2011; van der Ham et al. 2008). It was a notable pioneer in late Holocene oxbow lakes in the Drava valley at the Hungarian Croatian border (Gałka et al. 2024).

Nymphoides peltata (S.G.Gmel.) Kuntze (gbif.org)

5.2.124 *Pilularia globulifera* L. Marsileaceae

Sporocarps of *Pilularia globulifera*

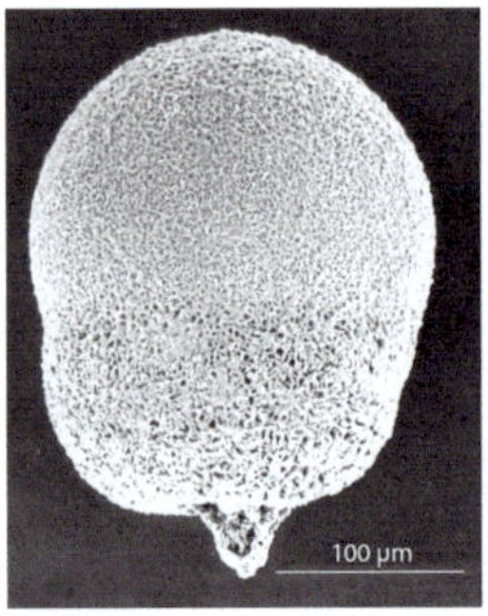

SEM of a *Pilularia globulifera* megaspore, modified from Stafford (1995).

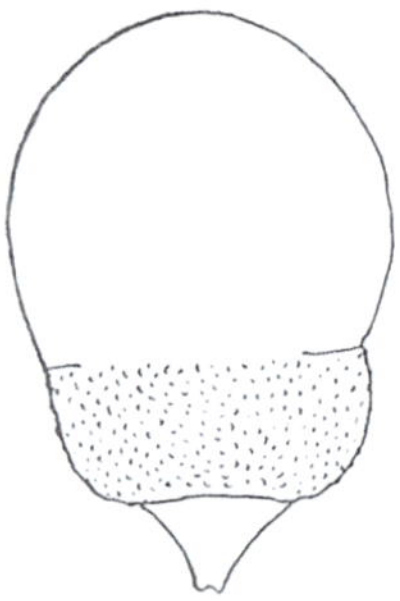

Drawing of a *Pilularia* megaspore modified from Watts (1959)

Morphology *Pilularia globulifera* sporocarps are spherical and divided into 4 chambers. Each chamber contains one or two sori bearing microspores and megaspores. The sporocarps are about 3 mm in diameter with a tough scabrous surface. The spores are released when the sporocarp decays. The megaspores are oval with a constriction giving them a pear-shaped appearance. Watts (1959) describes the megaspores 'as having the appearance of minute acorns with a constriction separating the spore into a smooth 'fruit' and a rough 'cupular' portion.

Life form and ecology *Pilularia globulifera* is an aquatic pteridophyte. Creeping rhizomes send up thin cylindrical fronds forming a grass-like mat. The spherical sporocarps are formed at the base of the fronds. *Pilularia* grows on peaty or sandy substrates on the shores of ponds and lakes, often being submerged, in nutrient-poor acid and dystrophic water.

Distribution *Pilularia globulifera* occurs in northern and western mainland Europe, in southernmost Scandinavia, and the Atlantic seaboards of Ireland and Iberia.

Quaternary history Megaspores have been recorded in the Gortian interglacial (Holsteinian) at Kilbeg in Ireland (Watts 1959). There are also a few Holocene records in Britain (Godwin 1975). These records are usually accompanied by microspores.

Pilularia globulifera L. (gbif.org)

5.2.125 *Montia fontana* L. Montiaceae

Morphology *Montia fontana* ssp. *fontana* seeds are round in outline, 1.0–1.2 mm in diameter, and usually shiny black. The radicle of the curved embryo projects slightly next to the micropyle. The cells of the seed coat are clearly delimited and domed, with slightly wavy margins, distinct from Caryophyllaceae. Even fossil fragments are readily identifiable. *Montia fontana* ssp. *amporitana* seeds are shiny and the marginal cells bear pointed tubercles. *Montia fontana* ssp. *chondrosperma* seeds are dull black and are coarsely tuberculate (Schou et al. 2023).

Life form and ecology *Montia fontana* is an annual to perennial herb with stems up to 50 cm long. It grows as mats, sometimes floating, in shallow water such as in slow streams and lake shores. It forms mounds in montane bryophyte springs and grows as wide cushions on wet ground. It prefers slightly acidic to neutral substrates.

Distribution *Montia fontana* has a boreo-temperate distribution throughout northwest Europe from southern Spain to northern Norway. It also occurs in central and east Asia, Oceania, Africa, and in the Americas.

Quaternary history *Montia fontana* seeds have been found in Quaternary sediments (e.g. Godwin 1975; Burrows 1974; Wood and Walker 2008).

Montia fontana L. (gbif.org)

5.2.126 ***Nuphar lutea*** (L.) Sm. Nymphaeaceae

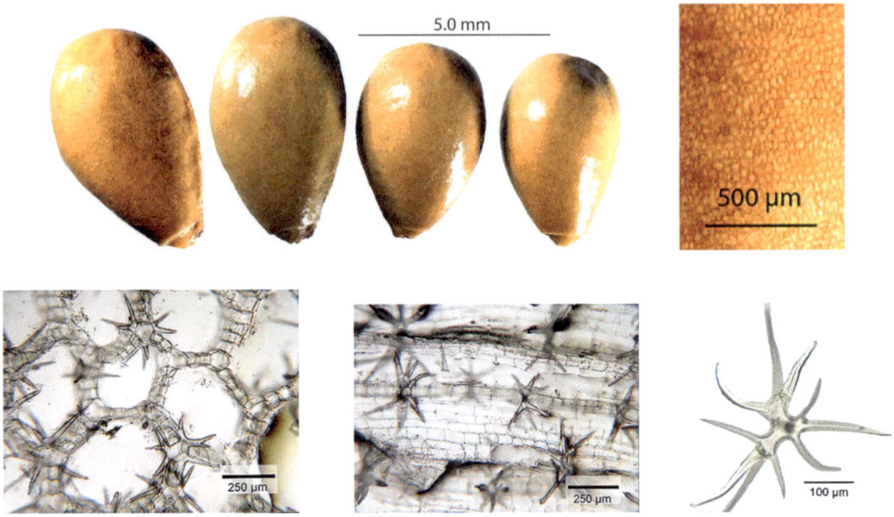

Nuphar lutea seeds; close-up of surface cells; sclereids in the leaf and stem, and a sclereid isolated by KOH digestion. Sclereid photographs are extracted from Marrotte et al. (2012)

Morphology The large seeds of *Nuphar lutea*, 4.0–5.8 mm long, are ovoid, smooth, and with an apical point with a collar at the point of attachment. This may be lost in fossil material, leaving a hole. The seeds are bluntly keeled, with flatter faces towards the keel. The cells of the surface are distinct, small, and isodiametric, and broken seeds can be identified. Sclereids (sometimes called trichosclereids or star cells) are also found as macrofossils. They are supporting cells within the tissues to ensure air circulation. They are usually star-shaped with a minutely scabrous surface. They survive acetolysis during pollen preparation and can be counted on pollen slides (Marrotte et al. 2012), (see also *Nymphaea alba*). They may be more representative of Nymphaeaceae presence than the seeds, which, if found at all, are usually in small numbers or as fragments (see Birks (1973) for *N. variegatum*).

Life form and ecology *Nuphar lutea,* yellow waterlily, is a robust rhizomatous perennial with submerged and floating leaves that spread up to 2 m on the water surface. It grows in mildly acidic to basic nutrient-enriched shallow water up to 5 m deep in shallow lakes, ponds, canals, and slow streams. Seeds are dispersed either in the floating capsule segments or by short-time floatation across the water surface before sinking to the sediment (Ridley 1930; Hart and Cox 1995), or by being transported by waterfowl and fish (Ridley 1930; Preston and Croft 1997).

Distribution *Nuphar lutea* occurs throughout temperate, boreal, and subarctic Europe and west Asia. Related species occur in North America.

Quaternary history *Nuphar lutea* seeds have been frequently recorded in Quaternary sediments (e.g. Hannon and Gaillard 1997; Godwin 1975). Philippsen (2013) shows that *Nuphar* has a freshwater reservoir effect and should not be used for AMS radiocarbon dating.

Nuphar lutea (L.) Sibth. & Sm. (gbif.org)

5.2.127 *Nuphar pumila* (Timm.) DC. Nymphaeaceae

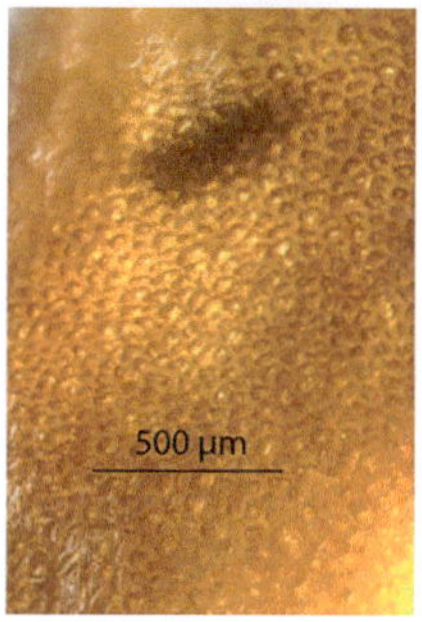

Morphology The seeds of *Nuphar pumila* are obovate, with a collar at the basal point of attachment. They are about 3.5–4.0 mm long and 2.2–2.6 mm wide. The surface cells are small, distinct, and isodiametric.

Life form and ecology *Nuphar pumila* is a floating-leaved water lily, smaller than *N. lutea*. It grows on mud, silt and peaty substrates in mesotrophic lakes in water up to 2.5 m deep. It is intolerant of eutrophication. The seeds sink on release and do not survive prolonged drying out or digestion by animals. Those that are enclosed in floating fruit segments are dispersed by water currents (Preston and Croft 1997).

Distribution *Nuphar pumila* is a circumboreal species, frequent in northern Europe, particularly Scandinavia. In the UK, it is centred on Scotland. It is known from The Alps and is rare in northern Spain. It occurs throughout Asia and in eastern North America.

Quaternary history There are no direct records of *Nuphar pumila* seeds, but its relict distribution in Britain may imply that it has survived there during the Holocene (Preston and Croft 1997).

Nuphar pumila (Timm) DC. (gbif.org)

5.2.128 ***Nymphaea alba*** L. Nymphaeaceae

Nymphaea alba seeds

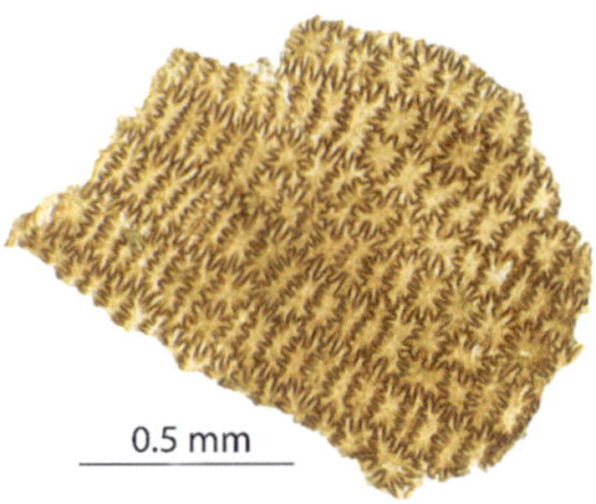

Surface cells, stereomicroscope

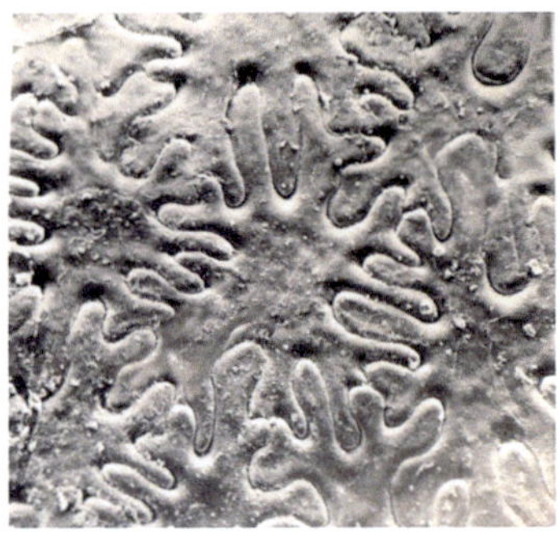

Scanning electron micrograph of surface cells. Modified from Schoch et al. (1988)

Sclereids from the petiole, stereomicroscope

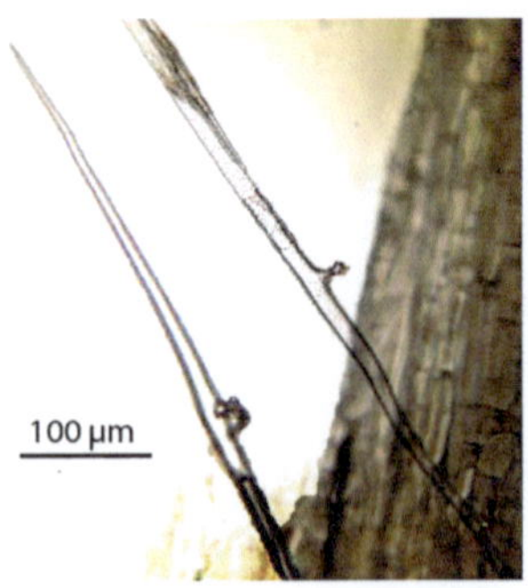

Individual petiole sclereids of *N. odorata* (modified from Marrotte et al. 2012)

Morphology *Nymphaea alba* seeds are large and obovate, about 2.4–3.0 mm long and 1.5–1.8 mm wide. The surface cell pattern is unique to *Nymphaea*, consisting of interlocking 'jig-saw' cells. Even small fragments can be recognised. The sclereids are supporting cells found throughout the plant, enabling aeration of the tissues. In the petiole, they are elongated, and elsewhere they are usually branched or star-shaped, as in *Nuphar*. They are destroyed by acetolysis during pollen sample preparation (cf. *Nuphar*) but they can be found as macrofossils (Marrotte et al. 2012).

Life form and ecology *Nymphaea alba* is the familiar white waterlily, a large rhizomatous plant with floating leaves. It grows on a range of fine-grained lake sediment in oligotrophic to eutrophic water 0.5 to 1.5 m deep in sheltered bays of lakes or in slow rivers. It is intolerant of disturbance (e.g. by boat traffic) and exposure to wind, and it decreases with excessive eutrophication. The seeds can float for 1–3 days and are dispersed by water. They are fully digested when eaten by waterfowl and fish, and they cannot tolerate prolonged drying out. (See Preston and Croft (1997) and references therein).

Distribution *Nymphaea alba* is widespread in Europe, reaching 66°N in Norway, but it decreases in abundance southwards towards the Mediterranean and is rare in North Africa. It is common in western Asia but is scattered across the rest of Asia. It is naturalised in North America, where the closely related *N. odorata* is widespread. It is widely planted and can become a weed in southeast Australia and New Zealand.

Quaternary history *Nymphaea alba* joins the aquatic plant succession in new lakes after sufficient lake mud has been deposited to support its large rhizome. It recirculates nutrients, such as phosphorus that have been bound in the sediments and stimulates a productive early hydrosere (e.g. Birks 2000) accompanied by other large perennial macrophytes such as *Potamogeton* spp. Marrotte et al. (2012) found that the records of *Nymphaea* pollen and sclereids on pollen slides do not coincide, probably because *Nymphaea* sclereids are not preserved during pollen preparation. As macrofossils, the records of *Nymphaea* seeds and Nymphaeaceae sclereids (star cells) coincide very well in the Early Holocene, as at Kråkenes Lake in western Norway (Birks 2000). At two sites in southern Norway, '*Nymphaea* star cells' picked as macrofossils gave a better Holocene record of Nymphaeaceae presence and abundance than seeds, that are sparsely produced (Birks 1973), basal hair cells, or pollen, which were absent (Eide et al. 2006).

https://www.gbif.org/species/2882437

5.2.129 *Nymphaea tetragona* Georgi Nymphaeaceae

Morphology The ovoid seeds are 2–3 mm x 1.5–2 mm, smaller than *N. alba*. They are smooth but have the characteristic *Nymphaea* pattern of interlocking cells. However, these are not so easy to see as in *N. alba*.

Life form and ecology *Nymphaea tetragona* is a small rhizomatous nymphaeid with floating leaves. It can form dense monospecific stands in shallow (up to 2 m) acid or alkaline nutrient-rich water on muddy substrates in lakes, ponds, and slow-moving streams (Mossberg and Stenberg 2018). The seeds are dispersed by water or by attachment to waterfowl, or by simple decay of the fruit (e.g. Wiersema 1988).

Distribution In Europe, *Nymphaea tetragona* is almost confined to Finland, but it occurs across temperate Asia and in northwest North America, reaching the arctic. It is very rare in northern India.

Nymphaea tetragona Georgi (gbif.org)

5.2.130 *Epilobium hirsutum* L. Onagraceae

Morphology *Epilobium hirsutum* fruits are oblong to obovoid and acute at the base, with a dense covering of tubercles. They are 1.0–1.5 mm long. A pappus of hairs was attached at the top. The ventral side is flattened and has a longitudinal scar (see Berggren 1974). *Epilobium* species with similar fruits include *E. parviflorum, E. palustre, E. montanum, E. collinum, E. roseum, E. adnatum, E. lamyi,* and *E. obscurum* (Group 1, Berggren 1974).

Life form and ecology *Epilobium hirsutum* is a rhizomatous perennial up to 1.5 m tall. Plants occur in small to large clonal patches arising from the branching rhizomes. It grows beside streams and ponds, and in drier parts of fens. It typically grows in moist soils but can tolerate occasional flooding that covers the root system. It can tolerate low acidic to basic soil pH. Seeds are primarily dispersed by wind and the fluffy pappus allows them to float on water.

Distribution *Epilobium hirsutum* occurs throughout Europe reaching southern Sweden and Finland, temperate Asia, and parts of Africa. It is introduced in southern Norway and Finland, some areas of North America, and Australia, although it is not invasive. It is absent from Iceland and Greenland.

Quaternary history *Epilobium hirsutum* has been recorded in Quaternary sediments (e.g. Bittmann 2007; Field and Peglar 2010). The tuberculate seed coat remains quite recognisable through time.

Epilobium hirsutum L. (gbif.org)

5.2.131 *Epilobium palustre* L. Onagraceae

Morphology The fruits of *Epilobium palustre* are 1.8–2.0 mm long and narrowly obovoid with a distinct neck that bore the pappus, and a pointed base. The ventral side has a longitudinal scar (see Berggren 1974). The dorsal surface is finely pitted or slightly papillose (Group III, Berggren 1974).

Life form and ecology *Epilobium palustre* is a slender stoloniferous willowherb that inhabits acidic fens, flushes, and ditches, from lowland to subalpine elevations. The fruits are wind dispersed. The overwintering turions can be dispersed by wildfowl.

Distribution *Epilobium palustre* occurs widely across the temperate and boreal Northern Hemisphere, reaching southern Scandinavia, but it is absent from the Southern Hemisphere.

Quaternary history Although its fruits are distinctive, *Epilobium palustre* has not often been recorded from Holocene sediments (e.g. van Geel et al. 2020; Godwin 1975).

Epilobium palustre L. (gbif.org)

5.2.132 *Epilobium parviflorum* Schreb. Onagraceae

Morphology *Epilobium parviflorum* fruits are 0.8–1.0 mm long and obovate, being slightly wider at the rounded top. There is no neck (*cf. E. palustre*). The base is rounded. There is a ventral longitudinal scar. The dorsal surface is covered with dense, equally spaced papillae that are circular in outline (Group 1, Berggren 1974). The fruits are similar to but smaller than those of *E. hirsutum.*

Life form and ecology *Epilobium parviflorum* is a rhizomatous perennial willowherb up to 80 cm tall. It inhabits stream banks, fens, marshes, damp woods, and moist montane meadows.

Distribution *Epilobium parviflorum* occurs across central Europe to western Asia but hardly reaches as far north as Scandinavia. It is found in the Himalaya and Chinese mountains and Japan, and also the Canary Islands and the Azores. It is introduced in North America.

Quaternary history There are occasional records of *Epilobium parviflorum* in Late Glacial and Holocene sediments from Britain (Godwin 1975).

Epilobium parviflorum Schreb. (gbif.org)

5.2.133 ***Erythranthe guttata*** (DC.) G.L.Nesom (Syn. *Mimulus guttatus* Fisch. ex DC.) Phrymaceae (Scrophulariaceae)

Morphology The tiny seeds of *Erythranthe guttata* are about 0.5 mm long and 0.3 mm wide. The surface is reticulate with relatively large lumina with tall translucent walls.

Life form and ecology *Erythranthe guttata* is a rhizomatous and stoloniferous perennial up to 50 cm tall. It usually grows in large masses in water seeps on cliffs, in waterfall spray zones, beside slow rivers and streams, and in wet meadows and woodlands.

Distribution *Erythranthe guttata* is native to western North America, from Mexico to Alaska. It is introduced and often widespread in continental Europe, the Faroe Islands, New Zealand, and eastern North America.

Mimulus guttatus Fisch. ex DC. (gbif.org)

5.2.134 *Pedicularis palustris* L. Orobanchaceae

Morphology *Pedicularis palustris* seeds are 2.0–2.4 mm long and 1.0–1.4 mm wide. They are ovoid with a broad base narrowing to the top. The dorsal side is rounded. The ventral side is flatter with a deep furrow. The surface is pitted or reticulate with thick walls and rounded lumina. Seeds can float for up to 2 months before sinking (Praeger 1913).

Life form and ecology *Pedicularis palustris* is an annual or biennial herb up to 60 cm tall. It grows in a wide range of base-rich to acidic fens, wet heaths, and damp meadows.

Distribution In the British Isles *Pedicularis palustris* is most common in the north and west including Ireland, and in East Anglia. It is widespread throughout Europe, from arctic Scandinavia and southwards apart from Iberia and the Mediterranean, at elevations of up to 1200 m. It extends across boreal Asia to northern China. It also occurs in eastern Canada.

Pedicularis palustris L. (gbif.org)

5.2.135 *Callitriche* L. Plantaginaceae (Callitrichaceae)

Callitriche species produce schizocarp fruits consisting of four mericarps joined at the inner margins. The mericarps separate at maturity, and each contains a single seed. They are bean shaped, and 1–2 mm long. In fresh condition, the fruits can be identified to species or groups by their size, shape, the surface reticulum, and the presence and configuration of a lateral wing. Lansdown (2008) presents anatomical characters that can be diagnostic in fresh material, but which are rarely visible in subfossil specimens.

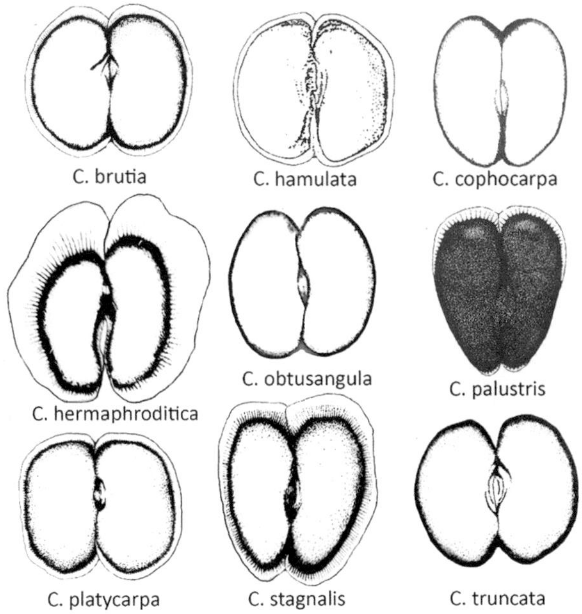

Drawings of paired *Callitriche* fruit mericarps. Extracted from Lansdown (2008) and Stace (2019). Not to scale

The mericarp walls have a characteristic reticulate pattern that is retained during preservation in lake sediments. It is often the only part that is preserved, and poorly preserved macrofossils may resemble highly characteristic net bags. These can be mistaken for animal remains. Fossil *Callitriche* fruits can seldom be identified to species, unless preservation is excellent and the wing, if any, is preserved. They fall into two indistinct groups separated by the size of the reticulum lumina. Large lumina occur in *C. cophocarpa* and *C. palustris*. The rest have smaller lumina. *C. truncata* has a very dense reticulum with tiny lumina.

Early Holocene fossil *Callitriche* fruits from Kråkenes Lake, western Norway

Callitriche species are slender elodeid aquatic plants rooted to the substrate on wet mud and in shallow still water often as pioneers in disturbed habitats. They grow in oligotrophic, mesotrophic, or eutrophic water and sometimes in brackish conditions. We do not consider all the European species but include most of those occurring in northwest Europe.

C. hermaphroditica and *C. truncata* have only submerged leaves, whereas the other species develop terminal rosettes floating on the water surface. In addition, all species except *C. hermaphroditica* and *C. brutia* have terrestrial forms that grow on damp mud beside water bodies or animal-disturbed puddles. They are difficult to identify, and flowers and fruits are usually needed and even pollen grains (see Lansdown 2008).

Callitriche fruits have often been recorded in Quaternary lake sediments. Various specific determinations have been made on the basis of the wing morphology and the size of the reticulum (e.g. Godwin 1975).

5.2.136 ***Callitriche brutia*** ssp ***brutia*** Petagna and ***Callitriche brutia*** ssp ***hamulata*** (Kütz. ex W.D.J.Koch) O.Bolòs & Vigo Plantaginaceae (Callitrichaceae)

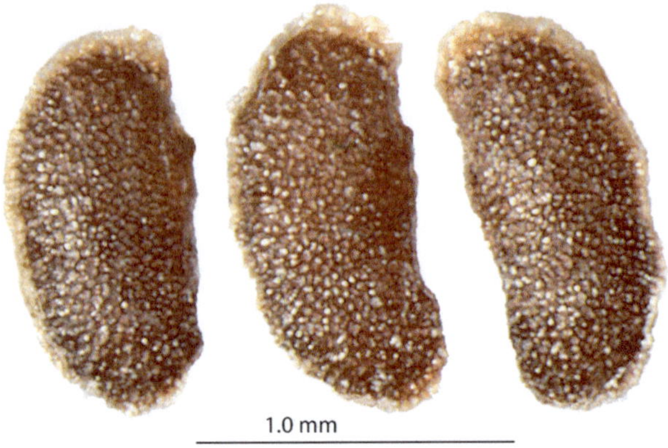

Callitriche brutia ssp *brutia*

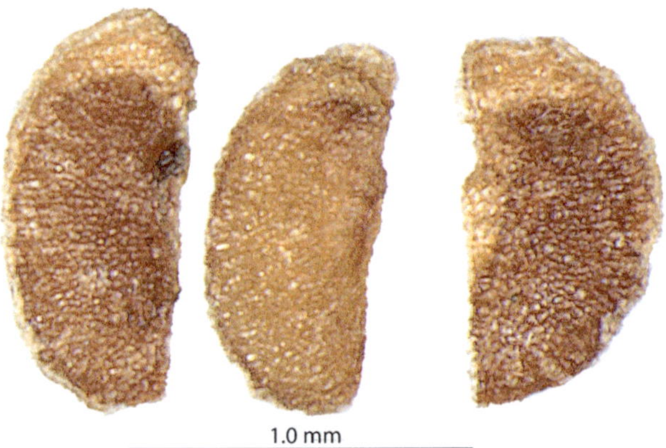

Callitriche brutia ssp *hamulata*

Morphology *C. hamulata* was redefined as *C. brutia* ssp. *hamulata* (Kütz. ex W.D.J.Koch) O.Bolòs & Vigo by Stace (Stace 2009). The fruits of both species are indistinguishable. They are up to 1.5 mm long and 0.5 mm wide. They are narrowly winged, the wing being slightly wider at the top. The reticulate pattern is well marked with small lumina.

Life form and ecology *C. brutia* ssp *brutia* grows in shallow water across a broad range of habitats including ephemeral ponds (including Mediterranean temporary ponds), wheel ruts, lakes, canals, and both slow and fast flowing rivers. Both species are calcifuge, growing in acid oligotrophic water. *C. brutia* ssp *hamulata* tends to occur in flowing water, while *C. brutia* is more prevalent in still water and more ephemeral habitats (Lansdown 2008).

Distribution *C. brutia* ssp *brutia* occurs throughout Europe extending north to Iceland and northern Scandinavia, south to the Mediterranean, and east to the Balkans. Beyond, it is found in Greenland, Morocco, Iran, and the Caucasus region. It is introduced in Australasia (Lansdown 2022) and North America (Lansdown 2008). *C. brutia* ssp *hamulata* occurs in Europe from The Netherlands to Iberia.

Callitriche brutia Petagna (gbif.org)

Callitriche brutia var. hamulata (Kütz. ex W.D.J.Koch) Lansdown (gbif.org)

5.2.137 ***Callitriche cophocarpa*** Sendtn. Plantaginaceae (Callitrichaceae)

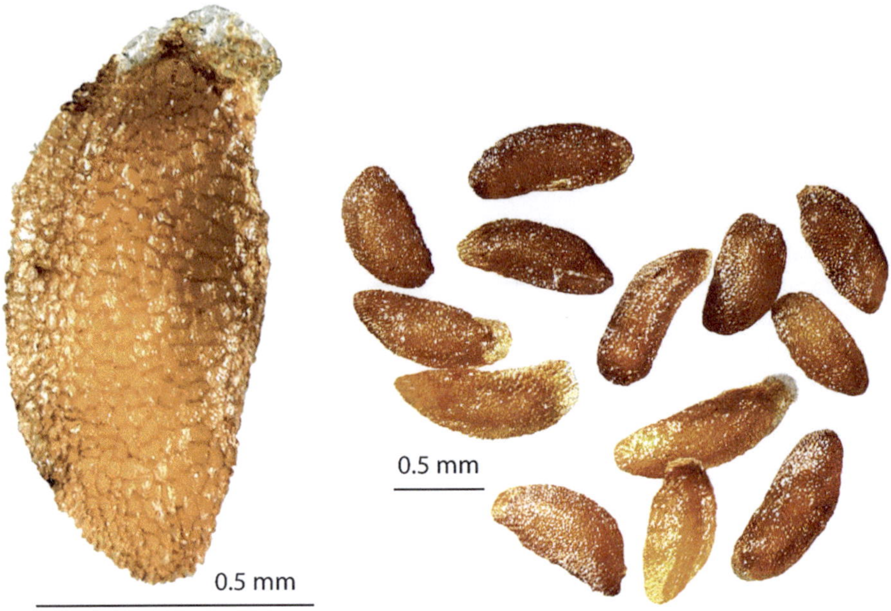

Morphology The mericarps of *Callitriche cophocarpa* are small, up to 1 mm long. They are wingless or very narrowly winged except at the top where the wing is more pronounced. The reticulum is distinct, with relatively large lumina.

Life form and ecology *Callitriche cophocarpa* grows in still water and in flowing streams and rivers, in marsh pools, ditches, ponds, and on damp soil (e.g. Mossberg and Stenberg 2018). It usually grows in mesotrophic water ranging from low to high conductivity (Lansdown 2008, 2011a).

Distribution *Calllitriche cophocarpa* has a continental European distribution from the Arctic to the Mediterranean. It extends across northwest Asia. It is naturally absent from Britain. It is thought to have become extinct in The Netherlands and Belgium (Lansdown 2008).

Callitriche cophocarpa Sendtn. (gbif.org)

5.2.138 *Callitriche hermaphroditica* L. Plantaginaceae (Callitrichaceae)

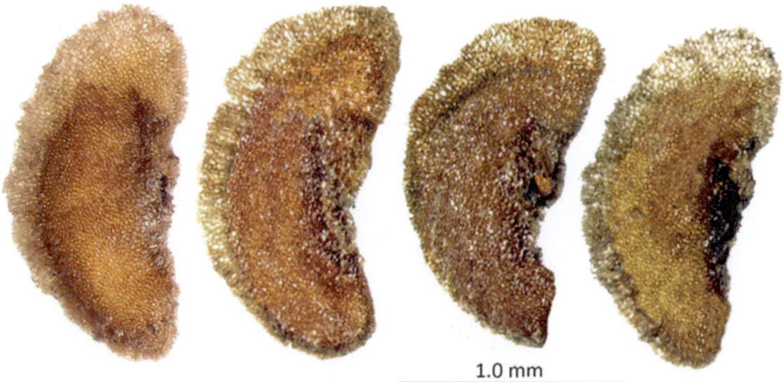

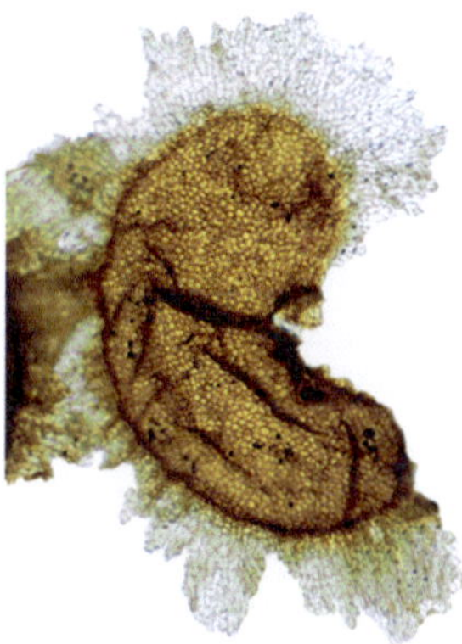

Photograph of a well-preserved fossil by B. Brandon Curry

Morphology The mericarps of *Callitriche hermaphroditica* are 1.2–2.4 mm long and bear a pronounced wing which is broader at the top. The reticulum is distinct with small lumina.

Life form and ecology *Callitriche hermaphroditica* grows on clay and silty substrates in still water in lakes, canals, and in slow flowing rivers, and brackish bays. It prefers neutral to basic, mesotrophic to eutrophic water but does not tolerate strong eutrophication or competition from larger species. It can fruit prolifically.

Distribution *Callitriche hermaphroditica* has a circumboreal distribution. In Europe it is a northern species (primarily north of 53°N) extending south to northern Britain, northern Germany, and Romania. There are two recognised sub-species. C. *hermaphroditica* ssp. *macrocarpa* occurs throughout northern Scandinavia and east to Tibet and across North America; C. *hermaphroditica* ssp. *hermaphroditica* is more southerly (Lansdown 2008).

Callitriche hermaphroditica L. (gbif.org)

5.2.139 *Callitriche obtusangula* Le Gall
Plantaginaceae (Callitrichaceae)

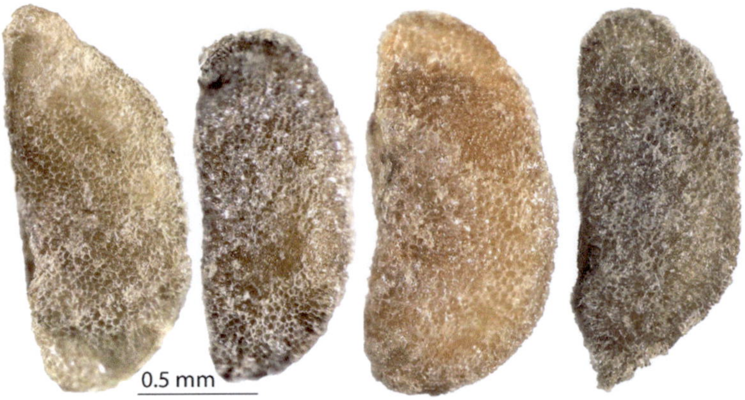

Morphology The mericarps of *Callitriche obtusangula* are about 1.5 mm long. They lack any wing. The reticulum is fine with rather indistinct lumina.

Life form and ecology *Callitriche obtusangula* is primarily a lowland species, typically found in eutrophic-mesotrophic and alkaline rivers, fens, and ditches growing on a wide range of substrates. It can tolerate water with pH 4–6, as well as brackish water associated with coastal grazing marshes (Lansdown 2008). In still, non-flowing water it fruits freely.

Distribution *Callitriche obtusangula* is almost confined to Europe. It has a southerly distribution, reaching southern parts of the British Isles and Ireland, but rarely further north. It is absent in Scandinavia. It extends to the western Balkan region and around the Mediterranean. It is sparse in Iberia and North Africa.

Callitriche obtusangula Le Gall (gbif.org)

5.2.140 *Callitriche palustris* L. Plantaginaceae (Callitrichaceae)

Morphology The mericarps of *Callitriche palustris* are about 1 mm long. A lateral wing is absent or very narrow but is wider and more pronounced at the top. The fruits are usually blackish. The reticulum is strong with large lumina, sometimes tending to be aligned vertically.

Life form and ecology *Callitriche palustris* is a perennial elodeid that grows in a wide range of habitats including lakes and rivers where falling water levels expose bare mud substrates, and in disturbed muddy habitats such as wheel ruts and watering holes (Mossberg and Stenberg 2018; Roberts 2023). It prefers base-rich oligotrophic water. It reaches 2600 m elevation in The Alps, where it is more associated with permanent water (Lansdown 2008).

Distribution *Callitriche palustris* occurs across the Northern Hemisphere. It is widespread in Europe including Fennoscandia and Iceland. It is also found in Greenland, North America, Asia (including Japan and much of China) and in south-east Australia where it is a non-native (Lansdown 2022).

Callitriche palustris L. (gbif.org)

5.2.141 *Callitriche platycarpa* Kütz.
Plantaginaceae (Callitrichaceae)

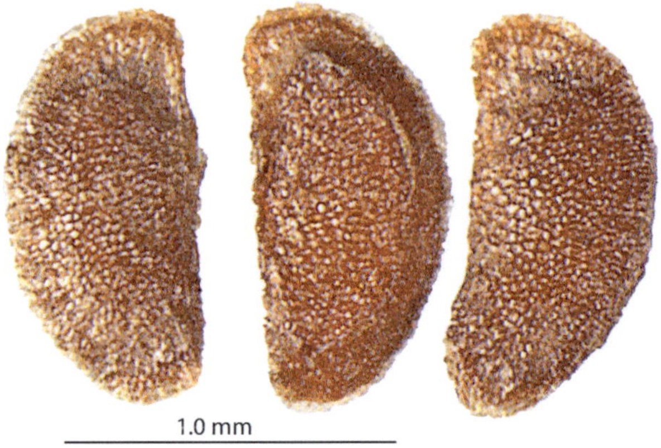

Morphology The mericarps of *Callitriche platycarpa* are 1.2–1.5 mm long. The lateral wings are very narrow and a bit wider at the top. The reticulum is fine with distinct lumina.

Life form and ecology *Callitriche platycarpa* is a perennial elodeid that grows in most lowland aquatic habitats and on damp mud. It is most frequent in mesotrophic to eutrophic water with a pH of 6–9. This species fruits less freely than other *Callitriche* taxa (Preston and Croft 1997).

Distribution *Callitriche platycarpa* is a northwest European species. It extends north to southern Sweden, south to southern France (and occasionally found in Spain), and east to the Czech Republic. It has recently been recorded in southern Norway (Rørslett 2022).

Callitriche platycarpa Kütz. (gbif.org)

5.2.142 ***Callitriche stagnalis*** Scop.
Plantaginaceae (Callitrichaceae)

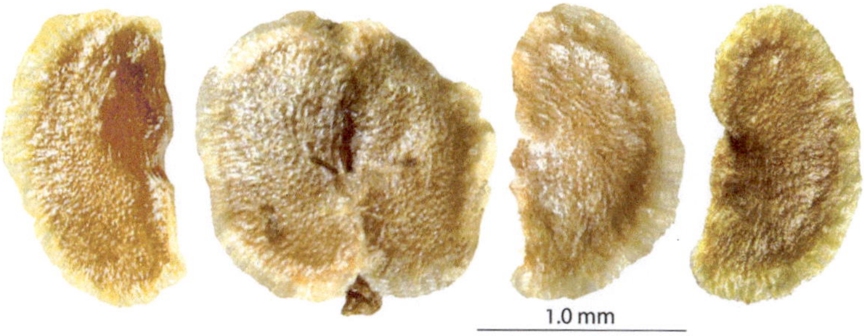

Morphology The mericarps of *Callitriche stagnalis* are 1.1–1.8 mm long. They have a prominent wing that is slightly wider at the top. The reticulum has small lumina, similar to *C. platycarpa.*

Life form and ecology *Callitriche stagnalis* is a perennial elodeid found on fine organic and inorganic substrates in ephemeral pools, wheel ruts, wet field corners (seasonally flooded ground), as well as in ponds, lakes, and slow-flowing rivers (Preston and Croft 1997). It grows in shallow mesotrophic-eutrophic water and occasionally in oligotrophic and acid conditions. Plants usually produce abundant fruit.

Distribution *Callitriche stagnalis* is found all over Europe extending eastwards into Belarus and western Russia and north to southern Scandinavia and Finland. It occurs as a non-native in North America (Philbrick et al. 1998) and Australasia (Lansdown 2022).

Callitriche stagnalis Scop. (gbif.org)

5.2.143 *Callitriche truncata* ssp. *occidentalis* (Rouy) Schotsman Plantaginaceae (Callitrichaceae)

Morphology The mericarps of *Callitriche truncata* are about 1 mm long with a rounded profile. *C. truncata* subsp. *truncata* has a narrow wing, but in *C. truncata* subsp. *occidentalis* (illustrated) the wing is absent. The fruits are dark, and the reticulum has tiny lumina, appearing as a simple cell pattern.

Life form and ecology *Callitriche truncata* ssp. *occidentalis* grows in brackish coastal lagoons and temporary pools and in mesotrophic-eutrophic lakes, canals, ditches and slow-flowing rivers where it can be a pioneer species (Barry and Wade 1986). C. *truncata* ssp. *truncata* is apparently more restricted to coastal lagoons and pools than subsp. *occidentalis*, but its ecology is currently poorly understood (Lansdown 2008).

Distribution *Callitriche truncata* ssp. *occidentali*s occurs along Mediterranean, Iberian, and French coasts. It also occurs in the northern Netherlands and in England (especially the Midlands region) and Ireland. C. *truncata* ssp. *truncata* is restricted to eastern Mediterranean coasts (see Lansdown 2008).

Callitriche truncata subsp. occidentalis (Rouy) Schotsman (gbif.org)

5.2.144 *Gratiola officinalis* L. Plantaginaceae

Morphology The seeds of *Gratiola officinalis* are very small, about 0.5 mm long. From a flat apex the seed narrows down to a pointed base. The dorsal side is convex, and the ventral side is flat. The surface is covered by a marked reticulum with relatively large lumina (see Bojňanský and Fargašová 2007).

Life form and ecology *Gratiola officinalis* is a perennial herb up to 30 cm tall. It grows along streams, irrigation channels, and in marshy ground, optimally in tall sedge beds on slightly acid to basic alluvial soil, avoiding the most acid sites.

Distribution *Gratiola officinalis* occurs across temperate Europe and western Asia, and from southern Denmark to the Mediterranean. It is very rare in Scandinavia and Britain.

Gratiola officinalis L. (gbif.org)

5.2.145 *Hippuris vulgaris* L. Plantaginaceae (Hippuridaceae)

Fresh fruits with exocarp

Fruits with exocarp removed, as in fossil material

Morphology The fresh fruit is a sessile or shortly stalked, dark green or blackish, ovoid drupe that is about 1.6–2.0 mm long and up to 1 mm wide with a round plug filling a hole at the top. The thin fleshy exocarp of the fruit is tightly appressed to the hard, slightly smaller, and highly distinctive endocarp that is similarly ovoid (or barrel-shaped), fairly smooth or rugose, and typically of a cream colour. A longitudinal 'seam' is sometimes visible. The emergence of the radicle during germination opens the plug. The plug is usually lost in fossil material, leaving a round opening at the top. Fresh fruits can float for several months (Praeger 1913).

Life form and ecology *Hipppuris vulgaris* is a rhizomatous perennial which has two growth forms, occurring as a long-stemmed, flaccid submerged elodeid and as a flowering and fruiting emergent, like a stiffly erect Christmas tree or more imaginatively like a horse-tail (its English common name is Mare's-tail). It grows in shallow (<2.5 m) water around lakes and ponds, beside sluggish rivers, and in swamp habitats. It prefers base-rich water and can also thrive in brackish lakes and freshwater tidal marshes. It grows in mesotrophic to eutrophic water but is negatively affected by severe eutrophication.

Distribution *Hippuris vulgaris* is widespread throughout the arctic, boreal, and temperate regions of the Northern Hemisphere, south to Myanmar and north to northern Ellesmere Island in the Canadian Arctic Archipelago, but it is absent from Greenland. It also occurs in Patagonia in southern South America. It is closely related *to H. lanceolata*, widely distributed in The Arctic, and *H. tetraphylla*, much rarer in Arctic regions (Flora of Svalbard. http://svalbardflora.no/index.php?id=661).

Quaternary history *Hippuris* fruits are commonly recorded in both glacial and interglacial Quaternary sediments (e.g. Bennike 1989). It is particularly characteristic of the last Glacial (Weichselian) period, and it occurs in lesser amounts throughout the Holocene (e.g. Godwin 1975). It seems to be well represented by its fruits when it is locally present.

Hippuris vulgaris L. (gbif.org)

5.2.146 *Littorella uniflora* (L.) Asch. Plantaginaceae

SEM of a *Littorella* fruit, modified from Arts and Heijden (1990)

Morphology The one-seeded nuts of *Littorella uniflora* are 1.5–1.8 mm long and 0.8–1.0 mm wide. The base is flat, and the apex is pointed, often with the remains of the style base. There is an obvious 'seam' or ridge down one side. The surface is pitted but appears rough and the cells are not distinct in fresh material where the pericarp is still present. The Scanning Electron Micrograph by Arts and Heijden (1990) shows the pitted surface as seen in fossil material, including details of the base.

Life form and ecology *Littorella uniflora* is a tufted perennial isoetid, spreading by stolons to form an underwater turf. Its narrow cylindrical pointed leaves do not have obvious air channels. It grows in shallow water at the margins of oligotrophic or mesotrophic lakes on gravel, sand, or other non-organic substrates, often in association with other isoetids. It can survive moderate cultural eutrophication and acidification but dies out when these are severe (Preston and Croft 1997). *Littorella* declines as sediment accumulates in lakes, or it becomes overgrown during eutrophication (e.g. Pedersen et al. 2006). When exposed by falling water-level or drying of shallow pools, it flowers and sets seeds. The seeds are sparsely produced but form a seed bank and are viable for up to 30 years (Arts and Heijden 1990).

Distribution *Littorella uniflora* is widespread across northwest Europe where it is declining, extending south to Spain and Portugal, Corsica, and Sardinia. It is common in the lowlands of Norway south of 65°N, and in southern Sweden, and rare in south Finland. It occurs throughout the British Isles, being particularly widespread in the north and west. It occurs in western Iceland, and is scattered in eastern Canada, especially on Prince Edward Island.

Quaternary history *Littorella uniflora* pollen is recorded far more frequently than its seeds. However, these have been found throughout the Late Glacial and the Holocene in the UK (Godwin 1975).

Littorella uniflora (L.) Ascherson (gbif.org)

5.2.147 ***Veronica*** L. Plantaginaceae

V. anagallis-aquatica L.

V. beccabunga L.

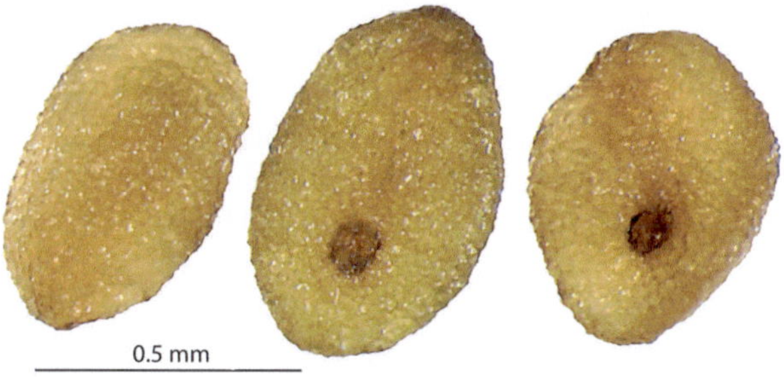

V. catenata Pennell

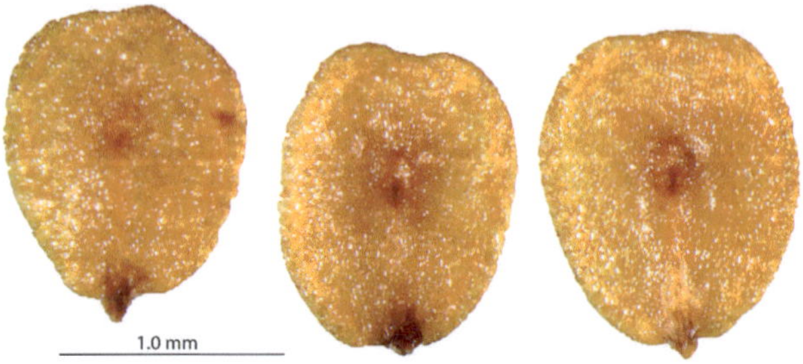

V. scutellata L.

Morphology The distinctive seeds of the genus *Veronica* have a similar structure. We illustrate seeds of four common wetland species. They are flattened, irregularly discoid, and sometimes concave. The ventral chalaza is visible as a dark spot and may project slightly. It is linked to the micropyle by a more or less obscure seam. *V. anagallis-aquatica, V. beccabunga,* and *V. catenata* are in Subgenus *Beccabunga* and have similar small seeds, about 0.5 mm in diameter. *V. scutellata* is in subgenus *Veronica* and has larger seeds, about 1.0 mm in diameter with a central chalaza. The surfaces are verrucate or minutely reticulate. Distinctive differences in the testa cell structure are seen under the SEM (e.g. Mazur et al. 2022 and references therein).

Life form and ecology *V. beccabunga* and *V. scutellata* are perennial stoloniferous herbs. *V. anagallis-aquatica* and *V. catenata* are annual or sometimes perennial herbs rooting from the stem nodes. All the species inhabit the margins of ponds, slow streams, ditches, springs, and wet meadows. They have a broad preference for mesotrophic to eutrophic, acid to alkaline water, except *V. scutellata* that prefers oligotrophic to mesotrophic, acid to neutral water. All fruit freely and the seeds are dispersed by water and animals. Detached plant parts can result in vegetative spread.

Distribution *V. anagallis-aquatica* is widespread across the temperate Northern Hemisphere, becoming scarce to the north, in southern Africa, South America, southeast Australia, and New Zealand.

V. beccabunga is widespread across temperate and boreal Europe, reaching central Asia. It becomes scattered in the boreal zone. It occurs around the Great lakes in the USA and is scattered in western USA and Canada.

V. catenata occurs across temperate Europe but is rare east of Poland and in western Asia. It is mostly absent from Fennoscandia. It occurs in temperate North America, and in southeast Australia and New Zealand.

V. scutellata is widespread throughout Europe reaching central Asia. It reaches The Arctic in Fennoscandia. It occurs across temperate and boreal North America, and in New Zealand.

Veronica anagallis-aquatica L. (gbif.org).

Veronica beccabunga L. (gbif.org).

Veronica catenata Pennell (gbif.org)

Veronica scutellata L. (gbif.org).

5.2.148 Poaceae L. (Syn. Gramineae)

Poaceae inflorescences have a common flower and fruit structure, distinct from other plants. The structure of the spikelet is shown in a). There are two glumes at the base of the spikelet. There may be one or more florets. Each of these is composed of an outer lemma and an inner palea, enclosing the fruit, the caryopsis or grain.

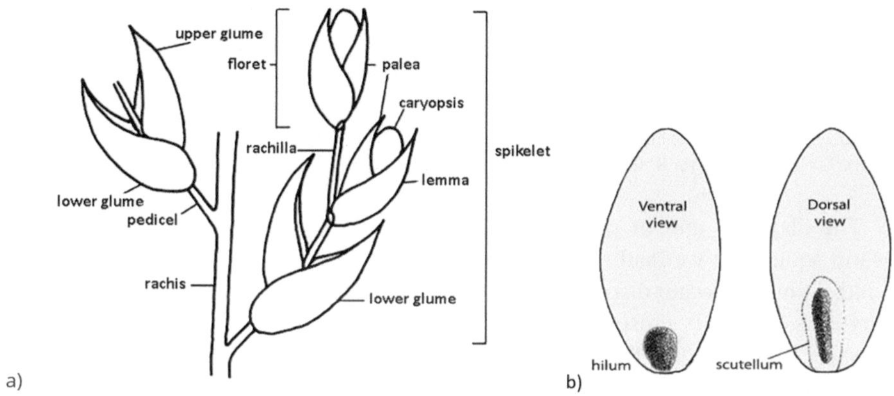

(a) Structure of a generalised grass spikelet (b) Structure of a generalised grass caryopsis. (Modified from Federal Noxious Weed Disseminules of the United States)

The distinctive structure of a grass caryopsis is shown in (b). The position of the embryo, the scutellum is on the dorsal side. The hilum, the attachment of the seed to the pericarp, is on the ventral side. The stigma is at the top, and the attachment to the floret is at the base. The seed enclosing the embryo is attached by its funicle to the pericarp at the hilum. If the pericarp is translucent, the hilum is usually visible as a dark area on the ventral side of the fruit. The position of the embryo is often apparent on the opposite side of the caryopsis in living material. If the pericarp is thicker, its structure usually reveals the positions of the embryo and the hilum. Fossilised caryopses generally do not show the position of the embryo, but the hilum remains as a distinct dark area.

The hilum is a distinctive feature of fossilised caryopses. It is often dark. It can be round, ellipsoid, or linear. It can be positioned right at the base of the caryopsis or further up the ventral side.

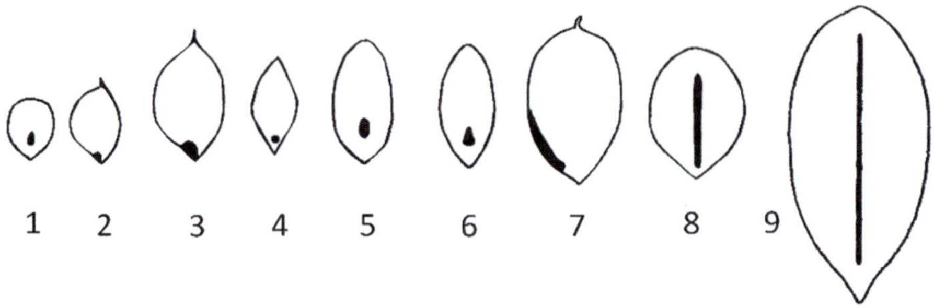

Drawings of hilums, modified from Körber-Grohne (1964): 1. *Catabrosa aquatica* 2. *Alopecurus aequalis* 3. *Alopecurus geniculatus* 4. *Poa palustris* 5. *Phagmites australis* 6. *Calamagrostis* 7. *Phalaris arundinacea* 8. *Glyceria maxima* 9. *Glyceria fluitans*

The dispersal unit of grasses is usually the floret, or sometimes the spikelet. Many aquatic or wetland species have hairs or awns on the lemma that assist in wind, animal, or water dispersal. The hairs allow the floret to float. Usually, only the caryopsis (grain) is preserved in sediments, the vegetative parts easily decaying. Poaceae caryopses are frequently recorded in Quaternary sediments, but they are seldom identified to species level. However, characters can be sufficiently distinctive that it is possible to identify many genera in fossil material. The illustrated identification keys for Poaceae fruits by Körber-Grohne (1964, 1991) are very useful for species identification; they are in German and English. Her work includes grasses identified from archaeological deposits in northern Germany and covers most of the grasses of the southern North Sea coast. Features useful for identification are size, shape, the shape of the hilum, the position of the hilum, and for *Glyceria*, the persistent forked stigma. The cellular structure of the caryopsis wall can be examined if preservation is good.

Poaceae caryopses from the large intestine of the 21,000-year-old Yukagir mammoth (modified from Van Geel et al. 2008). The hilums are clearly visible, and the different cell structures of the caryopses' walls are apparent.

5.2.149 *Alopecurus aequalis* Sobol., and *Alopecurus geniculatus* L. Poaceae

A. aequalis florets

A. geniculatus florets

A. aequalis caryopses

A. geniculatus caryopses

Morphology The dispersal units are the spikelets. The glumes are hairy and the palea is absent. The lemma of *A. aequalis* has a slender awn hardly exceeding the lemma. That of *A. geniculatus* has an awn about double the length of the lemma.

The caryopses are laterally flattened. The hilum is small and elliptical, positioned at the base. Caryopses of *A. aequalis* are ellipsoid, about 1 mm long. Caryopses of *A. geniculatus* are similar but larger, 1.2–1.7 mm long and 0.7–1.1 mm broad with "mottled brown depositions near the hilum "(Körber-Grohne 1964).

Life form and ecology *A. aequalis* is a tufted annual to perennial grass about 40 cm tall. It is characteristic of draw-down exposed muddy habitats. It also grows at the margin of ponds and ditches, and in wet meadows.

A. geniculatus is a tufted perennial grass about 50 cm tall, often with prostrate rooting culms that then characteristically ascend. It grows in wet meadows and grazing marshes on fertile soil, and beside ponds, ditches, and streams, sometimes forming floating mats or spreading by rhizomes on draw-down mud. It prefers fertile soils with pH 5.5–7-0.

Distribution *A. aequalis* has a circumboreal distribution and is widespread in temperate and boreal Europe, Scandinavia apart from the west coast of Norway, and Iceland, reaching the far north of Norway and the north of Spain. It is scattered in the UK, especially in Scotland, and is absent from Ireland, but it is widespread in Iceland.

A. geniculatus has a temperate-boreal distribution in Europe but is rare in Asia. It is widespread in northern Europe and Scandinavia, reaching the far north of Norway and the north of Spain. It is common in the British Isles but scattered in Ireland. It is introduced in the Southern Hemisphere and North America.

Alopecurus aequalis Sobol. (gbif.org)

Alopecurus geniculatus L. (gbif.org)

5.2.150 *Calamagrostis canescens* (Weber) Roth Poaceae

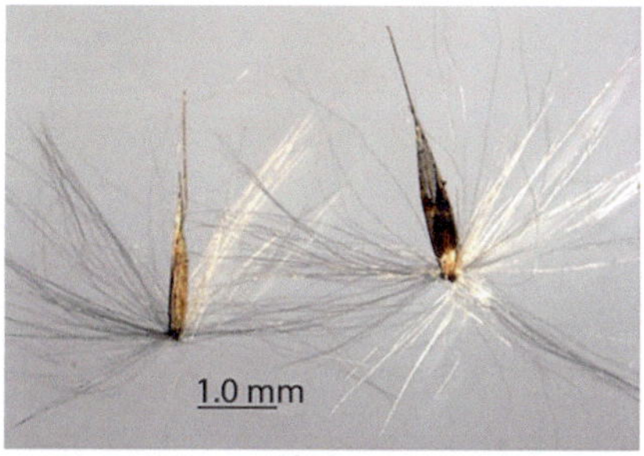

Calamagrostis canescens florets with basal hairs

Calamagrostis canescens caryopses. Right and left: ventral view showing the hilum. Centre: dorsal view showing the scutellum

Morphology The dispersal unit of *Calamagrostis canescens* is the floret. The long basal hairs are an adaptation to wind dispersal and floating on water. The small caryopses are about 0.8 mm long. The rounded hilum is relatively large and basal.

Life form and ecology *C. canescens* is a slender reed-like grass up to 120 cm tall. It grows in damp shady habitats, such as alder and willow fen woods, and also in ditches and fens. It grown on soils with a wide pH range (3.0–7.5).

Distribution *C. canescens* has a Euro-Siberian boreo-temperate distribution across continental northern Europe and central Asia. It occurs in Scandinavia east of the Scandes mountains in Sweden and Finland, and from Denmark to the northern treeline. It is rare in Norway and the UK and is absent from Ireland.

Quaternary history *Calamagrositis canescens* has been recorded in the late Holocene in the Bure Valley (Godwin 1975).

Calamagrostis canescens (Weber) Roth (gbif.org)

5.2.151 *Calamagrostis epigejos* (L.) Roth. Poaceae

Calamagrostis epigejos florets with basal hairs

C. epigejos caryopses. Left: lateral view, scutellum on left, hilum on right. Middle: dorsal view showing the scutellum. Right: ventral view showing the hilum

Morphology The dispersal unit of *Calamagrostis epigejos* is the floret. The numerous basal hairs are longer than the floret and aid wind dispersal and floating on water. The caryopses are elongated, about 1 mm long. The rather indistinct hilum does not reach the base and is elongated or trapezoidal, about $^1/_5$ the length of the caryopsis.

Life form and ecology *C. epigejos* is a strong reed-like grass up to 2 m tall. It has a wide habitat range. It often grows in damp woods, in ditches and fens, and dune slacks.

Distribution *C. epigejos* has a circumboreal distribution, rarely occurring north of the northern treeline. It less common across North America.

Calamagrostis epigejos (L.) Roth (gbif.org)

5.2.152 *Calamagrostis purpurea* (Trin.) Trin. Poaceae

Calamagrostis purpurea florets with basal hairs

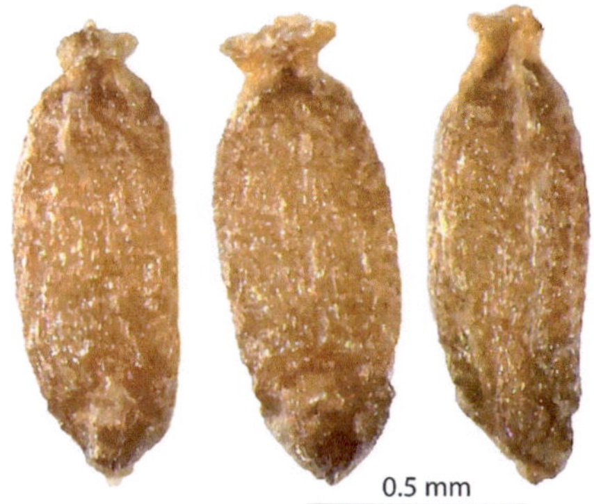

C. purpurea caryopses. Right: lateral view with the scutellum to the left. Middle: ventral view showing the hilum. Left: dorsal view showing the scutellum

Morphology The dispersal unit of *Calamagrostis purpurea* is the floret. The numerous basal hairs are longer than the florets and facilitate wind dispersal and floating on water. The caryopses are about 1 mm long and 0.3 mm wide. The rounded hilum is basal.

Life form and ecology *Calamagrostis. purpurea* grows in poor fens and wet woods and carr, and at the edges of ditches. It can form extensive stands in wet fens.

Distribution *Calamagrostis purpurea* has a circumboreal and circum arctic distribution. It occurs in Sweden, Norway, and Finland, but is absent in Denmark, Ireland, England, and Wales, and is very rare in Scotland.

Calamagrostis purpurea (Trin.) Trin.

5.2.153 ***Calamagrostis stricta*** (Timm) Koeler Poaceae

Calamagrostis stricta florets

Calamagrostis stricta caryopses: Left and middle; ventral side showing the hilum. Right: dorsal side showing the scutellum

Morphology The dispersal unit of *Calamagrostis stricta* is the floret. The numerous basal hairs are shorter than the floret. They aid dispersal by wind and floating on water. The caryopses are about 1.5 mm long and 0.5 mm wide. The oval hilum lies above the base of the caryopsis.

Life form and ecology *C. stricta* is a reed-like grass up to 1 m tall. It grows in wetlands, marshes, lake shores, ditches, fens etc.

Distribution *C. stricta* has a northern circumboreal distribution across Europe, Asia, and North America, including Svalbard, Iceland, and Greenland.

Calamagrostis stricta (Timm) Koeler (gbif.org)

5.2.154 *Catabrosa aquatica* (L.) P.Beauv. Poaceae

Morphology The dispersal unit of *Catabrosa aquatica* is the floret. The floret is glabrous. The entire-margined palea equals the lemma. The lemma apex is truncate with a ragged tip. The caryopsis is 1.5 mm long and 0.5–0.8 mm wide. It is ellipsoid and laterally flattened with a small elliptic hilum near the base and a marked cell pattern. The remains of the style are persistent.

Life form and ecology *Catabrosa aquatica* is a tall, reed-like grass characteristic of muddy banks of river and ditches, bordering still or slow-flowing water where it can form floating mats. It is readily grazed.

Distribution *Catabrosa aquatica* is widely distributed in the temperate circumboreal regions of the Northern Hemisphere.

Catabrosa aquatica (L.) P.Beauv. (gbif.org)

5.2.155 *Glyceria declinata* Bréb Poaceae

Morphology The florets of *Glyceria declinata* are 4–5 mm long, with strongly ribbed lemmas toothed at the apex. The palea is equal in length or lightly longer. The caryopses are 2–3 mm long, parallel sided, and with a rather broad hilum extending the length of the fruit. The surface cells form a marked rectangular pattern, similar to that of *G. fluitans*.

Life form and ecology *Glyceria declinata* is a loosely tufted grass lacking rhizomes, 10–60 cm tall and often prostrate, closely resembling *G. fluitans*. It grows in similar habitats on fertile soil in wet, cattle-disturbed meadows and beside streams, ditches, pools, and shallow muddy lake shores.

Distribution *Glyceria declinata* is widespread in mainland Europe, including the United Kingdom. It is rare in Ireland. Its northern limit lies in southern Norway and southwest Sweden. It is introduced in other parts of the world.

Glyceria declinata Bréb. (gbif.org)

5.2.156 *Glyceria fluitans* (L.) R.Br. Poaceae

Florets

Caryopses

Single caryopsis

Morphology The dispersal unit is the floret. The lemma and palea are equally long. The lemma is prominently nerved with a blunt tip. The caryopses are 2.5–3.0 mm long, much longer than wide. The hilum is narrow, running the length of the fruit. The scutellum is well marked. The persistent stigmas are typically recurved. The cells of the pericarp are polyhedric and well defined.

Life form and ecology *Glyceria fluitans* is a loosely tufted aquatic grass, often forming floating mats or decumbent in shallow water of lake shores or ditches. It may grow in wet depressions in water meadows and pastures. It prefers damp soils with pH 5.0–6.5 but has a wide tolerance. It is tolerant of disturbance and nutrient enrichment, for example by grazing animals. The florets are said to be dispersed by birds (Cope and Gray 2009) and by wind and water (Schou et al. 2023).

Distribution *Glyceria fluitans* is native to Europe, where it is widely distributed within the boreo-temperate zone south of the Arctic Circle, including Finland. It has become naturalized in northern Asia, the Americas, Australia, and New Zealand.

Quaternary history The caryopses of *Glyceria* often retain their hooked stigmas and are then instantly recognizable. "The fossil fruits are very tough, dark, and opaque with a long hilum, and the bifid style base is often still present. The cell pattern is only seen clearly if the cells of the pericarp have begun to disintegrate" (Dickson 1970 p.240; Körber-Grohne 1964). *Glyceria s*pecies can be distinguished on shape and size (Borrill 1956). *Glyceria* caryopses are occasionally found in lake sediments, e.g. in the late Holocene of England. However, fossils have rarely been identified below genus level.

Glyceria fluitans (L.) R.Br. (gbif.org)

5.2.157 *Glyceria lithuanica* (Gorski) Gorski Poaceae

Morphology The florets of *Glyceria lithuanica* are about 3 mm long, with a strongly ribbed and scabrous lemma. The caryopsis is shorter (2.5 mm) than *G. fluitans*, but equally wide, giving it a rounder appearance. It retains the recurved stigma characteristic of *Glyceria* spp., and the hilum is long. The cells of the pericarp are smaller than in *G. fluitans,* elongated, and not so well delimited.

Life form and ecology *Glyceria lithuanica* is a reed up to 120 cm tall with delicate hanging inflorescences. It occurs on damp or wet soils in shade, such as in wet woodland and stream margins.

Distribution *Glyceria lithuanica* has a boreal-nemoral Eurasian distribution from southern Scandinavia through Finland, northeast Europe, and Asia to Kamchatka. It is absent from the British Isles and northwest mainland Europe.

Glyceria lithuanica (Gorski) Gorski (gbif.org)

5.2.158 *Glyceria maxima* (Hartm.) Holmb. (Syn. *Glyceria aquatica* (L.) Wahlenb.) Poaceae

Morphology The florets of *Glyceria maxima* are 3–4 mm long, scarcely toothed at the apex. The caryopses are 1.5–2.0 mm long and ovate. The linear hilum runs the length of the caryopsis. The scutellum is well marked.

Life form and ecology *Glyceria maxima* is a robust rhizomatous reed, up to 2.5 m tall. It typically grows on waterlogged fertile soils of pH 6.0–8.0, beside freshwater ponds, lakes, canals, and ditches, and also in wet water-meadows that are flooded in winter. It can form dense stands and floating mats. It is highly competitive and invasive and is often considered to be a noxious weed outside its native range.

Distribution *G. maxima* is widespread in northwest Europe, extending to southern Sweden and southern Finland. It is introduced in southern Norway. In the Britain Isles it is rare in northern Scotland and in Ireland.

Quaternary history There is a Late Holocene record (zone VIII) from Apethorpe, UK (Godwin 1975).

Glyceria maxima (Hartm.) Holmb. (gbif.org)

5.2.159 ***Glyceria notata*** Chevall. (Syn. *Glyceria plicata* (Fries) Fries) Poaceae

Morphology The florets are about 4 mm long with ribbed lemmas with short scarious teeth. The paleas are the same length. The caryopses are about 2 mm long and 0.8–1.0 mm broad. The hilum is linear, as long as the caryopsis. The scutellum is well marked. The recurved styles are persistent.

Life form and ecology *G. notata* is similar to but larger than *G. fluitans*. It grows on damp or wet fertile calcareous soils, in similar habitats to *G. fluitans*; in wet grazed meadows and beside creeks and streams, pools and ditches, often on limestone.

Distribution *Glyceria notata* is widespread in Europe with its northern limit in southern Norway and Sweden. It is rare in Ireland. Its distribution is similar to *G. declinata*, but is more eastern, extending into western Siberia. It is introduced in Australasia, South America, Kamchatka, and eastern China.

Quaternary history Fruits tentatively assigned to *G. notata* (*plicata*) were found in the Kentish valley site at Frogholt, UK, at Late Bronze Age and Iron Age levels indicated by radiocarbon dates of 2860 and 2490 ^{14}C BP (Godwin 1975).

Glyceria notata Chevall. (gbif.org)

5.2.160 *Phalaris arundinacea* L. Poaceae

Morphology The dispersal unit is the whole floret. The smooth shiny lemma has short basal hairs that facilitate floating. The caryopses are ellipsoid and laterally flattened, about 1.5–1.8 mm long with a linear hilum one third the length of the caryopsis. The remains of the style often persist.

Life form and ecology *Phalaris arundinacea* is a reed up to 2 m tall. It grows on the edges of rivers and freshwater lakes, wetlands, and marshes, usually on drier ground than *Phragmites* and *Glyceria* spp. It also colonises disturbed wetland habitats and drier waste-ground such as roadsides. It thrives on most neutral to alkaline soils, especially on clay. It is tolerant of moderate intermittent flooding. Its deep roots allow it to survive in drier habitats. It spreads mainly by strong rhizomes, forming dense monocultures. It can produce prolific seeds that are dispersed by water and germinate in spring.

Distribution *Phalaris arundinacea* is very common throughout temperate lowland Europe. It is introduced and invasive throughout the rest of the world, particularly in temperate North America where vigorous ecotypes overcome native vegetation.

Phalaris arundinacea L. (gbif.org)

5.2.161 *Phragmites australis* ssp. *australis* (Cav.) Trin. ex Steud. Poaceae

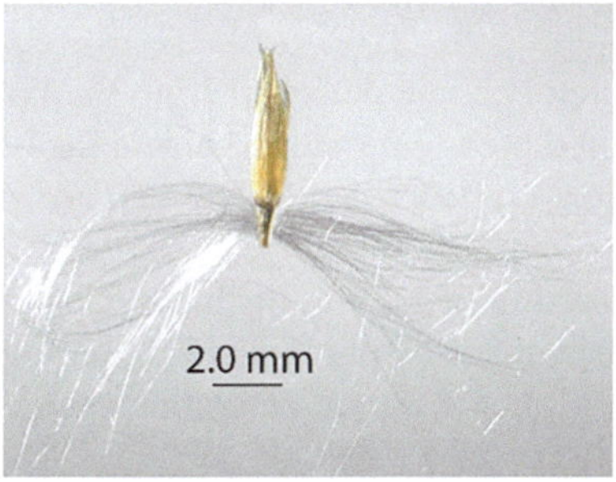

Floret with long silky hairs

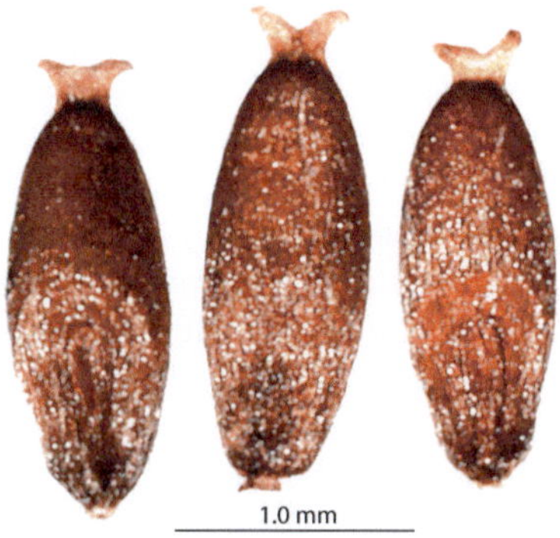

Caryopses showing the hilum on the outer two

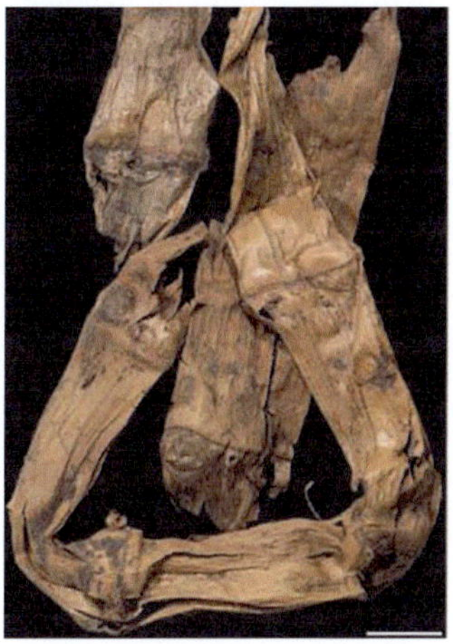

Phragmites rhizome

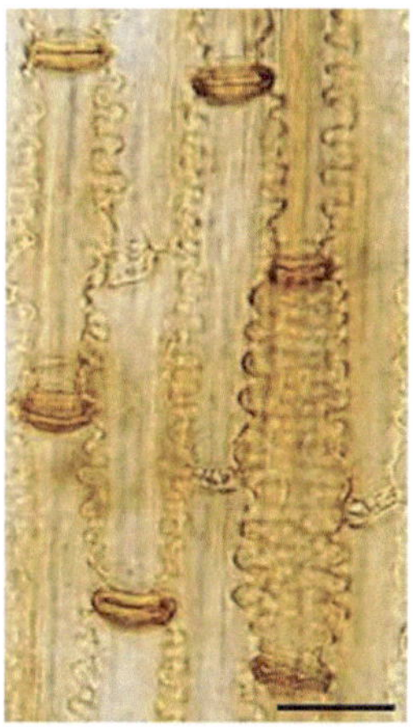

Surface cells of *Phragmites* rhizome. Modified from Mauquoy and van Geel (2007)

Morphology The fruits of *Phragmites australis* are typical grass caryopses. They are broadly ellipsoid, about 1.5 to 1.8 mm long. The hilum is distinct and elliptical, 2.3 x as long as broad, usually displaced from the mid-line and at least one hilum-length distant from the base of the fruit. These features are characteristic of *Phragmites*. The 2-pronged stigma shown in the photograph is lost on fossilization. The surface cell-pattern is indistinct.

The dispersal units are whole spikelets which become detached from the inflorescence. Dispersal is usually by wind, facilitated by 4–7 mm long hairs at the base of the floret, which also enable the floret to float on water. Birds and man are also dispersal agents.

The rhizome is pale and shiny when preserved in peat, with distinct nodes. The surface cells are long with wavy margins and distinct diaphragms (Mauquoy and van Geel 2007). Leaves are seldom found fossil, but they can be distinguished by the presence of stomata.

Life form and ecology *Phragmites* is the common reed that can form extensive stands (reed beds) in shallow water and muddy habitats, usually in freshwater, but also at the junction between saline and freshwater in upper saltmarsh. It can be up to 3.5 m tall. It grows best in alkaline (pH 5.5–7.5) nutrient-rich water but it can also tolerate acid fens and even flushed bogs. The fruiting inflorescences are silvery with the fine hairs at the bases of the spikelets. Fertile fruit production is infrequent, but vegetative reproduction is very effective. *Phragmites* facilitates hydroseres, colonizing lake mud under shallow water. The rhizomes can penetrate 40–100 cm depending on the water table. The dense plants trap sediment, thus raising the mud surface above the water level. As this proceeds, *Phragmites* is replaced by drier fen vegetation.

Distribution *Phragmites* occurs world-wide in temperate zones. It reaches northernmost Scandinavia, 71°N.

Quaternary history *Phragmites* is usually recognized from its rhizomes. These can form dense layers of *Phragmites* peat. It is recorded throughout the Holocene and becomes commoner after the mid-Holocene.

Phragmites communis Trin. (gbif.org)

5.2.162 *Poa palustris* L. Poaceae

Morphology The florets are narrow, and the lemmas have a tuft of woolly hair at the base. The caryopses are small, up to 1 mm long, with a pronounced round hilum near the base. The surface cell pattern is indistinct.

Life form and habitat *Poa palustris* is a loosely tufted grass up to 120 cm tall. It grows in a wide variety of wetland habitats on fertile soil, such as marshes, fens, ditches, *Salix* carr, and margins of ponds and streams.

Distribution *Poa palustris* has a circumboreal distribution. In Europe it is centred on northern Europe and southern Scandinavia, including southern Finland. It is rare in western Norway and north of the Arctic Circle. It is doubtfully native in the British Isles and was introduced in the early nineteenth century as a fodder plant (Cope and Gray 2009).

Poa palustris L. (gbif.org)

5.2.163 *Persicaria amphibia* (L.) Delarbre (Syn. *Polygonum amphibium* L.) Polygonaceae

Morphology The achenes of *Persicaria amphibia* are orbicular, biconvex to globose, 1.9–3.0 mm × 1.4–2.4 mm, and shiny with no discernible cell pattern.

Life form and ecology *Persicaria amphibia* has an aquatic form with floating leaves that grows in shallow water in lakes, ponds, canals, and slow-moving streams, where it can form dense stands. The terrestrial form with upright stems up to 60 cm tall grows at water margins on damp soil. It can also be a weed on rough ground. Dispersal is by water or by animals such as waterfowl that may ingest the seeds (Partridge 2001).

Distribution *Persicaria amphibia* occurs throughout the subarctic and temperate Northern Hemisphere and parts of Africa. It is introduced elsewhere where it can become a noxious weed.

Quaternary history *Persicaria amphibia* macrofossils are recorded occasionally from Holocene sediments (e.g. Godwin 1975; Willis 1992).

Persicaria amphibia (L.) Gray (gbif.org)

5.2.164 ***Persicaria hydropiper*** (L.) Delarbre (Syn. *Polygonum hydropiper* L.) Polygonaceae

Trigonous fruits

Lenticular fruits

Morphology The achenes of *Persicaria hydropiper* are triangular or lenticular in cross section and the cell pattern is minutely reticulate or verrucate. They are 2.2–3.2 mm long and 1.4–2.3 mm wide.

Life form and ecology *Persicaria hydropiper* is a hydrophilic annual up to 75 cm tall. It grows in shallow waters of lakes, ponds, ditches and in damp places by water. Dispersal is either by water or by animals such as waterfowl that may ingest or carry the seeds (Timson 1966).

Distribution *Persicaria hydropiper* is widespread throughout the boreal and temperate areas of Europe, North America, and Asia, and many parts of the Southern Hemisphere.

Quaternary history *Persicaria hydropiper* fruits have been recorded in Quaternary sediments (e.g. Holyoak 1983; Stachowicz-Rybka et al. 2009; Novenko et al. 2012; Godwin 1975).

Polygonum hydropiper L. (gbif.org)

5.2.165 *Rumex aquaticus* L. Polygonaceae

Achenes

Achene in perianth

Morphology The smooth shiny achenes of *Rumex aquaticus* are 3.0 mm long and about 1.5 mm wide. They are triangular in cross section with sharply defined edges. The achene is enclosed in the perianth, which can sometimes survive fossilisation, when it is useful for identification of the associated achenes (Godwin 1975; Stace 2019).

Life form and ecology *Rumex aquaticus* is a hydrophilic rhizomatous perennial up to 2 m tall. It grows in periodically wet, nutrient-rich soil, particularly alongside ditches, rivers, and lakes.

Distribution *Rumex aquaticus* has a circumpolar boreo-temperate distribution, occurring in continental lowland northern and central Europe and across temperate eastern Europe and Asia. It has one locality in the British Isles.

Quaternary history *Rumex aquaticus* macro-remains are occasionally recorded in Quaternary sediments (e.g. Collins et al. 1996; Wohlfarth et al. 2001; Stančikaitė et al. 2003; Godwin 1975). The achenes are similar to those of *Rumex hydrolapathum*.

https://www.gbif.org/species/2888992

5.2.166 ***Rumex hydrolapathum*** Huds. Polygonaceae

Achenes

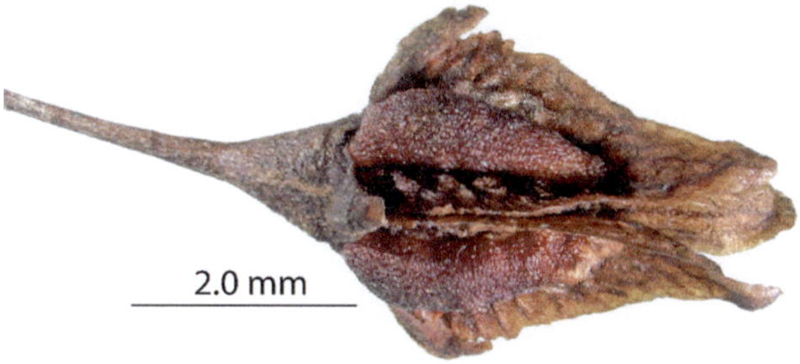

Achene in perianth

Morphology The achenes of *Rumex hydrolapathum* are triangular in cross section, with sharply defined edges and smooth surfaces. They are 3–4 mm long. The perianth surrounding them sometimes survives fossilisation, when it is useful for identification of the associated achenes (Godwin 1975; Stace 2019).

Life form and ecology *Rumex hydrolapathum* is a large rhizomatous perennial up to 2.5 m tall, often forming dense stands. It grows in damp ground alongside ditches, rivers, lakes, canals, and in marshes. The achenes or whole perianths are primarily dispersed by water, and sometimes by animals (Cappers 1993b; Soomers et al. 2013).

Distribution *Rumex hydrolapathum* occurs throughout temperate Europe, southern Scandinavia, and east into Russia. It reaches southern Scotland and is infrequent in Mediterranean regions.

Quaternary history *Rumex hydrolapathum* macro-remains have been recorded in Quaternary sediments (e.g. Wohlfarth et al. 2001; Bos and Urz 2003; Godwin 1975). The fruits are similar to those of *Rumex aquaticus*.

Rumex hydrolapathum (Scop.) Huds. (gbif.org)

5.2.167 *Rumex palustris* Sm. Polygonaceae

Morphology The achenes of *Rumex palustris* are sharply trigonous, smooth, pale brown, and at 1.8–2.0 mm long, notably smaller than other hydrophilic *Rumex* species.

Life form and ecology *Rumex palustris* is an annual or biennial herb up to 1 m tall. It typically grows in wet, nutrient-rich muddy areas that become dry in summer in marshes and beside ponds. Dispersal is primarily by water but can also occur via wind and animals (Godwin 1923; Cappers 1993b).

Distribution *Rumex palustris* occurs across temperate northwest Europe, and sparsely around the Mediterranean and in Iberia.

Quaternary history The achenes resemble other *Rumex* spp. and it is challenging to confidently identify them.

Rumex palustris Sm. (gbif.org)

5.2.168 ***Groenlandia densa*** (L.) Fourr. Potamogetonaceae

Morphology The delicate endocarps of *Groenlandia densa* are large, about 3 mm long and 2.5 mm wide. The curvature of the embryo is pronounced, resulting in a central depression leading to a marked groove along the position of the partition wall. The endocarps are ovoid and the ventral side is S-shaped. The sides are concave, and the endocarp is one-cell thick (Cappers 1993a). The narrow, rounded or roof-shaped lid reaches the beak, which is central or ventro-central. The style slopes dorsally. The stalk is ventral.

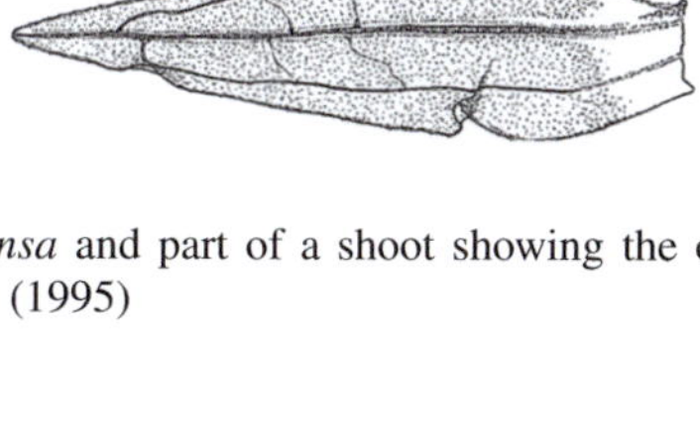

Leaf tips of *Groenlandia densa* and part of a shoot showing the opposite leaves. Drawings of *G. densa* leaves from Preston (1995)

The leaves of *G. densa* are linear, about 1.3 cm wide and faintly toothed. They taper to an obtuse or slightly emarginate tip (resembling the lingulate leaves of some *Callitriche* species). The main nerve is pronounced, but the 2–4 lateral nerves are inconspicuous.

Life form and ecology *Groenlandia densa* is a submerged rhizomatous pondweed with opposite leaves. It overwinters as leafy shoots and spreads vegetatively by detached branchlets (Preston 1995). It grows in shallow, clear, calcareous water in slow rivers, ditches, canals, ponds, and small lakes, often in sites at or close to spring-sources. It can also grow on acidic substrates such as peat and sand with calcifuge plants as long as the overlying water is sufficiently base-rich (Preston 1995). Given its need for highly transparent water, it typically declines with eutrophication. It is also threatened by hydrological changes associated with drought and water management practices that reduce water permanence (Šegota et al. 2019).

Distribution The centre of distribution of *Groenlandia densa* is temperate Europe. It extends north to Ireland and Scotland, south to southern Portugal, Greece, and Sicily, and east to the Baltic states It is declining due to habitat loss and is already extinct in Denmark, Norway, Sweden, Lithuania, and eastern Germany (Akhani 2014; Šegota et al. 2019). It is rare in southeast Europe and northwest Africa, and there are records from southwest Asia. It is probably introduced in North America and Australia (Akhani 2014).

Quaternary history *Groenlandia densa* endocarps are frequently recorded in Full- and Mid-Weichselian sediments in Britain and Ireland (Godwin (1975) and subsequent literature). For example, they were particularly abundant 30,000 years ago at Sourlie in Scotland (Bos et al. 2004). Although *G. densa* is not characteristic of cold climates, interstadial conditions provided shallow calcareous ponds as an ideal habitat. It is less frequently recorded in Holocene sediments.

Groenlandia densa (L.) Fourr. (gbif.org)

5.2.169 *Potamogeton* L. Potamogetonaceae

Potamogeton fruits found in sediments have usually lost their epicarp and mesocarp, and only the hard endocarp, or fruit stone, remains. It has a distinctive structure, as illustrated by Aalto (1970) (Fig. 5.1) and nearly all species can be identified in well-preserved fossil material. For comparison with fossils, removal of the outer fleshy layers from modern reference material is necessary. Fruits should be soaked or shortly heated in dilute KOH (5%) to soften the outer layers, which can then be carefully removed to reveal the endocarp.

Not all the characters are preserved in all fossil endocarps, but according to Aalto (1970) and Jessen (1955) the most constant and useful are: 1. The shape; the sides can be convex or flat or have a distinct central depression, 2. The size; measured from the top to the base, excluding the beak and the stalk, 3. The length of the lid compared with the length of the fruit, 4. The profile of the lid in top view; it can be rounded, roof-shaped, with or without a keel, and the keel can be smooth or crested, 5. The distance between the apex of the lid and the base of the beak, 6. The ventral area in side view can be straight, in which case the fruit is crescentic, convex, or

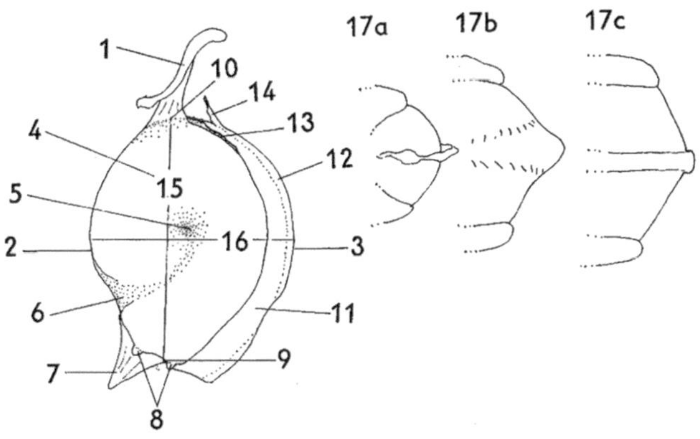

P. natans subfossil endocarp. 17a-c are top views of the lids of *P. alpinus, P. perfoliatus, and P. lucens*

1. beak (style and stigma)
2. ventral area
3. dorsal area
4. side
5. central depression
6. furrow marking partition wall of locule
7. stalk
8. basal warts and spines
9. base
10. top
11. lid
12. Keel of lid (or, in some species, crest or ridge)
13. Crispate margin of side
14. Tongue or projection of apex of lid
15. Length
16. breadth
17. lid profiles in top view
    a) rounded with crest
    b) keeled
    c) roof-shaped (with ridge)

Fig. 5.1 Diagram of an idealised *Potamogeton* endocarp illustrating diagnostic features. Modified from Aalto (1970)

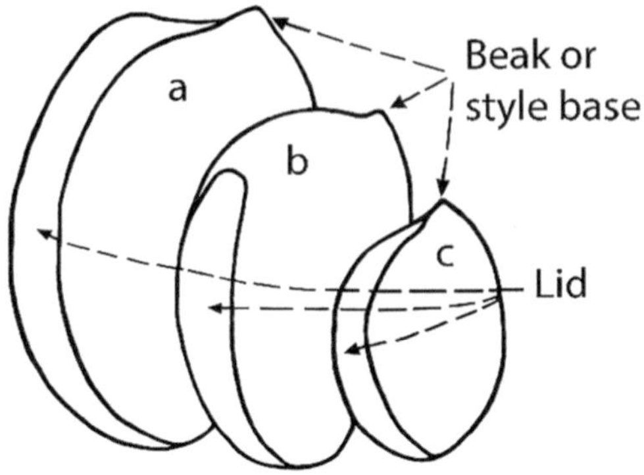

Fig. 5.2 Three types of *Potamogeton* endocarps. a) Large fruit with lid reaching the beak, b) *Stuckenia* type, with the lid with a rounded apex reaching two thirds up the back of the endocarp, c) small fruit, with lid sloping up to the beak and only just reaching it. Modified from Martin (1951)

S-shaped (convex above, concave below, as in Fig. 5.1), 7. The positions of the beak and stalk, 8. Spines or warts on the fruit. Projecting parts, such as warts, spines, the style base and stalk, and the crest and tongue of the lid can easily be lost during preservation. However, when they are present, they can be diagnostic of the species.

Potamogeton is divided into two subgenera. Subgenus *Coleogeton* (containing its own genus *Stuckenia*) has distinctive fruits (Figs. 5.2b and 5.4, 6–10). They have a convex ventral profile, smooth rounded sides, and rounded lids with rounded or subacute apices that reach half to two thirds up the length of the fruit. Subgenus *Eupotamogeton* (containing the genus *Potamogeton)* has fruits more typical of Figs. 5.1 and 5.2a, c. They can be broadly divided into three size categories, large (>2. 5–4.0 mm long), medium (2.0–2.5 mm long), and small (<2.0 mm long) (Aalto 1970). Aalto (1970) (Fig. 5.3) and Jessen (1955) provide an overview of the variation in length of *Potamogeton* endocarps.

The overview illustration of most of the northwest European species by Jessen (1955) is useful for narrowing down possible identifications (Fig. 5.4). More features are included in similar drawings by Madalski (1949), although this was inaccessible by us. Aalto (1970) and Katz et al. (1965) provide excellent illustrations of the endocarps, including the range of variation. Aalto (1970) also emphasises and illustrates the distinctive cell structure of the endocarp walls. Cappers (1993a) illustrates seventeen species from The Netherlands with scanning electron micrographs, also concentrating on cellular structure.

Aalto (1970) provides a comprehensive key to endocarps. Jessen (1955) provides a simpler and more accessible key. Martin (1951) gives an illustrated key to North American species that uses different characters, including those seen in vertical sections of the endocarps.

Aalto (1970) discusses some difficulties that may be encountered in identifying fossil endocarps to species level. She divides the species into those that are 'easily' identified, and those that are 'difficult' to identify. Easily identified species are *P. natans, P. praelongus, P. alpinus, P. crispus, P. obtusifolius, S. pectinatus,* and *Groenlandia densa.* Causes of difficulty for the rest are: 1. There is often a large range of variation in recent material, and variability of fossil endocarps can be even greater, rendering them unidentifiable. 2. Erosion of fossil endocarps, causing the

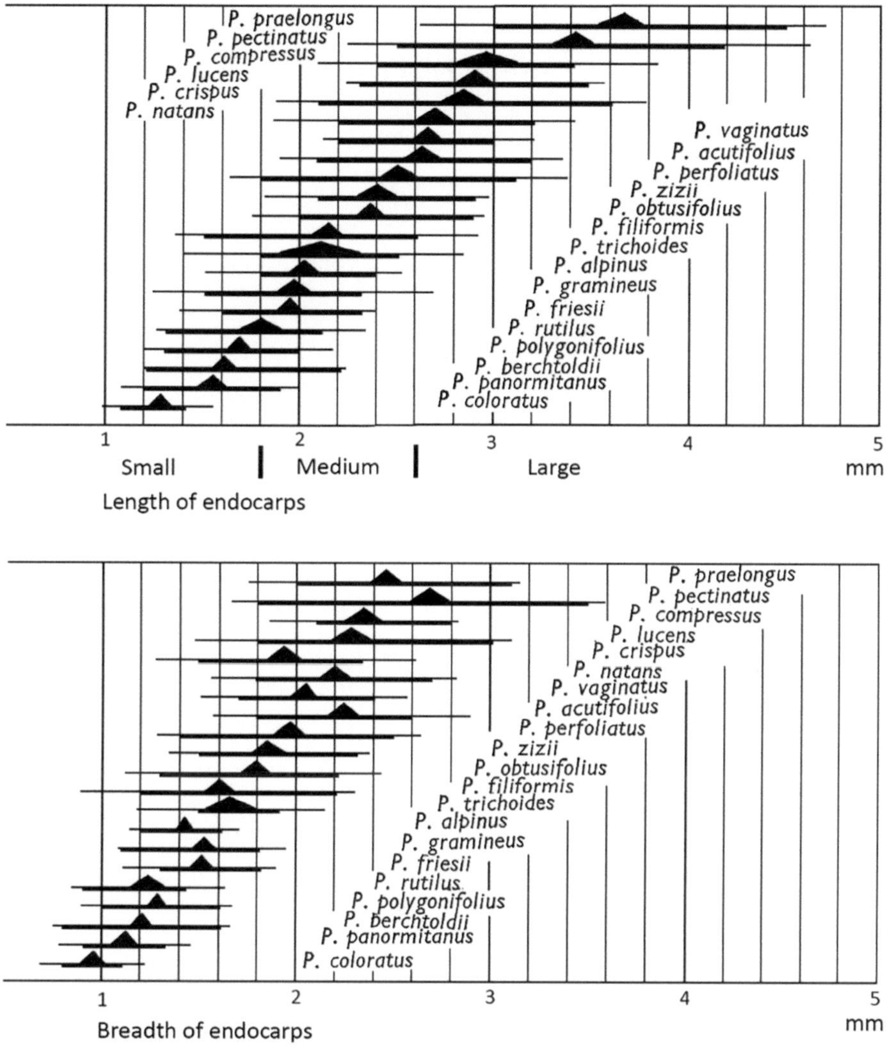

Fig. 5.3 Length and breadth of endocarps (mm) showing the mean, 99% standard deviation (base of the triangle), the range between minimum and maximum measurements (thick lines) and ± 3 standard deviations of the population (thin lines). Modified from Aalto (1970). Note: *P. panormitanus* is now *P. pusillus*

loss of projecting parts, such as beak, stem, warts, crests on lids, and even the loss of the lid. This may render small endocarps such as *P. berchtoldii, P. pusillus, P. friesii,* and *P. rutilus* indistinguishable. They can only be referred to a type. 3. Similarity of closely related species can cause difficulty, such as between *S. filiformis* and *S. vaginata*, between *P. gramineus, P. x angustifolius*, and *P. lucens,* and between *P. compressus* and *P. acutifolius*. 4. Hybridisation is common in *Potamogeton* (e.g.

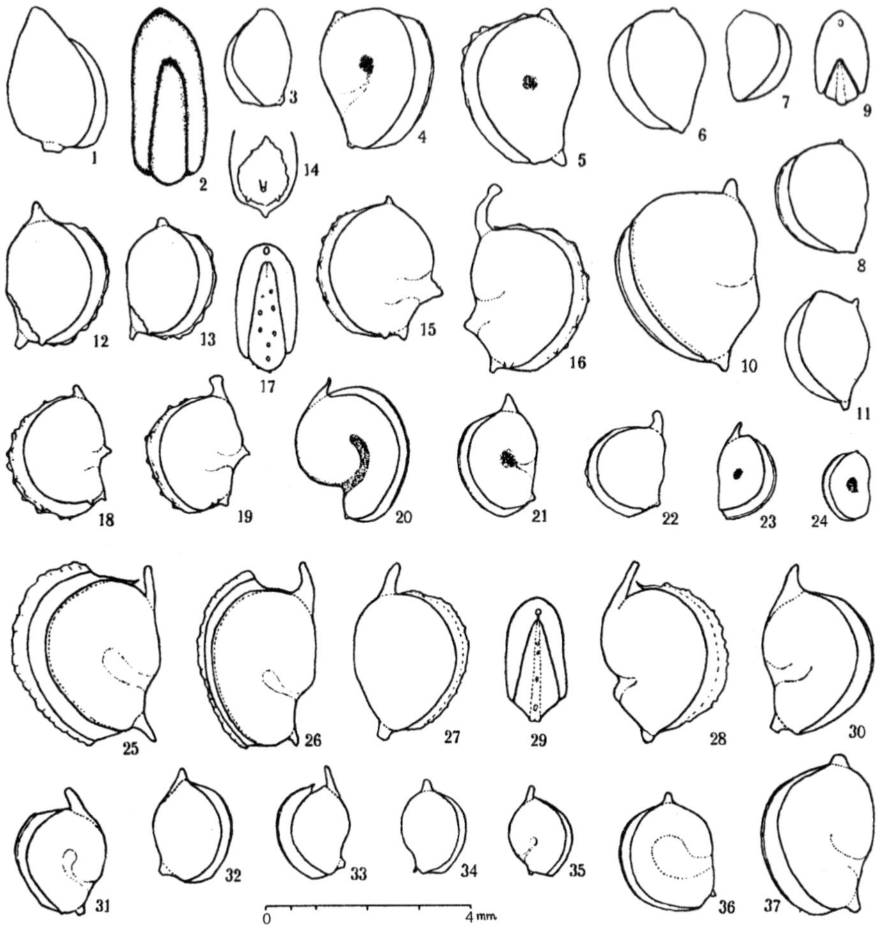

Fig. 5.4 Overview of *Potamogeton* endocarps, mostly in lateral view, by Jessen (1955). 1. *P. crispus* lateral view, 2. *P. crispus* dorsal view; 3. *P. rutilus*; 4, 5. *P. natans*; 6, 7. *P. filiformis*; 8. *P. vaginatus*; 9. *P. vaginatus* dorsal view; 10. *P. pectinatus*; 11. *P. alpinus*; 12, 13. *P. obtusifolius*, 14. *P. obtusifolius* dorsal view; 15, 16. *P. acutifolius*; 17. *P. acutifolius* dorsal view; 18, 19. *P. trichodes*; 20. *P. densus*; 21, 22. *P. gramineus*; 23. *P. polygonifolius*; 24. *P. coloratus*; 25, 26. *P. praelongus*; 27, 28. *P. compressus*; 29. *P. compressus* dorsal view; 30, 31. *P. perfoliatus*; 32, 33. *P. friesii*; 34 *P. pusillus*; 35. *P. panormitanus*; 36. *P. x zizii*; 37. *P. lucens*, From Jessen (1955). Note: *P. filiformis* is now *Stuckenia filiformis, P. pectinatus* is now *S. pectinata, P. vaginatus* is now *S. vaginata, P. densus* is now *Groenlandia densa, P. panormitanus* is now *P. pusillus, P. x zizii* is now *P. x angustifolius*

Preston 1995; Stace 2019; Wiegleb and Kaplan 1998; Schou et al. 2023) leading to atypical endocarps.

Occasionally, leaf remains are preserved in lake sediments. Leaf tip shape and venation of some of the narrow-leaved submerged species can sometimes be quite characteristic, but on the whole, they are difficult to differentiate (Fig. 5.5; Preston

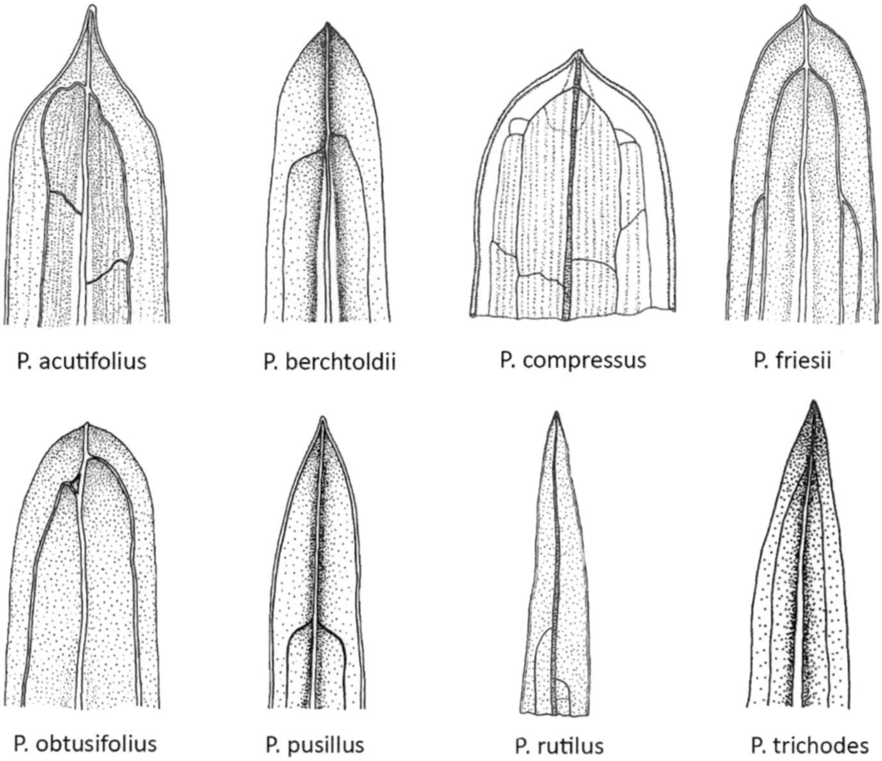

Fig. 5.5 Drawings of leaf tips of narrow-leaved submerged elodeid *Potamogeton* species compiled from Preston (1995)

1995). Leaf tips are well illustrated for British species by Preston (1995) and for Fennoscandian species by Mossberg and Stenberg (2018). Drawings by Preston (1995) of leaf tips of the narrow-leaved submerged species are shown in Fig. 5.5. We provide photographs and/or drawings of leaf tips and leaf fragments where appropriate. Well-preserved leaf tip material may be identified to species (e.g. *Potamogeton acutifolius* and *Potamogeton trichoides*) or at least to an aggregate group. For example, the relatively wide and blunt leaves of *Potamogeton friesii* and *Potamogeton obtusifolius* can often be relatively confidently combined as an aggregate group, with the same being true of the typically narrower and slightly pointier leaves of *Potamogeton pusillus* and *Potamogeton berchtoldii*. In some cases known species distributions can inform and help to refine identifications. For example, *P. trichoides* does not overlap geographically with *Potamogeton rutilus* in most areas, making finely tapering and highly pointed *Potamogeton* leaves (Fig. 5.5) easier to assign to a single species. Sometimes, layers of *Potamogeton natans* leaves are found in anoxic pond sediments.

5.2.170 ***Potamogeton acutifolius*** Link ex Roem. & Schult. Potamogetonaceae

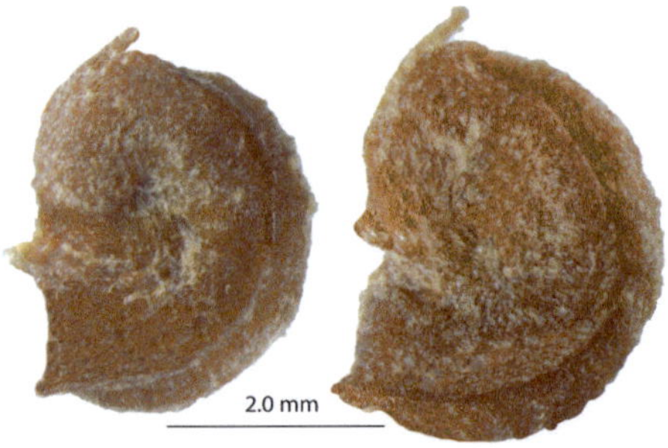

Morphology The endocarps of *Potamogeton acutifolius* are large, about 2.6 mm long and 2.3 mm broad, nearly as broad as long. The size is rather variable. The ventral side is flat and may have a protruding boss at the position of the partition wall. The sides are flat, often with a central depression and a prominent wart at the base. The rounded lid reaches nearly to the beak and can be covered in scattered warts (Aalto 1970), but this is not obvious in our material. The beak is ventral. The stalk is inconspicuous.

The submerged leaves of *Potamogeton acutifolius* are 2–5 mm wide with pointed tips. There are three main veins and a strong marginal vein, all meeting at a distinctively drawn out mucronate tip: hence its common English name of sharp-leaved pondweed. Drawing modified from Preston (1995).

Life form and ecology *Potamogeton acutifolius* is a submerged pondweed typically found in shallow pools in generally calcareous, mesotrophic to meso-eutrophic fen, grazing marsh, and floodplain habitats. It also often grows in ditches, and under natural conditions, it can be speculated that it thrived in swampy lowland lake and river edge areas and especially in oxbows and beaver ponds. A detailed study of habitat preferences amongst English populations of *P. acutifolius* showed that the species has a pH range of 6.4–7.6 and two distinct bands of conductivity (239–355 and 611–804 $\mu s\ cm^{-1}$) (Newbold 2003). It is vulnerable to eutrophication and habitat destruction and appears to require regular habitat disturbance (e.g. ditch management) to avoid what appears to be unfavourable competition with other species.

Distribution *Potamogeton acutifolius* occurs in temperate Europe reaching central Sweden to the north and northern Italy, Serbia, and Romania to the south (Kaplan 2010). It is mostly absent from the Mediterranean area, although populations have recently been discovered in Turkey representing its southern-most occurrence (Aykurt et al. 2016). It is scattered across eastern central Asia. It is a rare and declining species across much of its range (Wiegleb et al. 1991; Preston and Croft 1997; Sand-Jensen et al. 2000).

Potamogeton acutifolius Link ex Roem. & Schult. (gbif.org)

5.2.171 *Potamogeton alpinus* Balb. Potamogetonaceae

Morphology The endocarps of *Potamogeton alpinus* are medium sized, about 2 mm long and 1.4 mm wide. The sides are rounded, sometimes with a central depression. The ventral side is convex. The lid is rounded, sometimes with a crest, and it reaches the top of the endocarp, but not the base of the beak. The large distance between the top of the beak and the lid is characteristic of this species. The gap between is usually straight and slanting. The beak is ventral, and the well-marked stalk is also ventral. In our photographs, the inner parts of the exocarp still adhere to the endocarp, which in the fossil state is smooth.

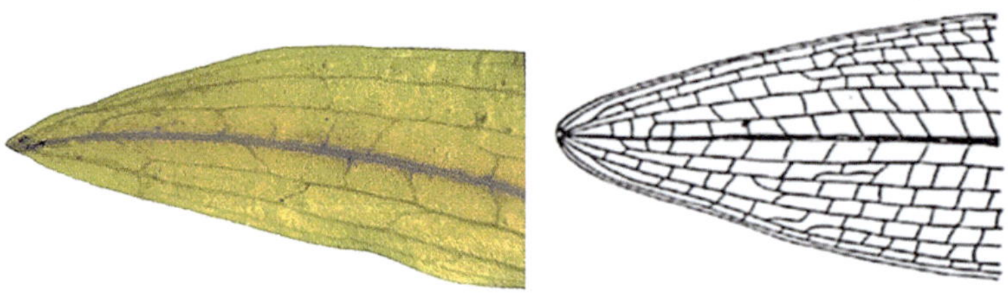

Potamogeton alpinus. Tip of a submerged leaf (photo modified from Schou et al. 2023; drawing modified from Raunkiær 1895–1899)

The submerged leaves of *P. alpinus* have rounded or subacute tips. The central vein is prominent, and the longitudinal veins are linked by a network of cross-veins.

Life form and ecology *Potamogeton alpinus* has floating and submerged leaves. It grows in lakes, rivers, ponds, canals, river backwaters and slow flowing rivers where it usually favours silt and peat substrates and relatively shallow (<1.5 m) water (Preston 1995). It is characteristic of cool, oligotrophic to mesotrophic water, and sometimes slightly acidic to highly alkaline waters (Wiegleb and Todeskino 1983). It is in decline over large areas of western Europe, primarily due to eutrophication (Sand-Jensen et al. 2000) and also to global warming which may reduce sexual reproduction (Robionek et al. 2018). *P. alpinus* propagates abundantly by seed but also produces tuber-like structures when under stress, for example during desiccation events (Brux et al. 1987).

Distribution *Potamogeton alpinus* is a circumboreal Northern Hemisphere species of arctic and boreal regions. It is widespread in northern Europe, but it is less common in central Europe and does not extend south of the Pyrenees (Baattrup-Pedersen et al. 2008). It is also found in Greenland, Canada, and northern USA, and it is widespread in northwest Asia.

Quaternary history *Potamogeton alpinus* endocarps are readily recognisable. They have been recorded from lake sediments, particularly in the Late Glacial and in the Early Holocene (e.g. Gałka and Sznel (2013) and Gałka et al. (2014) in northern Poland; Mortensen et al. (2011) in Denmark; and Birks et al. (2000) in the Early Holocene at Kråkenes Lake, western Norway).

Potamogeton alpinus Balb. (gbif.org)

5.2.172 ***Potamogeton x angustifolius*** J.Presl. (Syn. ***Potamogeton x zizii*** W.D.J.Koch ex Roth) Potamogetonaceae

Morphology The endocarps of *Potamogeton x angustifolius* are medium sized, 2.0–3.0 mm long and 1.8–2.0 mm wide. Two collections are illustrated. The endocarps are D-shaped with a flat or slightly convex ventral side indented at the position of the partition wall. The sides are flat. The lid is narrow and roof shaped and reaches the beak. In side view its top is hidden behind the projecting side walls. The beak and stalk are ventral. *P. x angustifolius* is usually regarded as a hybrid between *P. gramineus* and *P. lucens* and is said to fruit abundantly (Aalto 1970).

Leaf tip of a submerged leaf

Life form and ecology *Potamogeton x angustifolius* is a robust rhizomatous pondweed with submerged, and often, floating leaves. It fruits freely, but it is uncertain if the fruits are viable (Preston 1995). It overwinters and reproduces vegetatively by rhizomes and plant fragments. It has been found as a pioneer in a reservoir in Czech Republic more than 50 km from its nearest occurrence (Krahulec and Lepš 1993). It grows in lakes in water at least 0.5 m deep that is base-rich and mesotrophic (Preston and Croft 1997).

Distribution *Potamogeton x angustifolius* occurs in temperate Europe, including Scandinavia, the British Isles and Ireland, across Asia to Japan, and in North America. A new record in Japan was verified by DNA comparison (Shutoh et al. 2020).

Potamogeton zizii W.D.J.Koch ex Roth (gbif.org)

Potamogeton angustifolius J.Presl (gbif.org)

5.2.173 ***Potamogeton berchtoldii*** Fieber Potamogetonaceae

Morphology *Potamogeton berchtoldii* endocarps are small, about 1.5 mm long and 1.2 mm wide. The base of the endocarp is relatively broad. The sides are slightly convex with a shallow central depression. The ventral side is S-shaped to convex. The rounded lid reaches the base of the beak. The base of the lid is relatively broad, reaching below the base of the endocarp. The beak is central and decurrent into the ventral area. The stalk is ventral and prominent.

Potamogeton berchtoldii leaf tips.

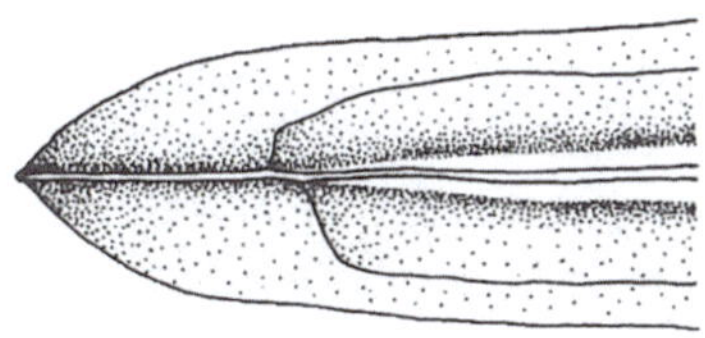

Drawing of a leaf tip modified from Preston (1995)

The leaves are narrow with acute or blunt-pointed tips, but these and the width of the leaves are variable. They lack a marginal vein (cf. *P. pusillus*, Preston 1995), but the narrow-leaved species are difficult to differentiate on leaf tips (see Preston 1995).

Life form and ecology *Potamogeton berchtoldii* is a small, fine-leaved, submerged pondweed. It has a very broad physical and chemical tolerance. It lives in lakes, slow flowing rivers, ponds, and ditches, in base-rich or base-poor, mesotrophic to eutrophic waters where it usually roots in silty substrates. It is rarely found in oligotrophic, brown-water habitats. In brackish and eutrophic conditions, it is thought to give way to the very similar *Potamogeton pusillus* (Preston and Croft 1997). It overwinters as turions, but also fruits freely.

Distribution *Potamogeton berchtoldii* is widespread across the temperate and boreal Northern Hemisphere. It is found throughout most of Europe, in Iceland and southwest Greenland and south into the Mediterranean region, becoming rare in the Iberian Peninsula and in the Arctic. *P. berchtoldii* also occurs in North America, and across Asia, including China and Japan. Unlike the superficially similar *P. pusillus,* it is rare in Africa (Kaplan and Štěpánek 2003).

Quaternary history *Potamogeton berchtoldii* endocarps are difficult to separate from similar small endocarps. However, they have been identified in the Late Glacial and Holocene of Britain (Godwin 1975; Bishop et al. 2019) and Sweden (Jacobson 1994). *P. berchtoldii* has increased recently in lakes suffering recent mild eutrophication (Bennion et al. 2018).

Potamogeton berchtoldii Fieber (gbif.org)

5.2.174 *Potamogeton coloratus* Hornem. Potamogetonaceae

Morphology The endocarps of *Potamogeton coloratus* are the smallest in our region, at about 1.2–1.5 mm long and 1 mm wide. They are oval and the ventral side is usually straight or it can be S-shaped. The sides are flat, with a large cavity in the centre. The lid usually reaches the beak and is rounded or roof-shaped with a keel. The beak position is ventral to central. The stalk is inconspicuous. The endocarps resemble those of *P. polygonifolius*, but they are smaller, with a straight or slightly S-shaped ventral side, a more pronounced central cavity, and a stouter beak.

Life form and ecology *Potamogeton coloratus* is a rhizomatous pondweed with narrow submerged leaves, but broader floating leaves. It is a calcicole typical of lime-rich and shallow (usually <1 m) ponds, ditches, karst springs and associated wetlands, lake edges, and slow flowing rivers and dykes. It often grows in association with charophytes in oligotrophic water. It is unable to utilise bicarbonate ions in photosynthesis, so is restricted to water containing dissolved CO_2 (Schou et al. 2023). Consequently, it declines with eutrophication. It can grow on peaty as well as clay-based substrates. It is vulnerable to physical disturbance and uprooting and so tends to prefer quiescent habitats with minimal water-level fluctuation (Lukács et al. 2020). *P. coloratus* generally overwinters as leafy shoots, but it also fruits freely. Its seeds are long-lived, retaining viability for up to 30 years and probably longer periods of time (Kaplan et al. 2014).

Distribution *Potamogeton coloratus* occurs throughout much of temperate Europe from the Mediterranean in the south to Scotland and Denmark and Bornholm in the north. It is also recorded in the Asian part of Turkey (Wiegleb 1989).

Quaternary history *P. coloratus* has been recorded from Quaternary interglacial and Holocene sediments in Britain (Godwin 1975; Brown et al. 2018). There are also records from the Early Holocene in Czech Republic (Petr et al. 2013) and the mid Holocene in Spain (Carrion and van Geel 1999).

Potamogeton coloratus Hornem. (gbif.org)

5.2.175 *Potamogeton compressus* L. Potamogetonaceae

Morphology The endocarps of *Potamogeton compressus* are large 2–3 mm long and about 2.0 mm wide. The ventral area is almost straight and often has a boss prolonging the partition wall. The sides are flat or slightly concave, tapering towards the stalk. The lid comes close to or reaches the beak. It is roof-shaped with a narrow crispate crest or rounded with scattered warts. The beak is more or less central and dorsally recurved. The conical stalk tapers gradually to the endocarp sides.

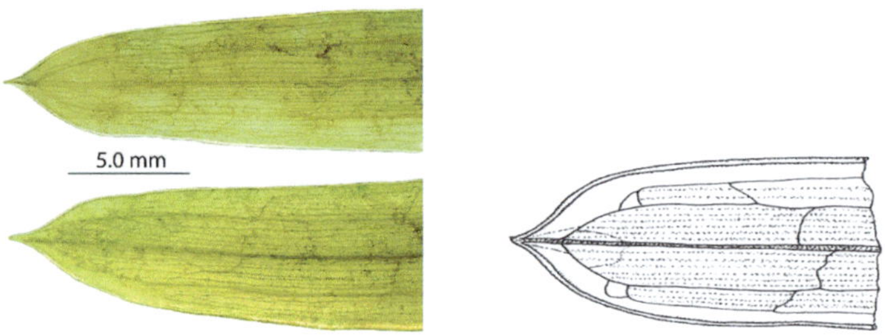

The leaves of *Potamogeton compressus* are 3–6 mm wide and mucronate. They have a strong central vein and two subsidiary veins linked by cross veins. (Drawing modified from Preston 1995).

Life form and ecology *P. compressus* is a perennial elodeid pondweed, characteristic of lowland still and slow flowing water, including lakes, ditches, river oxbows, canals, and sluggish rivers. It prefers moderately base-rich and mesotrophic to slightly eutrophic conditions and is sensitive to periodic habitat disturbance (e.g. scour in rivers and ditch management) (Birkinshaw et al. 2013). It grows typically on silt or sandy substrates (Nagasaka et al. 2002). It is a coloniser of open habitats and is intolerant of competition. It appears to reproduce primarily vegetatively by means of turions (Preston 1995).

Distribution *P. compressus* is distributed in temperate and boreal regions of Europe and Asia (Kaplan and Marhold 2012; Birkinshaw et al. 2013). In Europe it is absent from the extreme north of Sweden and Finland and is rare in Norway and Scotland. It is widespread in northwest continental Europe but is absent from the Mediterranean region (Birkinshaw et al. 2013). It extends through Asia across Mongolia, China, and Japan, although Kaplan and Marhold (2012) suggest that further sampling and detailed molecular and morphometric analyses are necessary to confidently assign plants to *P. compressus* (as opposed to the closely related *Potamogeton zosteriformis*) in this region. In North America, it is replaced by *P. zosteriformis*. Throughout much of its range *P. compressus* appears to be in decline due to eutrophication and habitat loss (e.g. by filling of backwaters and river regulation) (Schnittler and Gunther 1999; Birkinshaw et al. 2013).

Potamogeton compressus L. (gbif.org)

5.2.176 ***Potamogeton crispus*** L. Potamogetonaceae

Recent endocarps

Subfossil endocarps

Morphology The endocarps of *Potamogeton crispus* are large, around 3 mm long excluding the long beak and about 2.0–2.5 mm wide. They are ovoid, tapering to the beak, which is decurrent with the ventral side. The ventral side is convex. The sides are flat, with a central depression. The lid reaches half-way up the endocarp, its top hidden in side view by the protruding dorsal sides. It is keeled and crested, but this is often lost in fossil material. The beak is broad-based and decurrent with, and nearly as long as the endocarp. The stalk is inconspicuous, placed centrally in the flat base of the endocarp.

Potamogeton crispus turion upper side

Potamogeton crispus turion underside

Potamogeton crispus overwinters and spreads by turions. They are compressed thick horny leaves with spikey margins. Turion fragments can be preserved in sediments. A possible strategy for quantification is to count the number of teeth on each fragment.

Life form and ecology *Potamogeton crispus* is a submerged pondweed commonly found in lowland, meso-eutrophic to eutrophic waters, including lakes, ponds, rivers, canals and ditches, where it can grow to a depth of around 6–7 m. Mostly, *P. crispus* favours alkaline, calcareous water and it is also able to survive in slightly brackish water. In shallow lakes it is one of the last species to disappear under poorly illuminated, highly eutrophic conditions, where it frequently exhibits a canopy-forming strategy (Yu et al. 2018). It is versatile in terms of substrates being able to grow in the soft muds of lakes and ponds and across stone, gravel and sand-dominated habitats, especially in rivers. *P. crispus* fruits freely in shallow water. It persists throughout the winter as leafy shoots and plants. Turions, which generally form in late spring to early summer, are a key means of vegetative spread (Xie et al. 2015).

Distribution *P. crispus* is widely distributed in Europe, Asia, southern Africa, and Australia. It is widespread throughout most of Europe but becomes rare north of 60°N. It is introduced in the Americas and New Zealand where it is often invasive.

Potamogeton crispus L. (gbif.org)

5.2.177 ***Potamogeton epihydrus*** Raf. Potamogetonaceae

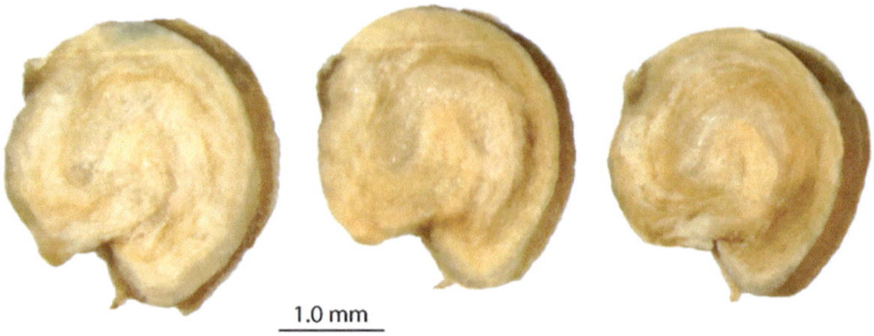

Morphology The endocarps of *Potamogeton epihydrus* are medium sized, about 3 mm long and 2.5 mm wide, appearing almost circular. The ventral side is strongly S-shaped with an indentation at the partition wall and a bulge above it. The embryo is tightly curved, (spiral), its position visible on the sides and resulting in an indentation. The ridged lid reaches the top of the endocarp and is relatively narrow. The inconspicuous beak is ventral, a short way down the ventral side. The small stalk is also ventral.

Life form and ecology *Potamogeton epihydrus* is a rhizomatous perennial pondweed with submerged and floating leaves. It reproduces mainly by overwintering stolons and buds, but it also fruits freely (Preston and Croft 1997). It grows in still or slow-moving base-poor, oligotrophic to mesotrophic water up to 1 m deep on peaty substrates.

Distribution The main occurrence of *Potamogeton epihydrus* is in North America, where it is widespread in oligotrophic waters from Alaska and Labrador to California and Tennessee. It has European outliers in a few small lakes in the Outer Hebrides, northwest Scotland, and in two English canals.

Potamogeton epihydrus Raf. (gbif.org)s.

5.2.178 *Potamogeton friesii* Rupr. Potamogetonaceae

Morphology The endocarps of *Potamogeton friesii* are small, about 2 mm long and 1.5 mm wide. They are a bit larger than the other three species in this group; *P. pusillus, P. berchtoldii,* and *P. rutlilus* (*Potamogeton pusillus* group). They are round to obovoid, the broadest point being above the middle. The ventral side is convex or almost straight. The sides are rounded, with a slight depression and rounded protuberances at the base. The lid is roof-shaped and usually reaches close to the base of the beak. The beak is ventral to centro-ventral and decurrent with the sides. The stalk is ventral.

Potamogeton friesii leaf tip

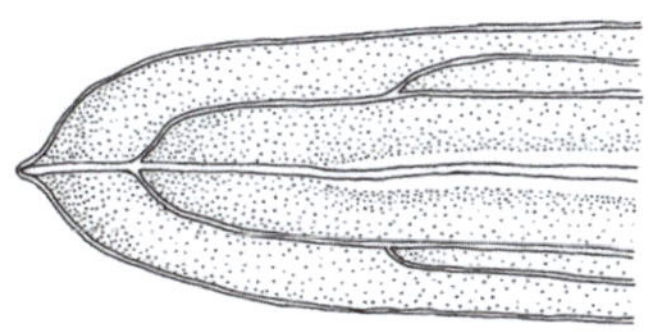

Drawing modified from Preston (1995)

The leaves of *Potamogeton friesii* are 1.5–3.5 mm wide (Preston 1995). The rounded tip is mucronate. There is a strong central vein and two side veins that curve to meet the central vein near the tip. Other side-veins develop further down the leaf. These characters may allow tentative separation from the leaf tips of other narrow-leaved species (Preston 1995).

Life form and ecology *Potamogeton friesii* is a fine leaved elodeid pondweed growing typically on silt substrates in lowland lakes and slow flowing rivers, canals, and ditches with calcareous water. It can survive in relatively eutrophic conditions

(Dynowski et al. 2019) but declines with excessive nutrient-enrichment (Preston and Croft 1997; Madgwick et al. 2011). *P. friesii* produces numerous robust over-wintering turions, but fruits relatively infrequently (Preston and Croft 1997).

Distribution *Potamogeton friesii* has a circumboreal distribution across the Northern Hemisphere (Wiegleb and Kaplan 1998). In Europe it is found north to the Arctic Circle. It is much less common in southern Europe and apparently absent from the Mediterranean area. It occurs across North America (Canada, northern USA) and Asia. Iran is its southern-most locality where it has only recently been found (Abbasi et al. 2015).

Potamogeton friesii Rupr. (gbif.org)

5.2.179 *Potamogeton gramineus* L. Potamogetonaceae

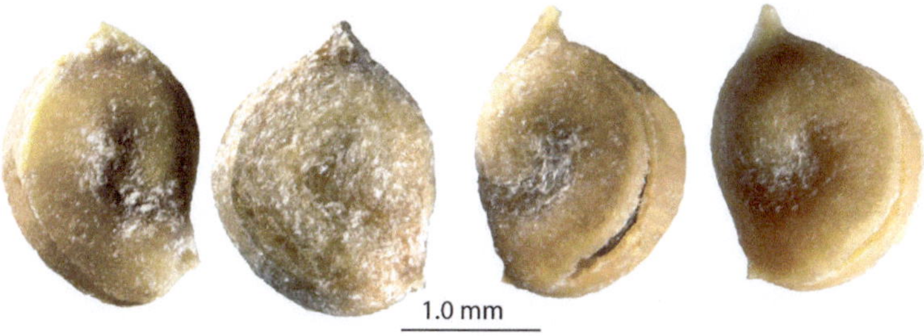

Morphology The endocarps of *Potamogeton gramineus* are small, about 2 mm long and relatively broad, at 1.5 mm wide. The ventral side can be S-shaped, straight, or convex. There is a shallow central depression and groove marking the internal partition wall. The sides are flat to somewhat rounded, with the dorsal edges projecting slightly beyond the lid. The roof-shaped lid almost reaches the top of the endocarp and the base of the beak. The beak is central, merging into the ventral side. The stalk is relatively high up the ventral side.

Potamogeton gramineus leaf tips

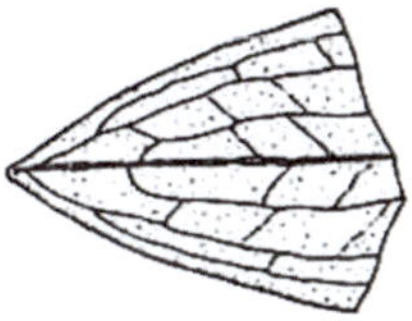

Drawing of a *Potamogeton gramineus* leaf tip, modified from Preston (1995)

Life form and ecology *P. gramineus* is a rhizomatous perennial pondweed with submerged and floating leaves. It dies back to winter buds formed at the ends of rhizomes (Spencer et al. 1994). It lacks a vegetative means of dispersal, but fruits freely (Preston and Croft 1997). It favours shallow clear water (usually <1.5 m, but sometimes deeper) and a range of substrates across a wide variety of habitats, mainly lakes, slow rivers, canals, and ditches. It grows in both base-poor and base-rich sites and in oligotrophic to meso-eutrophic waters, although it is absent from very oligotrophic sites. *P. gramineus* has been observed to decline in sites subject to acidification (Roelofs 1983) and nutrient-enrichment, being largely absent from eutrophic sites (Wiegleb et al. 1991; Preston and Croft 1997).

Distribution *Potamogeton gramineus* has a circumboreal distribution across the Northern Hemisphere. It occurs throughout much of Europe from Iceland and northernmost Scandinavia to central Europe, but it is exceptionally rare in the Mediterranean region, with only a few records (Raimondo 1996). It also occurs in boreal and arctic North America, Greenland, and across Asia extending east to Japan and south to the West Bengal area of India, where it has only recently been recorded (Chowdhury et al. 2015).

Quaternary history *Potamogeton gramineus* endocarps have been occasionally recorded from Quaternary deposits in the British Isles (Godwin 1975).

Potamogeton gramineus L. (gbif.org)

5.2.180 ***Potamogeton lucens*** L. Potamogetonaceae

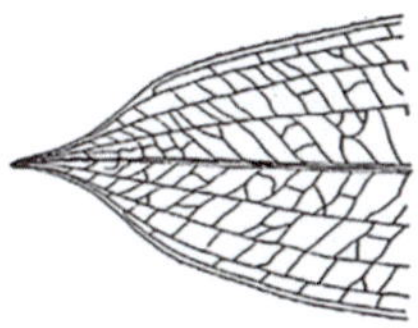

Potamogeton lucens leaf tip, drawing modified from Preston (1995)

Morphology The endocarps of *Potamogeton lucens* are large, about 3.0 mm long and 2.0 mm wide, but their size and shape are variable (Aalto 1970). The ventral side is straight, slightly convex, or slightly S-shaped. The sides are flat, often with a small central depression. The dorsal margins form raised ridges. The lid is roof-shaped with a prominent ridge and usually reaches the beak. The beak is central, inconspicuous, and decurrent with the ventral side. The stalk is usually broad and flat and situated high on the ventral side. The leaf tips are variable, mucronate or with a long excurrent midrib (Preston 1995).

Life form and ecology *Potamogeton lucens* is a large rhizomatous pondweed that grows in lowland, calcareous lakes, slow-flowing rivers, canals, and ditches. It prefers silty substrates and deep water. Plants can be extremely tall (up to 6 m) forming the freshwater equivalent of kelp forest. It lacks vegetative propagules and spreads from site to site primarily via seeds. It is associated with mesotrophic to mildly eutrophic water. It is highly vulnerable to eutrophication which has been responsible for its decline in many areas (Sand-Jensen et al. 2000).

Distribution *P. lucens* is widespread across temperate Europe extending north into Scandinavia and across the continent from Iberia to western Asia. Further east in Asia its occurrence is uncertain, but it reaches Japan (Shutoh et al. 2018). It is rare in North America where it is replaced by the similar *P. illinoensis* Morong. It also occurs in North Africa, where the related *P. schweinfurthii* A. Benn. is more common.

Potamogeton lucens L. (gbif.org)

5.2.181 ***Potamogeton natans*** L. Potamogetonaceae

Morphology The endocarps of *P. natans* are large, about 2.2–3.0 mm long and 2.0–3.0 mm wide. The sides are convex, giving a rounded shape, and with a central depression in the curve of the embryo. A furrow marks the position of the interior partition wall. Near the top, the sides project beyond the lid. There are lateral warts at the base and the ventral position of the stalk is obvious. The beak is usually ventral but can be more or less central. These projecting features may be lost on fossilisation. The ventral margin is S-shaped, convex at the top. The lid is broad, broader in the upper part, keeled or roof-shaped, and sometimes with a crest. Its apex is pointed, and often with a tongue-like projection. It reaches the top of the endocarp but does not usually reach the base of the beak.

Floating leaf of *P. natans* preserved in anoxic sediment from Bonfire Field Pond, Heydon, Norfolk, England. Photo: Eftesum Eftesum

The floating leaves of *P. natans* can sometimes be preserved in anoxic sediment, even surviving as almost whole leaves that can form distinctive layers in some situations. The leaves are up to 10 cm long and 4 cm wide and are recognisable in sediments by the distinctive elliptic or ovate spoon-shape and a characteristic vein pattern.

Life form and ecology *Potamogeton natans* is a rhizomatous perennial pondweed which has floating leaves that can cover large areas of still water, either as a monoculture or mixed with other nymphaeids and emergents. It overwinters as plants or as rhizomes (Preston and Croft 1997). It usually fruits freely and has long-lived (>150 years) seeds (Alderton et al. 2017). It readily colonises new water bodies, presumably by seeds dispersed by waterfowl. It has a wide ecological range, occurring in still-water habitats with pH ranging from 5.5 to 9.7 (mean 7.9, Pip (1987)), especially ponds and small lakes and the sheltered margins of larger lakes, and in ditches and river backwaters. It favours silt, sand, or peaty substrates, moderate water depths (most frequently <2 m) and can tolerate oligotrophic to eutrophic conditions. In acidic conditions it frequently co-occurs with *Potamogeton polygonifolius*, which outcompetes it as acidity increases. *P. natans* utilises aerial and dissolved CO_2 in photosynthesis (Bodner 1994).

Distribution *Potamogeton natans* has a circumpolar boreo-temperate distribution through Europe, Asia, and North America. It is widespread in Europe extending south from Iceland and northernmost Scandinavia to southern Spain and Sicily. It is less frequent south of The Alps.

Quaternary history *Potamogeton natans* is the most commonly recorded *Potamogeton* species in Holocene lake-sediments.

Potamogeton natans L. (gbif.org)

5.2.182 ***Potamogeton nodosus*** Poir. Potamogetonaceae

Morphology The endocarps of *Potamogeton nodosus* are medium to large, about 2.2–3.5 mm long and 2.0 mm wide. The ventral side is straight, and the sides are indented. The lid does not reach the beak, and it is heavily crested with lateral ridges. The strong decurrent beak is ventral. The thick stalk is ventral. Some of our material lacks a conspicuous crest.

Life form and ecology *Potamogeton nodosus* is a robust rhizomatous perennial pondweed with submerged and floating leaves. It appears to be restricted to slow flowing rivers in some parts of Europe including England (Preston and Croft 1997) and the Danube Delta (Schneider-Binder 2018) but grows in a much wider range of habitats elsewhere including shallow lakes, ponds, and ditches on soft silty substrates. It favours calcareous water and can tolerate moderately eutrophic conditions. It depends on dissolved CO_2 for photosynthesis and cannot utilise bicarbonate ions (Schou et al. 2023). *P. nodosus* regenerates from tubers and also fruits freely over much of its range, but not everywhere, with fruits rarely seen in English populations, for example (Preston 1995).

Distribution This species is common in western Europe as far north as England, Estonia, Poland, and The Netherlands. It is widely distributed in Africa (Kaplan and Symoens 2005), in the USA and southern Canada, and also temperate and tropical Asia extending to the Philippines and New Guinea.

Potamogeton nodosus Poir. (gbif.org)

5.2.183 ***Potamogeton obtusifolius*** Mert. & Koch Potamogetonaceae

Morphology The endocarps of *Potamogeton obtusifolius* are medium sized, 2.5 mm long and about 2.0 mm wide. The ventral side is convex, or slightly S-shaped, or more or less straight. The sides are slightly rounded or flat with an obscure central depression. There is an acute wart at the base. The lid is rounded with a crest and does not reach the base of the beak. The crest can be much more pronounced than in our specimens (see Aalto 1970; Cappers 1993a, b). Jessen's (1955) illustration suggests a less pronounced crest and also shows the wavy ridge that can surround the stalk. The beak is central and decurrent into the ventral side. The stalk is ventral. Aalto (1970) mentions that a 'microcarp' form of *P. obtusifolius* endocarps has been described, which lacks obvious crests and warts and closely resembles *P. berchtoldii*. Four of our five collections from the UK and Scandinavia would fall into this category and none is as extremely crested as those described by Aalto (1970). This suggests that the 'microcarp' form might be rather common in northern Europe.

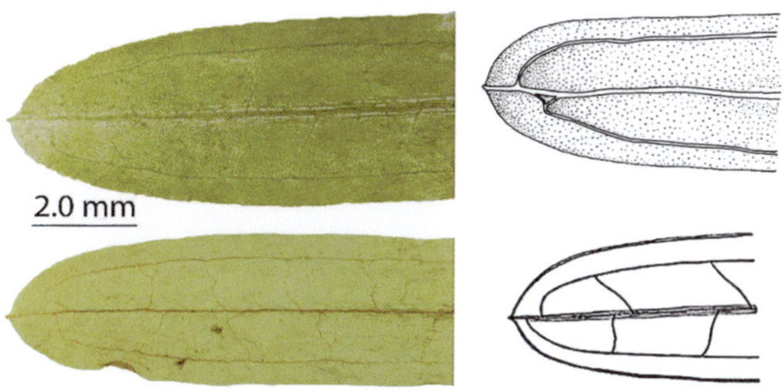

Drawings of leaf tips: upper from Preston (1995) and lower from Raunkiær (1895–1899)

Potamogeton obtusifolius leaves are 2.5–3.5 mm wide with a characteristic, blunt, obscurely mucronate apex. There are two parallel side veins that converge with the central vein just short of the apex. Lateral veins are rather obscure. There are fine teeth on the margin.

Life form and ecology *Potamogeton obtusifolius* is a submerged, non-rhizomatous, slender pondweed that overwinters by turions. However, it also fruits freely (Maberly 1993; Preston and Croft 1997). It is found in lowland lakes and slow-flowing rivers, canals, and ditches where it grows in shallow water (generally <1.5 m) on silty and sometimes peaty substrates. It grows in circum-neutral to slightly acidic, mesotrophic to slightly eutrophic water. Like the morphologically quite similar *P. friesii*, it declines with excessive eutrophication (Seddon 1972).

Distribution *P. obtusifolius* has a northern temperate and boreal circumpolar distribution. In Europe it is most common in central Europe, It is absent from the Mediterranean region and present, but scattered north of the Arctic Circle. It is widespread in Canada, but is apparently absent from Greenland. It occurs across Asia including Russia, China, and Japan, with its most southerly location in Myanmar (Potamogeton obtusifolius Mert. & W.D.J.Koch | Plants of the World Online | Kew Science).

Quaternary history *Potamogeton obtusifolius* fruits have been recorded in many Holocene sites in the British Isles and Ireland and also other parts of north-west Europe such as Denmark and Poland.

Potamogeton obtusifolius Mert. & W.D.J.Koch (gbif.org)

5.2.184 *Potamogeton perfoliatus* L. Potamogetonaceae

Morphology *Potamogeton perfoliatus* endocarps are medium to large, 2.0–3.0 mm long and 1.5–2.5 mm broad. Their shape is variable. The sides are usually flat with a slight central depression and a shallow groove at the position of the internal partition wall. The sides taper towards the top and to the base where they terminate in distinct warts. The ventral side is usually S-shaped but can be flat or irregular. The broad-based beak is centrally placed, erect or dorsally tilted, and the stalk is ventral. The lid is relatively broad, more or less keeled, and occasionally with a crest. It reaches to the top but does not usually quite reach the beak.

Life form and ecology *Potamogeton perfoliatus* is a robust, submerged, perennial pondweed. It lacks turions and overwinters as rhizomes. It spreads via rhizomes that creep horizontally through the sediment (Ozimek et al. 1976; Adamec 2018b). *P. perfoliatus* fruits freely. It grows in a wide range of freshwater habitats including lakes, rivers, and canals. It usually grows in water deeper than 1 m and up to 5 m on silty and sandy substrates away from the shoreline. It has a wide trophic tolerance and most frequently occurs in base-rich mesotrophic to mildly eutrophic water (Preston and Croft 1997).

Distribution This species has a circumpolar distribution occurring all over Europe including Iceland and Scandinavia to the north and the Mediterranean region to the south, reaching North Africa. It is widespread across Asia east to Japan and south to India and Indonesia. It occurs in Greenland, and in eastern Canada and USA. It is introduced in eastern Australia and New Zealand.

Quaternary history *Potamogeton perfoliatus* endocarps have been recorded in Holocene sediments in the British Isles (e.g. Godwin 1975) and in Ireland (Mitchell 1952), Denmark (Bennike et al. 2020), and Czech Republic (Kołaczek et al. 2015).

Potamogeton perfoliatus L. (gbif.org)

5.2.185 ***Potamogeton polygonifolius*** Pourr. Potamogetonaceae

Morphology The endocarps of *Potamogeton polygonifolius* are small, about 1.5 mm long and 1.2 mm wide. They are almost round. The ventral side is convex, and the sides are flat with a pronounced central cavity. The dorsal margins of the sides are thickened and extend beyond the top of the lid. The lid is roof-shaped with a narrow, low, distinct ridge (Aalto 1970). It scarcely reaches the top of the endocarp or the beak but extends below the base of the endocarp. The short stigma is ventral to central, and decurrent with the ventral side. It is tilted dorsally. The stalk is tiny and not apparent in fossil material. Endocarps of *Potamogeton coloratus* are similar, but smaller, and with a more pronounced central cavity.

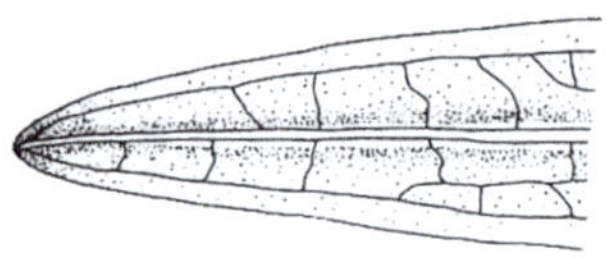

Left. The floating leaves of *P. polygonifolius* are opaque, elliptic to ovate with a rounded tip. The veins all converge at the tip. The subsidiary veins are small, at right angles to the main veins, thus forming a reticulate pattern. Right. The submerged leaves are delicate and narrower, the central vein being bordered by lacunae (drawing modified from Preston 1995).

Life form and ecology *Potamogeton polygonifolius* is a rhizomatous perennial pondweed with floating and submerged leaves. It fruits freely and lacks specialised vegetative propagules (Preston and Croft 1997). It grows on peaty substrates in shallow (<1 m), base-poor, oligotrophic to mesotrophic water at the edges of lakes and in ponds, streams, ditches, peat cuttings, and flushes. It also grows in sub-terrestrial habitats in damp boggy ground and in oligotrophic *Sphagnum* flushes and bogs associated with heaths and moorlands. It sometimes occurs in calcareous habitats, but only when oligotrophic conditions prevail (Preston and Croft 1997) as it is unable to photosynthesise using bicarbonate ions (Maberly and Spence 1983). It declines in response to severe acidification, and also to eutrophication (Thiébaut et al. 2002), which threatens it in marginal parts of its range, especially central Europe (Zalewska-Galosz et al. 2012). It has also suffered habitat loss by drainage for agriculture. It can survive at relatively high elevations.

Distribution *Potamogeton polygonifolius* has an amphi-atlantic distribution. It is most frequent in western Europe north to southern Sweden and lowland Norway, where it reaches north of the Arctic Circle, and south to the Iberian Peninsula and rarely in North Africa. It occurs in the Azores and Madeira (Kaplan et al. 2013). In Canada it is restricted to Nova Scotia and Newfoundland.

Quaternary history *Potamogeton polygonifolius* has been frequently recorded in Holocene sediments in the British Isles, especially from the north and west (Godwin 1975) and in Ireland (Jessen 1949).

Potamogeton polygonifolius Pourr. (gbif.org)

5.2.186 ***Potamogeton pusillus*** L. (Syn. *P. panormitanus* Biv.) Potamogetonaceae

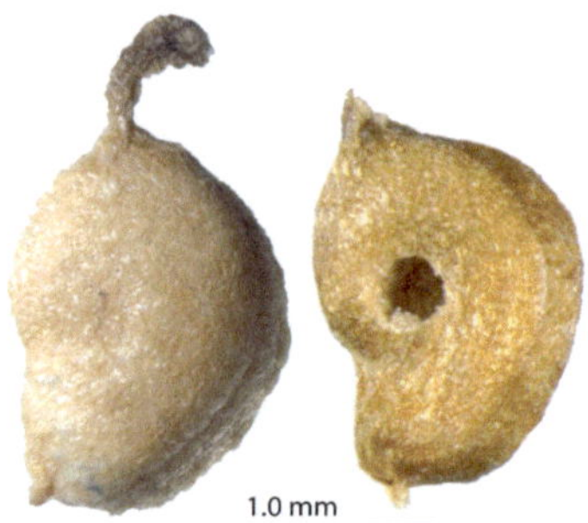

Morphology The endocarps of *Potamogeton pusillus* are small, about 1.5 mm long and 0.9–1.3 mm wide. They taper towards the base and the ventral area is S-shaped or convex. The sides are convex sometimes with a slight depression and groove at the partition wall. With wear, a hole may develop where the embryo curved within the endocarp. The rounded lid seldom reaches the beak. The beak is central and dorsally tilted. The small stalk is ventral to centro-ventral. The endocarps are often difficult to distinguish from other species of narrow-leaved submerged *Potamogeton (P. berchtoldii, P. trichodes, P. friesii., P. rutilus)* (Aalto 1970).

Potamogeton pusillus leaf tip

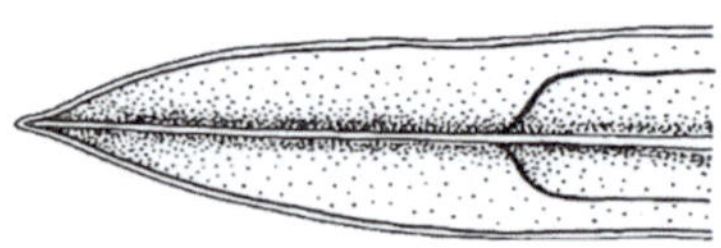

Drawing modified from Preston (1995)

The leaves of *P. pusillus* are about 1.0–1.5 mm wide with an apiculate tip and a marginal vein. Lateral veins are distinct. These characters may differentiate them from the very similar *Potamogeton berchtoldii,* but usually they and other narrow-leaves species cannot be differentiated based on leaf tips (see Preston 1995). Leaf tips of *Potamogeton pusillus* or *P. berchtoldii* were recorded in recent lake sediments by Salgado et al. (2010), Bennion et al. (2018), and Walton et al. (2021b).

Life form and ecology *Potamogeton pusillus* is a slender elodeid pondweed that overwinters by turions. Although it fruits freely, turions are the main method of reproduction (Preston 1995). It occurs in lakes, ponds and slow flowing rivers, canals and ditches where it prefers silty substrates. It is a lowland species of shallow (<1.5 m), calcareous, mesotrophic-eutrophic water and is one of the last species to disappear from shallow lakes subject to progressive nutrient-enrichment (Sayer et al. 2010a, b). It is also tolerant of slightly brackish water (Preston and Croft 1997). It is a pioneer species in new or disturbed water bodies (Preston and Croft 1997).

Distribution *Potamogeton pusillus* occurs on every continent except Australia (Kaplan and Štěpánek 2003) and Antarctica. It has a temperate-boreal distribution in the Northern Hemisphere including Iceland. It is widespread in Europe from southern Scandinavia to the Mediterranean and reaches the northern coast of Norway. It extends across Asia and is also widespread in USA, southern Canada, and central America. It occurs in South America and Africa (Preston and Croft 1997; Wiegleb and Kaplan 1998).

Quaternary history *P. pusillus* belongs in the group (*P. pusillus, P. berchtoldii, P. friesii, P. rutilus*) that Aalto (1970) considers are sometimes difficult to determine to species as fossils. Nevertheless, Godwin (1975) lists several localities in the British Isles where *P. pusillus* has been identified, although often with a proviso or as a type. *P. pusillus* is most frequently grouped with *P. berchtoldii,* e.g. by Bennion et al. (2018). Other records for example, are from Ireland (Jessen 1949; Mitchell 1952), Denmark (Bennike et al. 2020), northeast Poland (Gałka and Sznel 2013; Gałka et al. 2014), and western Russia (Sapelko et al. 2022).

Potamogeton pusillus L. (gbif.org)

5.2.187 *Potamogeton rutilus* Wolfg. Potamogetonaceae

Morphology The endocarps of *Potamogeton rutilus* are small, about 1.8–2.0 mm long and 1.5 mm wide. They are typically narrowly crescentic. The ventral side is flat, with an indent at the partition wall. The sides are convex, sloping to the top. The rounded lid reaches the top of the endocarp and sometimes the beak. The beak is ventral to central, with a broad base tapering upwards. The small stalk is distinct, ventrally placed.

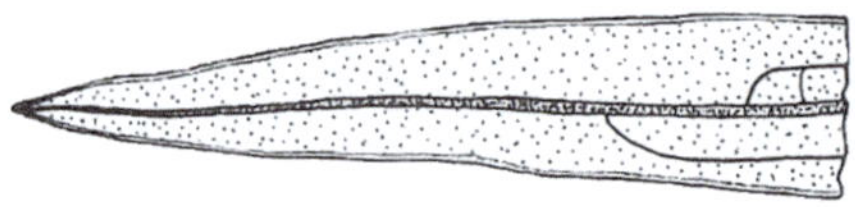

Drawing of a leaf tip of *Potamogeton rutilus* modified from Preston (1995). The leaves of *P. rutilus* narrow gradually to an acute apex

Life form and ecology *Potamogeton rutilus* is a submerged, fine-leaved pondweed that lacks rhizomes. Reproduction is usually by bud-like turions in the leaf axils, as seeds seem to be produced in low quantities (Preston and Croft 1997; Wallace 2005), thus limiting its ability to disperse between sites. *P. rutilus* grows in lowland, mildly basic, mesotrophic, and naturally eutrophic lakes, or their adjoining streams (Preston and Croft 1997; Mossberg and Stenberg 2018). It grows in relatively deep water (1–2 m) on clay, silt, or lake-mud substrates. In Scotland it tends to occur in coastal areas in limestone and sandstone catchments and machair lochs. *P. rutilus* appears to be sensitive to eutrophication and the resulting competition with other more robust submerged macrophytes (e.g. Preston and Croft 1997; Bobrov et al. 2018).

Distribution *Potamogeton rutilus* is endemic to northern Europe, north into the Arctic Circle and south to Germany and Poland. In the British Isles it is confined to northern Scotland. Recently, utilising molecular analysis, Bobrov et al. (2018) showed that *P. rutilus* occurs in west Siberia and central Yakutia (east Siberia, Asian Russia). It may be one of the rarest *Potamogeton* species worldwide (Preston and Croft 1997) and given its local distribution and scarcity in many of the areas where it is known, Bobrov et al. (2018) concluded that *P. rutilus* should be classified as Vulnerable in regional and global Red Lists.

Potamogeton rutilus Wolfg. (gbif.org)

5.2.188 *Potamogeton trichoides* Cham. & Schltdl. Potamogetonaceae

Morphology *Potamogeton trichoides* endocarps are medium to small, 2.0–2.5 mm long and about 1.5–1.9 mm wide. They are roundish to oval in shape. The ventral side is almost straight with a central boss continuing from the partition wall. The sides are flat to slightly convex and terminate in a prominent tubercle at the base. The lid usually reaches the base of the beak and is rounded with scattered warts (Aalto 1970), not prominent in our material. The beak is central and the stalk is inconspicuous. Aalto (1970) discusses the variability of *P. trichoides* endocarps, which can havc morc pronounccd tubcrclcs and largcr sizc.

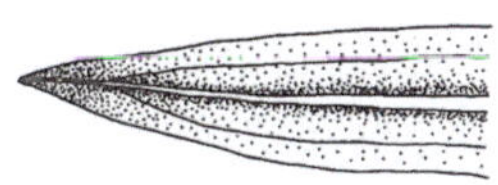

Drawing of a *Potamogeton trichoides* leaf tip, modified from Preston (1995)

The very narrow leaves of *P. trichoides* taper to an acute apex. They lack an obvious marginal vein and have a prominent midrib (Preston 1995). They can be identified to some level of certainty when preserved in lake sediment.

Life form and ecology *Potamogeton trichoides* is a very slender, much branched, non-rhizomatous pondweed. It regenerates in spring from overwintering turions. (Van Wijk and Trompenaars 1985). Fruits are sparingly produced and are important for dispersal rather than population maintenance. *P. trichoides* is a frequent colonist following disturbance (Wade and Edwards 1980) suggesting that seeds are long-lived and important for its long-term persistence. It grows in lakes, ponds, ditches, and slow flowing river sections in shallow water up to 50 cm deep where it can form very thick growths. It favours meso-eutrophic to eutrophic, base-rich water and soft substrates and tends to increase with mild eutrophication (Preston and Croft 1997).

Distribution *Potamogeton trichoides* is found across temperate Europe, western Asia, and Africa. Its northern limit is in northern England and southern Sweden (where it is very rare), and it is absent from Ireland, Norway, and Finland.

Quaternary history The Holocene record of *Potamogeton trichoides* is sparse in the British Isles, although it is recorded in many interglacial sequences (Godwin 1975). Other Holocene records come from southeast and northeast Poland (Kołaczek et al. 2015; Gałka 2014).

Potamogeton trichoides Cham. & Schltdl.

5.2.189 ***Stuckenia filiformis*** (Pers.) Börner (Syn. *Potamogeton filiformis* Pers.) Potamogetonaceae

Morphology The endocarps of *Stuckenia filiformis* are small to medium, about 2 mm long and 1.5 mm wide. The ventral side is almost straight, sometimes indented, and the sides are slightly convex. The lid is typical of *Stuckenia* (or subgenus *Coleogeton*). It reaches about three quarters up the back of the endocarp and is rounded at the top and sunken between the endocarp sides. The beak is central to centro-ventral and the stalk is indistinct.

Stuckenia filiformis leaf tip. Modified from Raunkiær (1895–1899)

The leaves of *S. filiformis* are very narrow with a blunt tip. There is a central vein and two faint lateral veins. They are recognisable when preserved in anaerobic sediments.

Life form and ecology *Stuckenia filiformis* is a perennial submerged pondweed with long rhizomes capable of colonising bare or sparsely vegetated areas. It over-winters and reproduces vegetatively by buds produced on the rhizomes, and it also fruits freely (Preston 1995). It usually grows in shallow (<1 m) water on minerogenic substrates, where competition from other macrophytes is minimal. It favours shallow lake margins with cold, clear, base-rich, mesotrophic to eutrophic water with pH from 7.5–10, or occasionally brackish water, such as sea-cliff ponds and machair lochs in Scotland (Preston and Croft 1997) and sea-shore habitats around the Baltic Sea (Mossberg and Stenberg 2018). It can also grow in slow rivers, streams, and ditches.

Distribution *Stuckenia filiformis* occurs across the Northern Hemisphere, reaching the Arctic in Europe, Siberia, Greenland, Canada, and Alaska. It also occurs in boreal and cool temperate regions through Europe, Asia, and North America. In Europe it is spread through the British Isles, Scandinavia, and especially around the Baltic coast (Mossberg and Stenberg 2018). Further south, it occurs in mountain areas, such as The Alps and Pyrenees and in the mountains of central Asia (Kaplan 2008).

Quaternary history *Stuckenia filiformis* endocarps are typically found in Late Glacial or Early Holocene sequences as part of the pioneer aquatic flora. They have been recorded in sites throughout Europe (UK, Ireland, Sweden, Denmark, Greenland, Svalbard, Poland, Czech Republic, Slovakia). For example, they were recorded in Bølling sediments in Denmark (Mortensen et al. 2011) and are typical of floras in Poland as a pioneer in early lake development (e.g. Gałka and Sznel 2013). *S. filiformis* has been widely recorded as a pioneer in the Full- and Late-Glacial of the British Isles, usually south of its present range (Godwin 1975). Late-Wisconsin records of S. *filiformis* in North America are also south of its current range (Warner 1988 and several subsequent records). In contrast, on Kongsøya, Svalbard, Late-Weichselian and Early Holocene records testify to a warmer climate than today as *S. filiformis* does not currently occur on Svalbard (Ingólfsson et al. 1995).

Potamogeton filiformis Pers. (gbif.org)

5.2.190 *Stuckenia pectinata* (L.) Börner (Syn. *Potamogeton pectinatus* L.) Potamogetonaceae

Morphology The endocarps of *Stuckenia pectinata* are large, about 3.5–4.0 mm long and 2.5 mm broad. They are crescentic or obovoid. The ventral side is usually straight. However, there is considerable variability in shape. The sides are flat. The lid is short, reaching about two thirds the way up the back of the endocarp. It is shorter than in the other two *Stuckenia* species (Sect. *Coleogeton*). It has a rounded top and a distinct rounded ridge down its length. The beak is ventral and not decurrent. The stalk is inconspicuous.

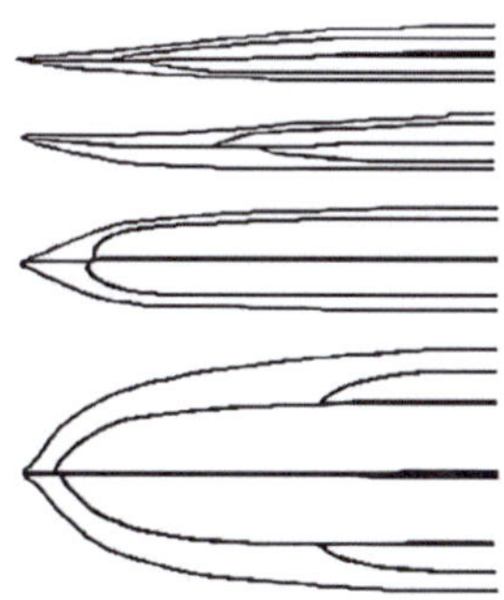

Drawings showing the variation in *S. pectinata* leaf tips, modified from Raunkiær (1885–1889)

Leaf tips of *S. pectinata* are sometimes found in sediments. They are acute to acuminate with very faint or invisible nerves. Stem nodes can also sometimes be found in sediments.

Life form and ecology *Stuckenia pectinata* is a submerged, slender to robust, rhizomatous, perennial pondweed. It overwinters as tubers on the rhizomes and leafy shoots (van Wijk 1989) and also fruits freely. Tubers and fruits are an important food source for wildfowl, which are probably the most important dispersal agent. It has a very wide habitat range including brackish water. It is frequent in base-rich, eutrophic water and is less common in base-poor acidic habitats (e.g. Birks 1973). It is tolerant of highly eutrophic conditions, and it is one of the last species to disappear as eutrophication becomes severe in shallow lakes (Sayer et al. 2010a, b). It grows across a wide range of habitats, including lakes, ponds, ditches, backwaters, slow rivers, and coastal lagoons. It prefers shallow (ca. 1.5 m) water and silt, sand, and peat substrates. It is often a pioneer in new waterbodies and increases with increased eutrophication.

Distribution *Stuckenia pectinata* is widely distributed in the world, being present on all continents except Antarctica. In Europe it is widespread in northwest Europe and the British Isles, but is less common in Scandinavia, where however it extends north to the Kola Peninsula. It is common in the southern Baltic Sea basin and extends to southern Spain and Crete.

Quaternary history *Stuckenia pectinata* endocarps have been recorded frequently in Late Glacial and Holocene sediment in the British Isles, in Poland, and in the southern North Sea. Occurrences in sediments have been recorded throughout the Northern Hemisphere. In sediments in North African lagoon lakes *S. pectinata* increased with recent eutrophication (Birks et al. 2001).

Potamogeton pectinatus L. (gbif.org)

5.2.191 ***Stuckenia vaginata*** (Magnin) Holub (Syn. *Potamogeton vaginatus* Turcz.) Potamogetonaceae

Morphology The endocarps of *Stuckenia vaginata* are medium to large, about 2.5 mm long and about 2.0 mm broad, intermediate between *S. pectinata* and *S. filiformis*. They are crescentic but longer than broad and widest near the top. The ventral side is straight or slightly indented. The sides are slightly convex, sloping towards the ventral area. They may have a small central depression. The lid is rounded with a distinct ridge. It is acute at the top, and reaches nearly to the top of the endocarp, further than in *S. pectinata* and *S. filiformis*. The beak and stalk are inconspicuous.

Life form and ecology *Stuckenia vaginata* is a rhizomatous submerged pondweed. It overwinters as tubers on the rhizomes, and it also fruits freely. The fruits are buoyant and easily dispersed by currents or animals. It grows on minerogenic substrates in fresh alkaline or brackish water 2–3 m deep, in deep lakes characterised by high conductivity, pH 8.5 (ranging from 6.7–9.7), low organic content, medium nitrogen values (mesotrophic), and low species diversity (Pip 1987).

Distribution *Stuckenia vaginata* has a circumboreal distribution. It occurs across boreal North America and central Asia (excluding China). Although widespread, it is often rare, occurring in only a few lakes. Its main occurrence in Europe is around the Gulf of Bothnia (Mossberg and Stenberg 2018) and in five lakes in northwest Norway (Artsdatabanken 2015).

Potamogeton vaginatus Turcz. (gbif.org)

5.2.192 *Hottonia palustris* L. Primulaceae

Morphology The angular seeds of *Hottonia palustris* are less than 1 mm across. They are characteristic of Primulaceae with relatively large cells unclearly arranged in rows originating from the attachment. The variety of seed shapes is due to their packed arrangement in the globular fruit.

Life form and ecology *Hottonia palustris* is a shallow-rooted perennial remaining green in winter. The seeds are abundantly produced and sink after release and germinate best on damp mud (Preston and Croft 1997). It prefers shallow water or the margins of small, sheltered ponds, often with fluctuating water levels, and can tolerate shading. It grows on organic and mineral substrates in oligotrophic to eutrophic, often calcareous water (Preston and Croft 1997; Lampinen and Metsäranta 2020). It is sometimes a pioneer in new water bodies. Seeds and detached plant parts are dispersed by water currents (Lampinen and Metsäranta 2020).

Distribution *Hottonia palustris* is almost confined to central Europe north to central Sweden and south to southern France and northern Italy. It occurs in southeast Britain and is absent from Iberia, the Mediterranean area and Norway, but has recently been recorded in southern Finland.

Hottonia palustris L. (gbif.org)

5.2.193 *Lysimachia thyrsiflora* L. Primulaceae

Morphology *Lysimachia thyrsiflora* seeds are globose or flattish and somewhat angular, with a pronounced hilum and sometimes a lateral rib. They are 1.0–1.6 mm in diameter. The pale brown surface is minutely pitted.

Life form and ecology *Lysimachia thyrsiflora* is an erect rhizomatous perennial up to 80 cm tall. It grows in wet marshes and meadows, swampy woods, and in shallow water by ditches, canals, and stoney lake margins.

Distribution *Lysimachia thyrsiflora* occurs across the boreal and temperate Northern Hemisphere. It reaches arctic Scandinavia but does not occur south of The Alps, in Ireland, or in Iberia. It is rare in western Europe.

Lysimachia thyrsiflora L. (gbif.org)

5.2.194 *Lysimachia vulgaris* L. Primulaceae

Morphology *Lysimachia vulgaris* seeds are mostly angular with a distinct border round the whole seed. The hilum is prominent on the ventral side, which is flattened as a result of the arrangement of the seeds in the capsule. The seeds are 1.5–1.8 mm in diameter. The pale brown surface is finely pitted.

Life form and ecology *Lysimachia vulgaris* is an erect rhizomatous perennial, 60–150 cm tall. It grows in shallow water beside lakes and rivers, in wet woods, fens, and swampy meadows.

Distribution *Lysimachia vulgaris* occurs across Europe and western Asia. It extends from southern Europe and North Africa to lowland Norway and southern Sweden and Finland. It is introduced in North America.

Lysimachia vulgaris L. (gbif.org)

5.2.195 *Caltha palustris* L. Ranunculaceae

Morphology Several dark brown seeds are contained in each follicle. They are approximately 2.5 mm long, ovoid, and usually with a constriction around one-third to one-half down the length of the seed delimiting a spongy floatation appendage. The surface pattern consists of small lengthwise ridges.

Life form and ecology *Caltha palustris* is a shortly rhizomatous perennial with cordate leaves up to 80 cm tall. It grows in damp and wet places such as small ponds, streams, fens, lakesides, and in wet woodlands where it can become luxuriant. It prefers slightly acidic mineral soil. Its seeds are primarily dispersed by water (van den Broek et al. 2005). *C. palustris* var. *radicans* is a smaller creeping plant (Lansdown and Pankhurst 2023).

Distribution *Caltha palustris* is native and widespread throughout the temperate and subarctic Northern Hemisphere.

Quaternary history *Caltha palustris* seeds have been recorded in Quaternary sediments (e.g. Hannon and Gaillard 1997; Yansa and Basinger 1999; Brown et al. 2018: Lapateva et al. 2023). They can be identified from their relatively large size and characteristic shape.

Caltha palustris L. (gbif.org)

5.2.196 ***Ranunculus*** L. Section ***Batrachium*** (DC.) Gray Ranunculaceae

The characteristic achenes of *R.* Sect. *Batrachium* are transversely ridged with longitudinally aligned ranks of cells. They all appear very similar, although *R. hederaceus, R. omiophyllus,* and *R. tripartitus* are somewhat more prominently ridged than the rest (G. Weigleb personal communication). It is probably not possible to separate the species in fossil material. We illustrate achenes of some of the European species here. Species delimitation of whole plants in Sect. *Batrachium* is complicated by a variety of factors. All taxa are characterized by extreme seasonal and aseasonal phenotypic plasticity (e.g. Cook 1966), allo- and autopolyploidization, breeding systems characterized by frequent hybridization, and persistence of genetically independent sterile lineages (see Wiegleb et al. 2017 and references therein).

Ranunculus aquatilis L. The left seed is split along the back line

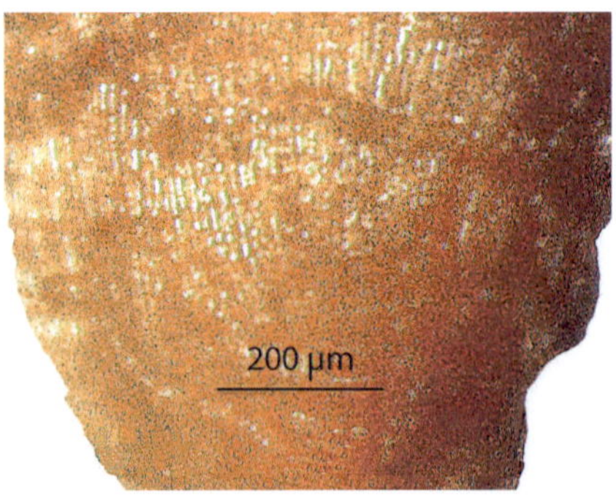

Ranunculus aquatilis cells

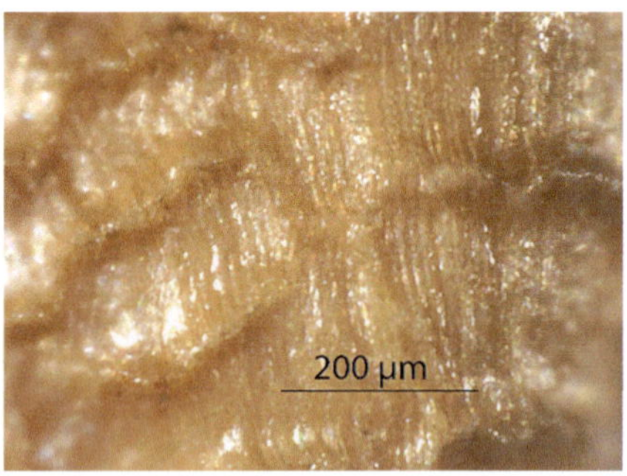

Ranunculus hederaceus cells

Ranunculus baudotii Godr.

Ranunculus circinatus Sibth.

Ranunculus confervoides (Fries) Fries

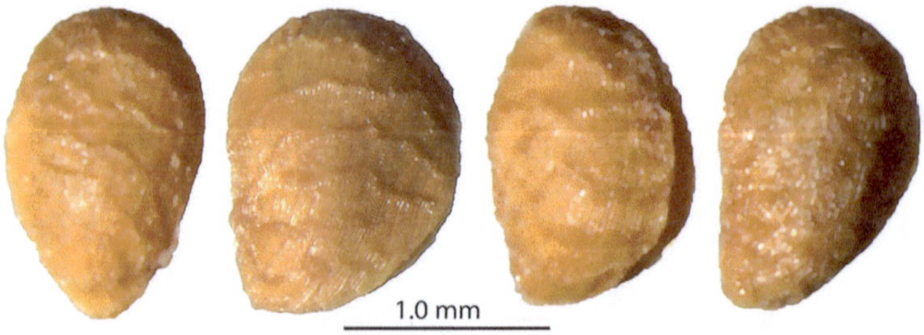

Ranunculus fluitans Lam.

Ranunculus hederaceus L. The second achene from left is a back view

Ranunculus omiophyllus Ten.

Ranunculus peltatus Schrank

Ranunculus trichophyllus Chaix. The beaks are lost in fossil material, and the achenes appear rounded. The long pedicels are also usually lost

Ranunculus tripartitus DC.

Morphology The achenes of *Ranunculus* Sect. *Batrachium* have a distinctive D shape, with one more or less straight side and the opposite side rounded. They are somewhat flattened from back to front and split easily into two longitudinal halves. In most species, the achenes are about 1.2–1.5 mm long, but range up to 1.5 mm in *R. aquatilis* and 1.8 mm or more in *R. peltatus* and *R. omiophyllus*, and down to

1.0–1.3 mm in *R. trichophyllus* and *R. confervoides*. The width of most species is about 0.9 mm but is 1.0–1.3 mm in *R. peltatus*. The cells of the pericarp are long and parallel to the longitudinal axis and are arranged in latitudinal rows. The raised ends of the cells form the distinct ridges, characteristic of the achenes. Even small fragments of the wall can be easily recognised. Although distinct to Section *Batrachium*, the achenes are variable and cannot be satisfactorily determined to species.

Life form and ecology *Ranunculus* Sect. *Batrachium* are aquatic annual or short-lived perennials, weakly competitive plants with rounded or lobed floating leaves, finely divided submerged leaves, or both type of leaf (heterophyllous). The flowers and fruit are borne above the water although the flowers can be self-pollinated under water. The fruits float well and germinate best after a period of drying. Wiegleb et al. (2017), Preston and Croft (1997), and Mossberg and Stenberg (2018) give good accounts of the ecology and distribution of the species. Many species are characteristic of shallow still or slow-flowing water, such as in ponds and ditches and river backwaters. They thrive where the banks are disturbed, e.g. by cattle poaching, or as pioneers in new water bodies. They can suffer from changes in land-use allowing the habitat to become overgrown, eutrophicated, or drained. Sometimes they can grow terrestrially on damp mud after water drawdown. *R. baudotii* is most frequently found in brackish water near the coast. *R. fluitans* grows in fast-flowing large rivers. Most species grow in mesotrophic base-rich clear water, but *R. omiophyllus* prefers acidic water over acid rocks or peat and *R. trichophyllus* grows in nutrient-poor sites flooded in winter but dry in summer.

Distribution *R. aquatilis* occurs across temperate Europe where it reaches southern Scandinavia and western European Russia. It is rare near the Mediterranean and has outliers in west Asia. Ranunculus aquatilis L. (gbif.org)

R. baudotii occurs across Europe and North Africa (Wiegleb et al. 2017) and is frequent around the coasts of northern Europe and the Mediterranean. Ranunculus baudotii Godr. (gbif.org)

R. circinatus occurs across temperate Europe and west Asia to Kazakhstan, and in North Africa. East Asian and boreal North American records are *R. subrigidus* (G. Weigleb personal communication). Its northern limit is the Gulf of Bothnia, and it is rare around the Mediterranean. Ranunculus circinatus Sibth. (gbif.org)

R. confervoides is an arctic-boreal species restricted to northern Europe, extending to the far north, and with outliers in the central Urals and in the eastern Alps. Ranunculus confervoides (Fries) Fries (gbif.org)

R. fluitans is endemic to temperate Europe, from the British Isles and Sweden to The Alps and Pyrenees and east to Belarus. There are scattered localities in northern Spain, southern Italy, Sardinia, and south Sweden. Ranunculus fluitans Lam. (gbif.org)

R. hederaceus is widespread in oceanic western Europe reaching south Sweden and Germany and south to Iberia. It may be introduced in eastern North America. Ranunculus hederaceus L. (gbif.org)

R. omiophyllus is endemic to westernmost Europe and North Africa, and it occurs in the British Isles, France, and Iberia. Sites in eastern Spain, southern Italy, Sicily, and Algeria are diploid. The northern plants are tetraploid (G. Weigleb personal communication). Ranunculus omiophyllus Ten. (gbif.org)

R. peltatus occurs throughout Europe to 60°N in Scandinavia and only at higher elevations in the Mediterranean area and Morocco. Sites in western Asia, e.g. Iran and the Altai mountains, may reflect bird migration routes (G. Weigleb personal communication). Ranunculus peltatus Schrank (gbif.org)

R. trichophyllus is a Holarctic species complex that is widespread across Eurasia and North America. It also occurs in North and South Africa, southeast Australia and Tasmania, and New Zealand. Ranunculus trichophyllus Chaix (gbif.org)

Ranunculus tripartitus is a rare species endemic to Europe and Morocco. It occurs mainly in western Europe from western UK to The Netherlands, with outliers in Sweden, north Germany, and possibly Greece. Ranunculus tripartitus DC. (gbif.org)

Other northwestern European species and numerous subspecies and hybrids are not included here (see Stace 2019; Schou et al. 2023).

Quaternary history Achenes of *Ranunculus* Sect. *Batrachium* are found frequently in Early Holocene lake sediments, Late-Glacial sediments, and earlier Quaternary sediments from glacial and interglacial periods. They often occur as pioneers in newly formed lakes (Bennike 1995) or in the earliest Holocene, e.g. in Greenland (Fredskild 1992). At Kråkenes Lake, western Norway, the peaks of *Ranunculus* Sect. *Batrachium* soon after deglaciation and again in the earliest Holocene were attributed to the relatively northern *R. confervoides*, whereas the separate later Holocene colonisation was by more temperate species (Birks 2000).

5.2.197 *Ranunculus flammula* L. Ranunculaceae

Morphology *Ranunculus flammula* achenes are about 1 mm long and rounded, often narrowed near the base. They easily split into halves (right photo). The positions of the style base and the stalk are obvious. The surface has a minutely and regularly pitted cell pattern, more obvious than in *R. reptans*.

Life form and habitat *Ranunculus flammula* is an erect or decumbent perennial herb, often rooting at the nodes. It is related to and sometimes combined with *R. reptans*, but *R. reptans* achenes are distinct, being D-shaped and with a less obvious cell pattern. It usually grows on wet mud, often seasonally inundated with oligotrophic or mesotrophic water, and where competition is limited e.g. by disturbance. It extends from the lowlands up to mid-elevations in the mountains. The achenes sink but may be dispersed by ingestion by grazing livestock. They form a persistent seed bank. See Preston and Croft (1997).

Distribution *Ranunculus flammula* is widespread in Europe and western Asia, becoming less frequent in the far north and round the Mediterranean. It occurs across North America except the southern USA and The Arctic.

Quaternary history *Ranunculus flammula* achenes have been frequently reported from both Holocene, Glacial, and Interglacial deposits in northwest Europe (e.g. Godwin 1975). They are often associated with human activity and archaeological sites.

Ranunculus flammula L. (gbif.org)

5.2.198 ***Ranunculus hyperboreus*** Rottb. Ranunculaceae

Morphology The distinctive whitish achenes of *Ranunculus hyperboreus* have one more or less straight side and the opposite side semicircular. They are about 1.2–1.3 mm long and about 1.1 mm wide. They are flattened and readily break into two halves along a pronounced division line (central photo). The surface is pale grey and consists of bubbly cells which form thickened margins, leaving a shallow hollow in the centre. These enable efficient floating.

Life form and ecology *Ranunculus hyperboreus* is a low-growing slender rhizomatous perennial. It spreads by rooting at the nodes and forms loose mats on open damp mud beside small streams and in tundra pools, in wheel tracks and road ditches, muddy beaches, and in cold springs, often on nitrogen-enriched rich soil associated with livestock and land use (Mossberg and Stenberg 2018). It is the only aquatic vascular plant found on Svalbard, where *R. hyperboreus* ssp. *arnelli* is common (The Flora of Svalbard https://archive.org/details/floraofsvalbard0000mni).

Distribution *Ranunculus hyperboreus* is circumpolar in the arctic and northern boreal zones. It occurs in the south Scandinavian mountains and northern boreal and arctic Fennoscandia, the Kola Peninsula, and across arctic Asia, on Svalbard, Iceland, and Greenland, and through arctic Canada and Alaska and in the Rocky Mountains.

Quaternary history *Ranunculus hyperboreus* achenes have been recorded in the British Isles during glacial periods but not in interglacials, including the Holocene (e.g. Godwin 1975; Birks 2008; Bos et al. 2004). It has also been recorded in cold periods in central Europe (e.g. Tralau 1963) and Norway. At Kråkenes Lake in western Norway, it was the only aquatic plant recorded during the Younger Dryas and rapidly disappeared in the Early Holocene (Birks 2000). It colonised lakes in Greenland at the start of the Holocene (e.g. Bennike 2000; Bennike and Björck 2000).

Ranunculus hyperboreus Rottb. (gbif.org)

5.2.199 *Ranunculus lingua* L. Ranunculaceae

Morphology The achenes of *Ranunculus lingua* are about 2.5 mm long excluding the beak and 2.0 mm wide. They are oval in side view and narrowly winged. The stalk and curved stigma are prominent. They are laterally flattened. Their surface is rough and appears granular.

Life form and ecology *Ranunculus lingua* is an erect and strongly rhizomatous perennial up to 120 cm tall. It is an emergent reed-like plant growing at the edges of ponds and ditches, in marshes, and sometimes by slow streams in base-rich water up to 0.5 m deep.

Distribution *Ranunculus lingua* has a temperate Euro-Siberian distribution covering Europe and central Asia. It is widespread through the British Isles, in southern Scandinavia, especially Sweden and Finland, and northwest Europe north of The Alps and the Pyrenees. Its abundance is being reduced by habitat loss.

Quaternary history *R. lingua* has been recorded in the British Isles in Quaternary warm stages, including the Allerød and the Holocene, although often tentatively as the achenes are difficult to distinguish from other large *Ranunculus* achenes, such as *R. repens* and *R. acris* (Godwin 1975). It was recorded as part of the infilling succession of an oxbow lake in northern Croatia in the late Holocene (Gałka et al. 2024).

Ranunculus lingua L. (gbif.org)

5.2.200 *Ranunculus reptans* L. Ranunculaceae

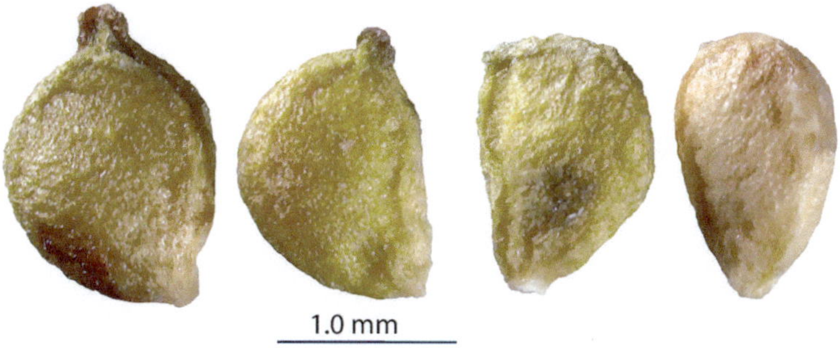

Morphology The achenes of *Ranunculus reptans* are D-shaped in side view, with a straight side opposite a semicirular side. They are small, 1.2–1.5 mm long and about 1 mm wide. The surface is minutely pitted. The style base and stalk are obvious. They readily split into two halves.

Life form and ecology *Ranunculus reptans* is a perennial creeping herb that often forms loose mats on periodically flooded gravelly or sandy shores of freshwater lakes and tarns and brackish bays. It grows in nutrient-poor water from high-water mark to 0.5 m deep.

Distribution *Ranunculus reptans* has a circumboreal and subarctic distribution across the Northern Hemisphere extending south in mountainous areas. It is rare in the British Isles but common in Fennoscandia and Iceland.

Quaternary history Mitchell (1952) recorded very small (1.2 × 0.9 mm) *Ranunculus* achenes in the Late Glacial of three sites in Ireland which he called *Ranunculus* cf. *reptans*.

Ranunculus reptans L. (gbif.org)

5.2.201 *Ranunculus sceleratus* L. Ranunculaceae

Morphology The achenes of *Ranunculus sceleratus* are about 1.0 mm long and 0.8 mm wide. They are asymmetrically oval, with one more or less straight side and one rounded side. The stigma base is usually persistent. They are flattened from front to back with a marked division between the two halves, which often splits. The surface consists of spongy cells that form a raised rim and a flattened centre similar to *R. hyperboreus*. This enables the fruits to float, and either be dispersed by water currents or by attachment to drinking animals.

Life form and ecology *Ranunculus sceleratus* is an annual that grows in shallow water or on wet, nutrient-rich mud at the edges of ponds, ditches, streams, or slow rivers, associated with disturbance by drinking livestock. It is salt-tolerant and often grows on grazed salt marshes. Its seeds are long-lived and form a seed bank and germinate after renewed disturbance.

Distribution *Ranunculus sceleratus* is a widespread plant with a circumpolar boreal distribution. It is a lowland temperate species and is not found in arctic biomes.

Quaternary history *Ranunculus sceleratus* achenes have been recovered frequently from Holocene deposits in the UK, often associated with archaeological sites and sometimes with brackish conditions (Godwin 1975; Brown et al. 2018).

Ranunculus sceleratus L. (gbif.org)

5.2.202 *Comarum palustre* L. Rosaceae

Morphology The achenes of *Comarum palustre* are characteristically semi-circular in side view, about 1.5 mm long and 1 mm broad. The style base is slightly bulbous, and often dark purple, whereas the rest of the achene is brown. The basal attachment point is usually obvious. The isodiametric cells are obscure. The achene often splits longitudinally showing the thick wall. The seeds can float for several months before sinking, allowing for wide dispersal (Praeger 1913).

Life form and ecology *Comarum palustre* is a robust rhizomatous perennial about 50 cm tall with pinnate leaves and pink flowers. It grows in acid to neutral shallow water on peaty substrates at lake margins, in littoral fen vegetation, and in pools in *Sphagnum*-dominated bogs.

Distribution *Comarum palustre* has a circumboreal distribution, particularly in northern and mountain regions. It occurs well north of the Arctic Circle.

Quaternary history Achenes of *Comarum palustre* are frequently found in Quaternary sediments. They often occur in fen sediments transitioning from lake to bog environments.

Comarum palustre L. (gbif.org)

5.2.203 ***Filipendula ulmaria*** (L.) Maxim. Rosaceae

Morphology The asymmetric achenes of *Filipendula ulmaria* are about 3 mm long. They are a characteristic shape, crescentic in side view with a persistent style and a projection from the hollow side. The fruit wall is composed of layers of long cells that criss-cross each other, a structure that is found in many Rosaceae.

Life form and ecology *Filipendula ulmaria* is a rhizomatous perennial up to 1.2 m tall. It grows in wet meadows, wet woods, marshes, fens and other terrestrial areas with high soil moisture. It typically grows in a range of soil types (sandy to heavy clay) that have a neutral to basic pH. Plants are clump-forming and can sometimes spread over large areas.

Distribution *Filipendula ulmaria* is common throughout Europe and temperate Asia, with boreal and montane tendencies. It is considered naturalised in eastern Canada, but it can be invasive elsewhere in eastern North America.

Quaternary history *F. ulmaria* fruits are occasionally found in Quaternary sediments, especially the Holocene (e.g. Godwin 1975; Burrows 1974; Bittmann 2007). However, the majority of its records are from fossil pollen (Godwin 1975).

Filipendula ulmaria (L.) Maxim. (gbif.org)

5.2.204 *Geum rivale* L. Rosaceae

Morphology The achenes of *Geum rivale* are narrowly oblong or elliptic and 3–4 mm long. They have a terminal hooked beak 4–9 mm long which is an adaptation for animal dispersal. Both achene and beak are hairy. The beak and hairs can be lost during preservation in sediments.

Life form and ecology *Geum rivale* is a rhizomatous perennial herb up to 30 cm tall. It grows on nutrient-rich soil in damp meadows with fluctuating water levels, wet deciduous woodlands, watersides, and in springs, in full sun or partial shades. The hooked achenes are an efficient means of animal dispersal.

Distribution *Geum rivale* is widespread across the whole of Europe, more confined to uplands and mountains in the south, across western Asia, and boreal and temperate North America. It grows as far north as northernmost Scandinavia and Russia, and in Iceland.

Quaternary history The achenes of *Geum rivale* are difficult to distinguish from those of *G. urbanum* according to Godwin (1975) and there are very few fossil records.

Geum rivale L. (gbif.org)

5.2.205 *Galium palustre* L. Rubiaceae

Morphology The mericarps of *Galium palustre* are borne in pairs. The nutlet is globose, 1.2–1.8 mm across, with a rugose surface. There is a groove on one side at the point of attachment. The nutlets can float for several months (Praeger 1913).

Life form and ecology *Galium palustre* is a straggling perennial herb ranging 60 cm to 1.5 m long. It grows in a wide range of wet places such as fens, marshes, and pond margins. It can tolerate drying soils and grows in oligotrophic to eutrophic conditions. The nutlets are primarily dispersed by water (Strykstra et al. 2002).

Distribution *Galium palustre* is found throughout Europe including Iceland and the Azores, and across western Siberia and eastern North America. It is introduced in western North America, northwest Australia, and parts of South America.

Quaternary history *Galium palustre* nutlets have been identified in late Holocene sediments (Koshkarova et al. 2006; Wasylikowa et al. 2012).

Galium palustre L. (gbif.org)

5.2.206 *Galium uliginosum* L. Rubiaceae

Morphology The mericarps of *Galium uliginosum* are borne in pairs. Each nutlet is semi-globose, about 1.0–1.5 mm long with a long attachment scar on the inner side. The surface is covered with small irregular projections.

Life form and ecology *Galium uliginosum* is a straggling perennial herb 60–100 cm long. It grows in wet places such fens and marshes and prefers alkaline soils. The nutlets are dispersed by water but can also be spread by animals (D'hondt et al. 2012).

Distribution *Galium uliginosum* is found throughout Europe including Iceland and across northern Asia.

Quaternary history *Galium uliginosum* nutlets have been recorded in Quaternary sediments (Bittmann 2007; Hunter 2012).

Galium uliginosum L. (gbif.org)

5.2.207 *Ruppia cirrhosa* (Petagna) Grande and *Ruppia maritima* L. Ruppiaceae

Ruppia cirrhosa

Ruppia maritima

Morphology The fruits of *Ruppia maritima* are black and have a distinctive obliquely ovoid shape with a persistent beak. They are 2.0–2.5 mm long. Like *Potamogeton* endocarps, they have a lid which may become separated in fossil material. The surface is covered by distinct, very small, cells. The fruits of *Ruppia cirrhosa* are similar to and probably indistinguishable from those of *R. maritima*. However, they are larger (2.8–3.4 mm long), narrower, and more tapering, and the cell pattern is less clearly marked.

Life form and ecology These two species are closely related, and plants can be problematic to identify. They are both submerged sea grasses, growing on soft mud or silt in brackish water such as in estuaries, tidal creeks and rivers, coastal lagoons, and pools on saltmarshes, and also in inland high-conductivity brackish lakes (Verhoeven 1979; Preston and Croft 1997; Mossberg and Stenberg 2018). *R. maritima* grows in water less than 0.5 m deep and is often exposed at low tide. The plants die if dried up, but the seeds survive desiccation. However, they do not survive long enough in sediments to form a seed bank (Strazisar et al. 2013). *R. cirrhosa* grows in water usually deeper than 0.5 m and tolerates more saline conditions. Both species regenerate in spring by seeds or from perennial rhizomes.

Distribution *Ruppia maritima* is found in coastal habitats around all continents except in the Arctic and Antarctica. It also occurs in inland brackish lakes on all continents. *Ruppia cirrhosa* has a similar distribution but is not so widely recorded and does not reach as far north. It occurs at west European and Mediterranean coasts, western North America only, southern Africa, and Australasia.

Quaternary history Most records of *Ruppia* fruits are from marine or brackish sediments associated with sea-level changes during the Holocene (e.g. Godwin 1975; Berglund et al. 2005) and with hydrological changes in inland lakes (e.g. Bradbury et al. 1989). *R. cirrhosa* was identified in the Early Holocene in the southern North Sea (Brown et al. 2018). *Ruppia* fruits should not be used for radiocarbon dating as the plants metabolise dissolved inorganic carbon (HCO_3^-; CO_2) and therefore can have a large hard water error. Rachal et al. (2024) demonstrated that the error can result in an age as much as 7500 years too old.

Ruppia maritima L. (gbif.org)

Ruppia cirrhosa (Petagna) Grande (gbif.org)

5.2.208 *Salix* L. spp. Salicaceae

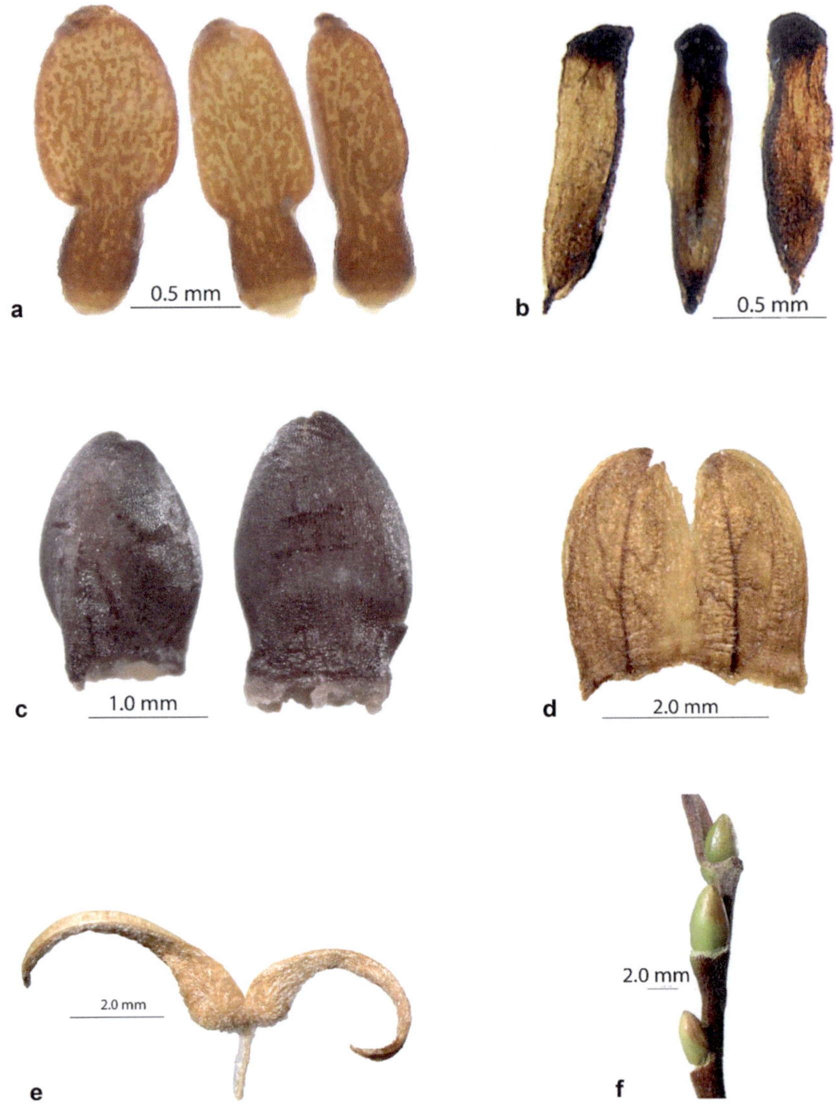

(**a**) Seeds of *S. caprea* (**b**) Fossil seeds from Abernethy Forest, Scotland (**c**) *Salix* buds (**d**) *Salix* bud scale (**e**) *Salix* capsule (**f**) *Salix* twig with buds

Morphology *Salix* seeds are about 1.2–1.7 mm long depending upon the species. They bear a plume of many fine white hairs for wind dispersal (Karrenberg et al. 2002; Sell and Murrell 1996–2018) which also allows them to float on the water surface. Fresh seeds a) have a transparent testa allowing the green cotyledons to show through. Fossil seeds b) are rather non-descript. Usually, one end is dark and blunt, and the other is pointed. The cells are rectangular but indistinct. *Salix* buds c) can be confused for seeds in fossil material. However, they have a broad base where they were attached to the twig and a notch at the top where the outer and inner budscales meet. They have a distinct angular cell pattern, characteristic of *Salix* cuticle. Buds can be small, about 1 mm long, or larger, up to 3 mm long. Sometimes, twigs with attached buds are found, f). *Salix* budscales d) have two characteristically rounded lobes and often tear in half. Like the buds, they can vary in size up to 5 mm long. *Salix* capsules e) have two valves that bend back when ripe to expose the seeds. Often, the valves are found singly, and broken capsules can be identified. Sometimes immature capsules are found. *Salix* buds and budscales are frequently found in sediments. The cuticle of twigs and bark is characteristic with straight-edged cells. They are often the only evidence for the presence of *Salix*, as the delicate seeds are not well preserved. Leaves are occasionally preserved and can sometimes be identified to species. Otherwise, it is not possible to separate hydrophilic species of *Salix* using macrofossil remains.

Life form and ecology *Salix* spp. are woody shrubs or trees. Arctic and alpine species are low-growing or creeping shrubs. Temperate and boreal species are shrubs or small trees and grow in soils that are either acid or alkaline, but which remain moderately damp throughout much of the year such as near ponds, lakes, rivers, and floodplains (Dickmann and Kuzovkina 2014). Seed dispersal is primarily by wind but can also occur by floating, the hairs trapping air (Johnson 2000; Karrenberg et al. 2002).

Distribution This genus occurs throughout the Northern Hemisphere.

Quaternary history *Salix* spp. macro-remains (seeds, budscales, leaves, etc.) are widely recorded in Quaternary sediments, e.g. Birks (1991, 2000); Bennike et al. (2006); Walton et al. (2021b).

https://www.gbif.org/species/3039576

5.2.209 *Azolla filiculoides* Lam. Salviniaceae

(**a**) Megaspore, stereomicroscope. (**b**) Megaspore, scanning electron microscope. The cap that normally covers the floats has been peeled back. (**c**) A megaspore of *A. filiculoides* surrounded by eight massulae of microspores attached by hooked glochidia to threads on the megaspore. From Hoxnian (Holsteinian) interglacial deposits at Hoxne, UK. Photomicrograph modified from West (1953)

Morphology The female spores are megaspores with a complex structure. The spore is 200–370 µm wide. The surface of the spore is papillate-reticulate. The upper part is covered by a collar bearing three floats (Large and Braggins 1993).

Life form and ecology *Azolla* is a water fern and the fronds float on the surface of ponds, lakes, canals, ditches and slow flowing rivers by means of air trapped between numerous, small, closely overlapping scale-like leaves. Small roots hang down in the water. The leaves have hollow structures that house the symbiotic cyanobacterium *Anabaena azollae*, which fixes atmospheric nitrogen. *Azolla* can grow extremely rapidly in eutrophic conditions and quickly forms a dense mat on the water surface. This excludes light for other aquatic plants. Decaying root and leaf matter below the mat creates an anaerobic environment which can reduce the quality of drinking water and make survival for other aquatic organisms impossible. *Azolla* can block waterways, canals, and irrigation systems, and can have a considerable economic impact. It can be controlled by the weevil *Stenopelmus rufinasus* Gyllenhal (McConnachie et al. 2004).

Distribution *A. filiculoides* is native to North and South America. Due to its ability to fix nitrogen, it has been widely used as a green fertilizer in rice cultivation. It has been introduced and has become invasive globally in temperate and tropical areas.

Quaternary history The explosive expansion of *Azolla* in the mid-Eocene 55 million years ago on the (then) freshwater surface of the Arctic Ocean has been proposed to have fixed enough carbon dioxide from the atmosphere to reduce the global greenhouse effect. Global cooling eventually led to the formation of ice sheets in Antarctica and to our current glacial-interglacial climate system. This hypothesis has been termed the *Azolla* event (Brinkhuis et al. 2006). *Azolla filiculoides* has been recorded as megaspores and microspore massulae in northwest European interglacials, especially the Holsteinian (Hoxnian) (MIS 11) interglacial (e.g. Godwin 1975; West 1953; Donner 1995), after which it became extinct in northwest Europe until its re-introduction in the late Holocene. In Egypt, increasing eutrophication of the Nile Delta lagoons has led to the extinction of *A. nilotica* and the spread of *A. filiculoides* (Birks 2002).

Azolla filiculoides Lam. (gbif.org)

5.2.210 *Limosella aquatica* L. Scrophulariaceae

Morphology The tiny seeds of *Limosella aquatica* are about 0.5 mm long. The basal attachment is usually prominent and darker, and the upper end is pointed and asymmetrical. The surface cells are laterally elongated and arranged in vertical rows, giving a ladder appearance. This is easier to see in fossilised seeds whose embryos have decayed.

Life form and ecology *Limosella aquatica* is a tufted annual isoetid with thin stolons forming a mat. The leaves are up to 10 cm long with spoon-shaped terminal blades surrounding the small flowers at the base. It produces many seeds which can form a seed bank for many years (Schou et al. 2023). It grows on sandy or muddy substrates in shallow, mildly mesotrophic and nutrient-rich water at the shores of ponds and lakes, streams, and rivers, that dry out in summer.

Distribution *Limosella aquatica* has a circumpolar-boreal distribution throughout much of the temperate Northern Hemisphere, with rare occurrences in South America and Africa. It is widespread in Scandinavia and northwest Europe and in western North America. It is scattered in the British Isles. Its habitat is threatened, and it has been extirpated from much of eastern North America.

Quaternary history Seeds of *Limosella aquatica* were recorded at Kråkenes Lake, western Norway (Birks 2000) where they occurred in abundance at the start of the Holocene. They initiated a remarkable isoetid succession before giving way to other isoetids.

Limosella aquatica L. (gbif.org)

5.2.211 *Scrophularia auriculata* L. Scrophulariaceae

Morphology *Scrophularia auriculata* seeds are 0.7–0.9 mm long, and ellipsoid with rounded or truncate ends. The longitudinal ribs are coarse and lumpy. The cells between are smooth with an irregular reticulate pattern. The seeds of *Scrophularia nodosa* are similar.

Life form and ecology *Scrophularia auriculata* is a perennial herb up to 150 cm tall. It prefers wet fertile soils in open or shaded habitats, such as in damp woodland, ditches and river banks, often in disturbed places. *S. nodosa* L. grows in similar but drier places.

Distribution *Scrophularia auriculata* is widespread throughout western Europe reaching its northern limit in southern Scandinavia. *S. nodosa* is widespread in Europe, extending into western Asia.

Scrophularia auriculata L. (gbif.org)

5.2.212 *Solanum dulcamara* L. Solanaceae

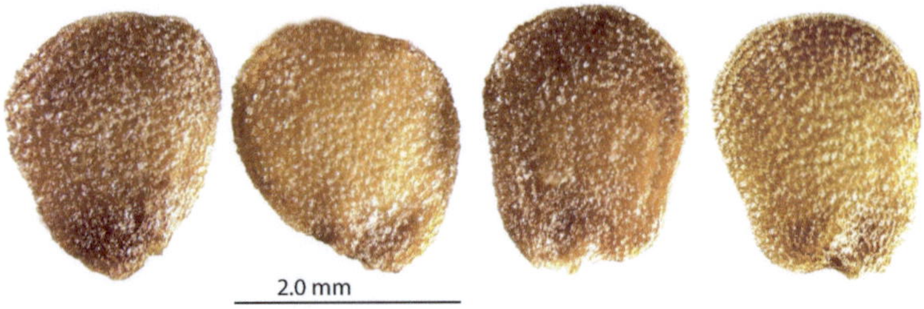

Morphology The yellowish or pale brown seeds are disc-shaped and flattened, approximately 2 mm across. The cells of the testa form a characteristic, distinct, irregularly reticulate pattern. The seeds are contained in a many-seeded berry. The berries are dispersed by animals.

Life form and ecology *Solanum dulcamara* is a perennial vine growing between 1–3 m tall. Although it grows in various habitats, including hedges, woods, and shingle beaches with very dry soil, it can often be found on damp soil in fens and fen carr on the banks of lakes, streams, and ponds.

Distribution *Solanum dulcamara* is native to much of Europe, temperate Asia, and North Africa. It has been introduced across North America where it can become invasive, and to parts of the Southern Hemisphere.

Quaternary history *Solanum dulcamara* has been recorded occasionally in Quaternary sediments (e.g. Godwin 1975; Jensen 1987; Waller and Early 2015; Gałka et al. 2024).

Solanum dulcamara L. (gbif.org)

5.2.213 *Sparganium angustifolium* Michx. Typhaceae (Sparganiaceae)

Morphology The achenes of *Sparganium angustifolium* are elongated, about 5–6 mm long with a long style and a constriction in the middle. The spongy exocarp facilitates floating for at least 6 months (Ridley 1930). It encloses a harder endocarp which is not readily separated from it. However, it is the part that becomes preserved in sediments. It is pear-shaped with a stalk, a few longitudinal lines, and a broad style base.

Life form and ecology *Sparganium angustifolium* is a floating-leaved rhizomatous perennial found in lakes, ponds, slow-flowing rivers, and ditches. It tolerates a wide variation in water-level, from exposed but still wet soil to depths of up to 2.5 m. It builds up a seed bank which, together with the rhizomes, helps it to survive dry periods (Rørslett and Mjelde 2021a). It is primarily associated with oligotrophic (sometimes mesotrophic) soft waters, including dystrophic sites, and it is tolerant of moderate acidification. The floating leaves use atmospheric CO_2 (Lindstrøm et al. 2004), and, like *Isoetes*, it can also utilize CO_2 from sediments via its roots (Lucassen et al. 2009). This species spreads vegetatively. It is dispersed by the floating fruits, which are sometimes eaten by waterfowl (Ridley 1930).

Distribution *S. angustifolium* is a northern species with a circumboreal distribution occurring mainly in northern parts of Europe and extending south to The Alps, Balkans and the Pyrenees (Cook 1961). Its main area is in northern Europe and Canada. It also occurs in eastern Asia in Kamchatka (Bobrov et al. 2023; Hultén and Fries 1986).

Sparganium angustifolium Michx. (gbif.org)

5.2.214 ***Sparganium emersum*** Rehmann
Typhaceae (Sparganiaceae)

Morphology The bodies of the spindle-shaped achenes of S*parganium emersum* are 4–5 mm long without a constriction. The exocarp is papery and the spongy base still adheres to the endocarp in our photograph. The tougher endocarps are 2.5–3.0 mm long, rounded, and not readily separated from the exocarp in fresh material. Several longitudinal lines are apparent.

Life form and ecology *Sparganium emersum* is a rhizomatous perennial, usually with upright stems bearing the inflorescences, and with floating and underwater leaves. It occurs in lowland, slow-flowing rivers, canals, ponds, lakes, and ditches, growing on fine silt and clay substrates. It can occupy slightly deeper water than *Sparganium erectum,* up to about 1 m. It is highly tolerant of physical disturbance due to deeply rooted rhizomes which allows it to survive boat traffic and river dredging (Preston and Croft 1997) and to survive in sites that undergo water-level change or varying current speeds (Rørslett and Mjelde 2021b). This species is highly tolerant of eutrophication and can be found across a broad nutrient range. *S. emersum* spreads locally by its rhizomes and is strongly competitive. It is dispersed further by the freely produced corky fruits floating on the water or being eaten by ducks (Ridley 1930). They can form a seed bank (Rørslett and Mjelde 2021b).

Distribution *Sparganium emersum* has a circumboreal distribution. It is present across large tracts of boreal Europe and Asia extending east to Japan (Kaneko and Jinguji 2012) and Kamchatka (Mochalova and Efimov 2022).

Sparganium emersum Rehmann (gbif.org)

5.2.215 ***Sparganium erectum*** L. Typhaceae (Sparganiaceae)

Morphology The achenes of *Sparganium erectum* are widest in the middle, round in cross section, and tapering to each end. They are about 6 mm long and 3 mm broad, with a pronounced beak. The shape varies and different shapes have been given subspecific status (Cook 1961) although these seem to overlap in their geographical or ecological ranges (Cook 1961, 1962). The outer exocarp is relatively tough and thick. The endocarps are elongated and irregularly angular in cross section.

Life form and ecology *Sparganium erectum* is a tough rhizomatous perennial burr reed 50–100 cm tall with variable morphology and fruit shape. It grows on waterlogged soil or in shallow water 10–20 cm deep beside ponds, small lakes, and slow flowing rivers, in ditches, and on ungrazed marshland. It has a broad tolerance of substrate type, pH, and trophic status (Cook 1962). The water-repellent fruits float for more than 12 months (Praeger 1913). The fruits are an important food for wildfowl, which eat them and excrete the hard resistant endocarps, thus dispersing them (Cook 1962). They may show enhanced germination after ingestion (Soons et al. 2008). However, survival of seedlings is poor (Cook 1962). Local spread is most frequently vegetative via rapid rhizome development (up to 2 m per year) and via uprooted plants (Cook 1961, 1962; Preston and Croft 1997).

Distribution *Sparganium erectum* occurs across boreal and temperate Eurasia, extending to northern Scandinavia and south into the Mediterranean region, including North Africa. In Asia it occurs in Tibet and the Himalayas and extends east to Japan and South Korea (Ota et al. 2016; Kim and Kim 2017). It occurs across North America, where it is more commonly known as *S. eurycarpum*. It is most likely introduced in Australia and New Zealand (Cook and Nicholls 1987).

Quaternary history *Sparganium erectum* has been recorded in British interglacials, the Late Glacial, and through the Holocene (Godwin 1975). However, endocarps of *Sparganium* are seldom identified beyond genus level.

Sparganium erectum L. (gbif.org)

5.2.216 *Sparganium glomeratum* (Laest. ex Beurl.) Beurl. Typhaceae (Sparganiacaeae)

Morphology The slender achenes of *Sparganium glomeratum* are about 4 mm long and 1.8 mm broad. They taper to the style and to the base. They have a slight central constriction.

Life form and ecology *Sparganium glomeratum* is a rhizomatous perennial with upright stems and leaves up to 40 cm tall. It grows, often in tight colonies, in nutrient-poor dystrophic to mesotrophic water up to 50 cm deep in small waterbodies and swampy woods on mineral or peat substrates (Mossberg and Stenberg 2018; Rørslett and Mjelde 2021c).

Distribution *Sparganium glomeratum* is widespread in southern Fennoscandia and in the boreal zone across the Northern Hemisphere. It is absent from Denmark and northwest Europe, including the British Isles.

Sparganium glomeratum (Laest. ex Beurl.) Beurl. (gbif.org)

5.2.217 *Sparganium gramineum* Georgi Typhaceae (Sparganiaceae)

Morphology The achenes of *Sparganium gramineum* are small, up to 3 mm long. They are cylindrical, sharply angled to the style and the base, and with a constriction in the middle.

Life form and ecology *Sparganium gramineum* is a robust rhizomatous perennial with 2–3 m long narrow (1.5–3.0 mm wide) floating leaves. It grows in oligotrophic or mildly mesotrophic water 1–2 m deep on a mud bottom. It spreads by rhizomes and by floating fruits (Mossberg and Stenberg 2018; Rørslett and Mjelde 2021d).

Distribution *Sparganium gramineum* occurs across boreal Europe and Asia. In Scandinavia, it is confined to the more continental climates of south Sweden and south Finland.

Sparganium gramineum Georgi (gbif.org)

5.2.218 ***Sparganium hyperboreum*** Laest. ex Burl. Typhaceae (Sparganiaceae)

Morphology The achenes of *Sparganium hyperboreum* are small, about 2.0–2.5 mm long and 1.5 mm broad, with a slight medial constriction and a few smooth longitudinal ribs. The spongy exocarp is tight-fitting and can be separated from the endocarps within. These are rounded with the remains of the stalk, and the style base is broad.

Life form and ecology *Sparganium hyperboreum* is an upright or floating long-rhizomatous telmatophyte, growing in cool, shallow water (0.2–1.0 m). In deeper water it is submerged. It grows in tarns, fen pools, bog hollows, and in man-made ditches and reservoirs on a peaty or silty soil. It prefers slightly acidic to neutral clear water often with a fluctuating water level, and it is able to colonise new artificial water bodies (Belyakov et al. 2022).

Distribution *Sparganium hyperboreum* has a circumpolar disjunct range in the boreal and tundra zones across Siberia and North America. More southerly locations are at higher elevations. It is widespread across Fennoscandia apart from in Denmark, southern Sweden, and Finland. It is absent from Britain and Ireland.

Sparganium hyperboreum Laest. ex Beurl. (gbif.org)

5.2.219 ***Sparganium natans*** L. Typhacaeae (Sparganiaceae)

Leaf tips

Morphology The achenes of *Sparganium natans* are ovoid and cylindrical in cross section with no constriction and are about 3 mm long excluding the beak. The endocarps are about 2–2.5 mm long, rounded with a pointed remnant stalk and a broader style base. They have a few longitudinal lines originating at the base. The leaves are long and narrow with rounded tips.

Life form and ecology *Sparganium natans* is a rhizomatous perennial which has submerged and long floating leaves. It grows in quiescent habitats including lake edges, drainage ditches, slow flowing rivers, and ponds. It favours peat substrates and base-rich, mesotrophic, shallow (10–50 cm depth) water, and naturally fluctuating water levels (Belyakov and Philippov 2018). It is intolerant of eutrophication (Preston and Croft 1997). *S. natans* fruits freely in shallow water. The fruits can float for many months (Praeger 1913) and are its main means of dispersal.

Distribution *Sparganium natans* has a circumboreal distribution across the temperate Northern Hemisphere. It is widespread in Europe (but absent from parts of the Mediterranean region) and North America and throughout Russia and Siberia into western Asia (Preston and Croft 1997; Kaplan et al. 2015).

Sparganium natans L. (gbif.org)

5.2.220 ***Typha latifolia*** L., ***Typha angustifolia*** L., ***Typha domingensis*** Pers. Typhaceae

Typha latifolia

Typha angustifolia

Typha domingensis

Morphology Fruits of *Typha* cannot be distinguished to species. They are small and cylindrical, about 1 mm long and 0.3 mm wide. There is a collar at the top surrounding the stigma base, and the bottom is pointed and usually black. The rest of the fruits are a characteristic yellow. The surface is slightly roughened. Seeds of *Juncus* appear similar in size and colour but have a different morphology. Seeds can remain viable up to 70–100 years in suitable soil conditions. The perianth of female flowers consists of many fine hairs and fruits are released in autumn as fluff. This is dispersed by wind and it also floats well in water, being blown towards the lake shores (Birks 1973). When wet, the hairs close, the pericarp splits and the achene is released and sinks, resulting in large monotypic stands around a lake.

Life form and ecology *Typha* spp. are robust rhizomatous reeds usually forming dense stands in shallow water at the edges of lakes and backwaters of slow rivers, in ditches, and irrigation canals. Due to their rapid growth rate and rhizomatous spread, *Typha* spp. are capable of rapidly colonizing and becoming invasive in wetland habitats that have been disturbed by nutrient or silt inputs (Bansal et al. 2019). All species prefer mesotrophic, nutrient- and base-rich water, growing on organic mud, peat, sand, and gravel.

T. latifolia grows in shallow water up to 15 cm deep, whereas *T. angustifolia* and *T. domingensis* can survive in water 1–1.5 m deep and can also tolerate greater salinity.

Distribution *T. latifolia* is a cosmopolitan plant, occurring through most temperature zones in Europe, North America, and Asia, and many subtropical areas. It reaches as far north as the Arctic Circle. *T. angustifolia* has a similar distribution but is more southern, only reaching south Scandinavia, and southeast Canada. *Typha domingensis* is a warm temperate and pantropical species found in southern Europe, Africa, southeast Asia, India, and Australia, and in the southern USA and central and south America. It can be a subdominant associate in mangrove ecosystems. *Typha* spp. are considered to be weed species in southern USA (Bansal et al. 2019).

Quaternary history The tetrads of *Typha latifolia* pollen are distinct and widely recorded. The monads of *T. angustifolia* and *T. domingensis* are hardly distinguishable from *Sparganium* pollen. *Typha* fruits are commonly recorded in Quaternary sediments, but the species cannot be distinguished. (See e.g. Godwin 1975).

Typha latifolia L. (gbif.org)

Typha angustifolia L. (gbif.org)

Typha domingensis Pers. (gbif.org)

5.2.221 ***Sphagnum*** L. spp. Sphagnaceae

Sphagnum recurvum agg. P. Beauv.
Syn. S. fallax (H.Klinggr,) H.Klinggr.

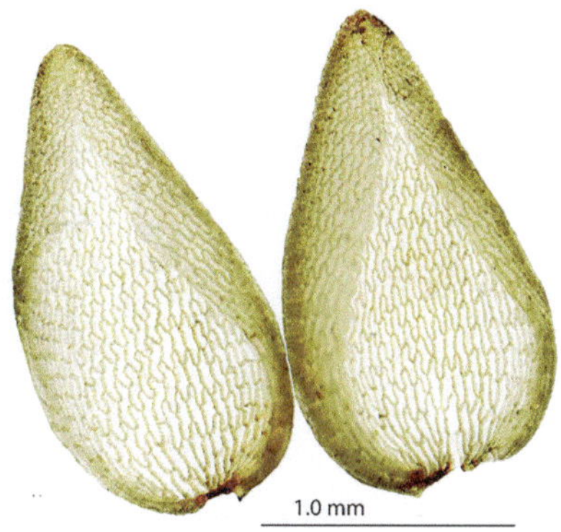

Sphagnum palustre L.

Morphology *Sphagnum* leaves are one cell thick. The cells are differentiated into dead hyaline cells with spiral thickenings and oval or circular pores and green capillary cells containing chloroplasts. These form a characteristic lacy network. There are about 380 species worldwide and in the British Isles 38 species are recognised by Blockeel et al. (2021). Mauquoy and van Geel (2007) have informative photographs showing the different cell structure found in some typical species. Identification keys are provided by Smith (1978) and Hill (1992). Capsules and their discoid lids may also be found in sediments.

Life form and ecology *Sphagnum* species have an apical growing capitulum and dead stems below. They form mats or hummocks in wet and typically oligotrophic habitats with a low pH. The hyaline cells absorb and retain water like a sponge. As the lower parts of the stems die, decay is hindered by the waterlogged anaerobic conditions and by phenolic compounds in the cell walls. As a result, they accumulate as peat in bogs and fens beneath the *Sphagnum* carpets or hummocks. Hence *Sphagnum* is often called 'bog moss'.

Quaternary history *Sphagnum* can be a major component of peat, especially in ombrotrophic bogs. Leaves and stems are also found in lake sediments, originating either from plants growing in the marginal fen or shallow water, or from inwash from the catchment. Curiously, leaves are frequently found in minerogenic sediments of Late-Glacial age (e.g. Birks 2000). The recent increase in leaves in oligotrophic lakes testifies to acidification by acid rain and the replacement of aquatic vegetation by submerged *Sphagnum* (Renberg and Wallin 1985). Leaves are seldom identified to species as diagnostic characters are often absent, although there are numerous specific records from British Quaternary deposits (Dickson 1973).

5.2.222 *Viola palustris* L. Violaceae

Morphology *Viola palustris* seeds are about 1.5 mm long and 1.0 mm wide. They are obovate with rounded tops marked by a small depression with a raised centre. The base is pointed with a scar where the small aril was positioned. The surface is smooth, and the cells are indistinctly isodiametric and faintly papillose, except at the top of the aril, where they are elongated along the length of the seed, forming a 'seam'. In fossil condition, the seeds often break into characteristic halves along the seam and are often pale brown.

Life form and ecology *Viola palustris* is a perennial herb with a dwarf rosette producing creeping rhizomes or thin stolons growing amongst low vegetation. The arillate seeds are adapted for myrmecory, dispersal by ants. *V. palustris* grows in damp habitats in mesotrophic bogs, fens, marshes, wet heaths, and swampy woods, frequently with *Sphagnum* spp., and often bordering oligotrophic pools, lakes, or streams.

Distribution *Viola palustris* is widely distributed through northwest Europe, including Scandinavia, Iceland, and the British Isles, and in The Alps and Spain in suitable habitats. It extends across boreal Asia to Mongolia, and across Canada, southern Alaska, and northwest North America.

Quaternary history *Viola palustris* seeds have been widely recorded in British Holocene deposits (Godwin 1975). Seeds are frequently recorded in Quaternary sediments and are usually attributed to '*Viola*', but the species can be separated with good preservation and care.

Viola palustris L. (gbif.org)

5.2.223 *Zannichellia palustris* L. Zannichelliaceae

Morphology The highly distinctive fruits of *Zannichellia palustris* are bean-shaped, 2–3 mm long and less than 1 mm wide. The lid occupies about half the width and is adorned with blunt spikes. It sometimes splits away from the rest of the fruit. The long style may not be preserved in sediments. *Z. palustris* produces abundant seeds and is an 'over-represented' species in macrofossil studies (Clarke et al. 2014).

Life form and ecology *Zannichellia palustris* is a submerged pondweed with narrow linear leaves. It is annual or perennial, spreading by rhizomes (Preston and Croft 1997). It grows in a broad range of habitats including lakes, ponds, ditches, rivers, canals, and coastal marshes and lagoons. It usually favours very shallow (often <0.5 m) calcareous freshwater and in some cases brackish environments. It tolerates desiccation and seeds can survive dry for at least 3 months (Preston and Croft 1997). It thrives in eutrophic conditions in lakes and is often one of the final species to persist prior to macrophyte loss (Sayer et al. 2010a, b; Wang et al. 2024).

Distribution *Zannichellia palustris* has a cosmopolitan distribution. It is found throughout Europe from Iceland and the Scandinavian coasts, especially round the Baltic Sea in the north to the southern edge of the Mediterranean basin. It is widespread in North America south of the Arctic, and in South Africa (Cholo and Foden 2006), but it is less abundant in the tropics. It also occurs in Asia, and it is declining in some Asian countries such as Japan (Tanaka et al. 2006). It is found in south eastern Australia and in New Zealand. *Zannichellia obtusifolia*, a similar Mediterranean species of brackish or high conductivity fresh water, has been recently discovered in eastern England (Adams 2017; BSBI News 2018).

Quaternary history *Zannichellia palustris* seeds are widely recorded in British Quaternary sediments but less commonly in the Holocene (Godwin 1975). They have been recovered from Early-Holocene sediments in the southern North Sea (Brown et al. 2018) and late-Holocene sediments in central Europe (Gałka et al. 2024).

Zannichellia palustris L. (gbif.org)

5.2.224 *Zostera marina* L. Zosteraceae

Morphology *Zostera marina* seeds are elongate-ellipsoid and circular in cross section, about 3.5 mm long and about 1.5 mm wide. They have 16–20 raised longitudinal ribs (Shin and Choi 1998) with smaller cells in between.

Life form and ecology *Zostera marina* is an annual or perennial seagrass. It forms 'meadows' characteristic of muddy and sandy intertidal and subtidal environments including estuaries, sea loughs, brackish coastal lagoons and sheltered coastal bays and inlets where exposure to wave action is low to moderate (Gundersen et al. 2017). These are important habitats for marine animals. Under ideal light conditions *Z. marina* can survive down to a depth of around 20 m (Boutahar et al. 2020), but it is usually found in shallower (<5 m) water. It is threatened by eutrophication which reduces light penetration, and by physical disturbance of rooting substrates. Other threats reducing its extent include coastal development, water quality degradation, mechanical damage (e.g. bottom trawling, mooring, and anchoring), alien species invasion, and climate change (e.g. warming, ocean acidification). At least 21% of the extent of seagrass habitat has disappeared from European Mediterranean coasts since 1869 (Boutahar et al. 2020). Progammes of seagrass reestablishment are being undertaken in many countries.

Distribution *Zostera marina* is a widespread Northern Hemisphere temperate species. It occurs round the coasts of Europe to 70°N in Norway and in Iceland (about 64°N) and extends south to the Mediterranean and Black Seas (Olesen et al. 2015; Boutahar et al. 2020). It grows along the Atlantic and Pacific coastlines of North America from Newfoundland and Alaska and south to Baja California and Mexico, and North Carolina. It follows the Pacific coasts of eastern Asia especially in Korea and Japan (Shin and Choi 1998).

Quaternary history *Zostera marina* seeds have been recovered from brackish-water sediments in the southern North Sea (Brown et al. 2018) and Tunisia (Birks et al. 2001).

Zostera marina L. (gbif.org)

5.2.225 *Zostera noltii* Hornem. Zosteraceae

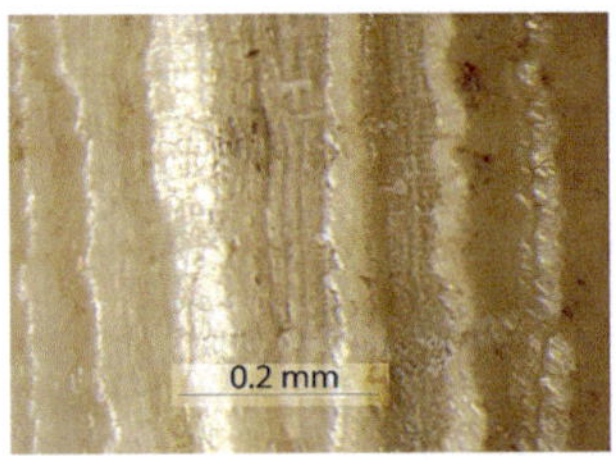

Morphology *Zostera noltii* seeds are ovate-ellipsoid, about 2.2 mm long and about 1 mm wide (Loque and Meinesz 1988). They are smaller than *Z. marina* seeds. However, they are ribbed in a similar way, the raised ribs having smaller longitudinal cells between them.

Life form and ecology Similar to *Zostera marina, Zostera noltii* is a rhizomatous meadow-forming seagrass. It requires saline or brackish water and muddy and detritus-rich sandy substrates associated with the intertidal and shallow subtidal zones of sea shores, estuaries, sea loughs, and brackish coastal lagoons. It tolerates high light levels (as at low tide), allowing it to survive better than *Z. marina* in upper shore and intertidal zones (Vermaat and Verhagen 1996), generally inshore of *Z. marina* (Short et al. 2007). While it usually grows in shallow water 1–3 m deep, it can survive in clear water to a depth of about 10 m (Short et al. 2010). Ripe seeds sink and are buried in the sediment, but germination has not been observed in nature. Dispersal is by detachment and floating of spathes bearing ripe fruits, by herbivorous fish, or by rhizome fragmentation (Loque and Meinesz 1988). *Z. noltii* is adversely affected by high nutrient levels and cloudy water and is threatened by coastal development and pollution (Short et al. 2010).

Distribution *Zostera noltii* is widely distributed in Europe on the Atlantic, Mediterranean, and Black Sea coasts. It extends northwards to the Baltic Sea, south to northwest Africa, the Canary Islands, and the Cape Verde archipelago, and eastwards to the Black, Caspian, and Aral Seas (Short et al. 2010). It is the only seagrass species found in the Caspian and Aral seas (Short et al. 2007).

Zostera noltii Hornem. (gbif.org)

Appendices

Appendix 1. Some Books and Papers on Macrofossil Methods and Identification

Some General Papers and Books

Birks HH (1980) Plant macrofossils in Quaternary lake sediments. Archiv für Hydrobiologie 15: 1–60

Birks HH (2001) Plant macrofossils. In: Smol JP, Birks HJB, Last, WM (eds) Tracking Environmental Change using Lake Sediments. Volume 3: Terrestrial, Algal, and Siliceous Indicators. Kluwer Academic Publishers, Dordrecht. pp. 49–74

Birks HH, Birks HJB (2000) Future uses of pollen analysis must include plant macrofossils. Journal of Biogeography 27: 31–35

Birks HH (2013) Plant Macrofossil Introduction. In: Elias SA (ed) Encyclopedia of Quaternary Science. Second Edition. Volume 3. Elsevier BV, pp. 593–612

Birks HJB, Birks HH (1980) Quaternary Palaeoecology. Chapter 5. Edward Arnold, London. (reprinted 2005, Blackburn Press, New Jersey, USA)

Dickson CA (1970) The study of plant macrofossils in British Quaternary deposits. In: Walker D, West RG (eds) Studies in the vegetational history of the British Isles. Cambridge University Press, Cambridge, pp. 233–254

Godwin H (1975) History of the British Flora. Cambridge University Press, Cambridge.

Warner BG (1988) Methods in Quaternary Ecology #3. Plant macrofossils. Geoscience Canada 15: 121–129

Wasylikowa K (1986) Analysis of fossil fruits and seeds. In: Berglund BE (ed) Handbook of Holocene Palaeoecology and Palaeohydrology. John Wiley & Sons Ltd. Chichester, pp. 571–590

H. H. Birks et al., *Seeds and Fruits of Aquatic Plants of Northern Europe - A Handbook*, Developments in Paleoenvironmental Research 23,
https://doi.org/10.1007/978-3-032-04170-8

Watts WA (1978) Plant macrofossils and Quaternary paleoecology. In: Walker D, Guppy JC (eds) Biology and Quaternary Environments. Australian Academy of Science, Canberra, pp 53–67

Some Macrofossil Identification Manuals

Anderberg A-L (1994) Atlas of Seeds. Part 4. Resedaceae - Umbelliferae. Swedish Museum of Natural History, Stockholm. 281 pp

Beijerinck W (1976) Zadenatlas der Nederlandsche Flora. Backhuys & Meesters, Amsterdam. 316 pp. In Dutch

Berggren G (1964) Atlas of seeds. Part 2. Cyperaceae. Swedish Natural Science Research Council, Stockholm. 68 pp

Berggren G (1981) Atlas of Seeds. Part 3. Salicaceae - Cruciferae. Swedish Museum of Natural History, Stockholm. 259 pp

Bertsch K (1941) Fruchte und Samen. Ein Bestimmungsbuch zur Pflanzendunde der vorgeschichtlichen Zeit. Ferdinand Enke, Stuttgart

Bojňanský V, Fargašová A (2007) Atlas of seeds and fruits of Central and East-European flora. Springer

Cappers RTJ, Bekker RM, Jans JEA (2012) Digitale Zadenatlas van Nederland—Digital Seed Atlas of The Netherlands. second edition. Groningen Archaeological studies 4. Barkhuis & Groningen University Library. www.gas.ub.rug.nl

Katz NJ, Katz SV, Kipiani MG (1965) Atlas and keys of fruits and seeds occurring in the Quaternary deposits of the USSR. Nauka, Moscow. 365 pp. In Russian

Körber-Grohne U (1964) Bestimmungsschlussel für subfossile Juncus-samen und Gramineen-fruchte. Probleme der Kustenforschung im Sudlichen Nordseegebiet 7, August Lax, Hildesheim

Lévesque PEM, Dinel H, Larouche A (1988) Guide to the identification of plant macrofossils in Canadian peatlands. Publication 1817. Agriculture Canada, Canadian Government Publishing Centre Supply and Services, Canada, Ottawa

Martin JB, Barkley WD (1961) Seed Identification Manual. California University Press, Berkeley. Reprinted: 2000 by The Blackburn Press, ISBN 1-930665-03-2

Sabato D, Peña-Chocarro L (2021) Seeds and fruits from the Mediterranean Basin. Maris Nostri Novus Atlas, Doce Galles Ediciones, Spain

Schoch WH, Pawlik B, Schweingruber FH (1988) Botanische makroreste. P. Haupt, Bern & Stuttgart. German, English, French

Velichkevich FY, Zastawniak E (2006) Atlas of the Pleistocene vascular plant macrofossils of Central and Eastern Europe. Part 1: Pteridophytes and monocotyledons. Part 2: Herbaceous dicotyledons. W. Szafer Institute of Botany, Polish Academy of Sciences, Kraków

Mosses

Fossil mosses are identified using modern moss floras

Dickson JH (1973) Bryophytes of the Pleistocene. Cambridge University Press, Cambridge

Janssens JA (1990) Methods in Quaternary Ecology 11. Bryophytes. Geosciences Canada 17: 13–24

Wood

Barefoot AC, Hankins FW (1982) Identification of modern and Tertiary woods. Clarendon Press, Oxford

Grosser D (1977) Die Hölzer Mitteleuropas. Springer, Berlin

Jane FW (1956) The Structure of Wood. A. and C. Black, London

Mork E (1966) Vedanatomi. With an identification key for microscopic wood sections. 2nd ed. Forlag Johan Grundt Tanum, Oslo. In Norwegian

Schweingruber FH (1990) Anatomie europäischer Hölzer; Anatomy of European woods. Verlag Paul Haupt, Bern & Stuttgart. In German and English

Cuticles, Bud Scales, and Leaves

Grosse-Brauchmann G (1972) Über pflanzliche makrofossilien mitteleuropäischer Torfe - I. Gewebereste Krautiger Pflanzen und ihre Merkmale. Telma 2: 19–55. In German

Grosse-Brauchmann G (1973) Über pflanzliche makrofossilien mitteleuropäischer Torfe - II. Weitere Reste (Früchte und Samen, Moose, U.A.) und ihre Bestimmungsmöglichkeiten. Telma 4: 51–117. In German

Grosse-Brauchmann G (1986) Analysis of vegetative plant macrofossils. In: Berglund BE (ed) Holocene Handbook of Palaeoecology and Palaeohydrology. John Wiley & Sons, Chichester, pp. 591–618

Katz NJ, Katz SV, Skovejeva EI (1977) Atlas of Plant Remains in Peat-soil. Nedra, Moscow. In Russian

Tomlinson P (1985) An aid to the identification of fossil buds, bud-scales and catkin-bracts of British trees and shrubs. Circaea 3: 45–130

Appendix 2. A Sampling Protocol for AMS Radiocarbon Dating by H.H. Birks and A.F. Lotter

In contrast to conventional radiocarbon decay-counting, AMS radiocarbon dating requires sample sizes that are substantially smaller. Organic carbon comprises about half the weight of a sample. Generally, a sample size with at least 0.2 mg organic carbon is needed for AMS radiocarbon dating. This varies between laboratories, so it is wise to check with your laboratory before proceeding. Besides the advantage of allowing a better stratigraphic resolution (i.e. slices of 0.5–1 cm instead of e.g. bulk sediment samples using 5–10 cm of the core) we must realise that, when dealing with such small samples, a small amount of contamination will lead to a substantial deviation from the 'true' radiocarbon age.

Separation of Terrestrial Plant Macrofossils from the Sediment

1. Cut the wet sediment core into suitable slices. Avoid sampling slumps or turbidites as they may include older reworked material. Alternatively, select samples of interest, e.g. from archaeological sites.
2. Do not store these samples more than a few weeks in the refrigerator. If you cannot proceed immediately, it is best to deep-freeze the samples.
3. If the sediment will not break down immediately, soak it under water in a clean beaker with a lid. If necessary, sediment breakdown can be assisted by adding a small spoon of tetra-sodium diphosphate-10-hydrate crystals $Na_4P_2O_710H_2O$, a.k.a. sodium pyrophosphate, or Calgon water softener. Shake to dissolve the powder and stand overnight or as long as necessary in the refrigerator. If the sediment is calcareous, treatment with 10% HCl may be desirable. If it is humified peat, treatment with 10% KOH may be desirable.
4. Sieve the sample through a 250 μm mesh (NB! if macrofossil analysis is to be done on the sample(s), sieve through a 125 m mesh to retain small seeds etc.). It may be useful to use a selection of mesh sizes. If macrofossil material is sparse, save the sediments >125 μm in addition to the larger fractions.
5. Store the sieved material under water in clean beakers with lids, both clearly labelled, in the refrigerator. If you cannot proceed immediately, deep-freeze the samples.
6. Disperse a little of the material in water in a clean shallow (petri) dish. Under a stereo-microscope, magnification ca. 12x, pick out the terrestrial plant fragments with soft (entomological) forceps. It may be useful to separate the different types, e.g. mosses, leaves, seeds, etc. for easier identification, or selection of one sort of material. The material for dating should be identified and that of terrestrial origin should be selected.

Selection of Material for Dating

1. Remove any sediment or other material adhering to the selected macrofossil material with a small brush or forceps.
2. Under the microscope, pick out the remains once more, carefully avoiding sediment and any dust, hairs etc. and place them in clean tap water in a new clean container. This is a washing process.
3. Pick out the remains again and place them in a dry, dust-free box or glass. Let the sample dry, covered, at room temperature.

4. Weigh the sample to ensure you have enough material for dating, as required by your dating laboratory. Although some laboratories can use as little as 2 mg dry material, to produce sufficient carbon it is usually best to submit at least 6 mg. For $\delta^{13}C$ measurements, 10 mg is required.
5. Control the dry sample once more for dust or clothing hairs. Wool hairs for example will make the sample 'younger'; synthetic oil-derived hairs will make the sample 'older'. With such small samples it is very important to do this.
6. Sterilize a glass vial at 600 °C for 3 minutes. Cool it. Place the clean sample in it. Label both the vial and the lid. Glass containers are preferable, as plastic suffers from static electricity which makes the fragments jump about.
7. The dry samples can be stored as necessary in a dark cold-room until being sent to the laboratory.

Steps 1–4 can be carried out in a laminated airflow chamber to reduce the chances of contamination by air-borne dust and clothing hairs.

Important: As AMS samples are very small, small amounts of fungal or bacterial biomass may easily contaminate the samples. Therefore, care should be taken to avoid this at all stages in the preparation; by proceeding rapidly, and by storing the samples in a dark refrigerator between stages. If delay is unavoidable, the material should be deep frozen. Samples should never be left in water at room temperature, as fungal growth can be extremely rapid. Samples that have been stored in this way, or in organic liquids (e.g. alcohol, glycerol etc.) should not be used for dating or $\delta^{13}C$ measurements. Working conditions should be as clean as possible, and care should be taken to remove all dust, hairs, etc. from containers before use, and to control the samples for these contaminants as outlined above. With larger fragments, such as wood, do not handle them with your fingers.

Before sending off the samples for dating, note the following data in your lab journal:

- site name
- core ID, segment, date of coring; sample ID
- core depth of sample
- date of macrofossil processing, treatment
- what plant material is used for dating (e.g. 5 Pinus needles, 20 Betula fruits, etc.)
- dry weight of submitted material
- expected age

When reporting the radiocarbon ages in publications, the following information should be given:

- sample depth or code
- dating lab sample number
- material dated
- radiocarbon age ± 1sd
- calibrated age (as cal years BP, with reference to the program and calibration dataset used)
- $\delta^{13}C$ if measured

Radiocarbon dates are not exact equivalents of calendar ages due to past fluctuations in the concentration of atmospheric ^{14}C. For calculating calendar ages of samples and accumulation rates for biotic and abiotic variables we suggest converting radiocarbon dates to calendar ages using a calibration curve and a program such as OxCal 4.4 (ORAU - OxCal, C. Bronk Ramsey) or CALIB (CALIB 14C Calibration Program, Stuiver and Reimer 1993). The latest calibration curve is IntCal20 (Reimer et al. 2020).

Reimer P, and 41 others (2020 The IntCal20 Northern Hemisphere Radiocarbon Age Calibration Curve (0–55 cal kBP). Radiocarbon 62: 725–757.

Stuiver M, Reimer PJ, (1993) CALIB Radiocarbon Calibration. Radiocarbon 35: 215–230.

Glossary

^{14}C The radioactive isotope of carbon, also known as radiocarbon

Achene A small dry indehiscent fruit where the pericarp is not fused to the seed

Alien A plant believed on good evidence to have been introduced by man and now naturalized

Aquatic plant Plant that completes its life cycle with all its vegetative parts submersed in or supported by water

Allochthonous Sediment that is derived from material that is produced away from the point of deposition and transported to the site (e.g. in the centre of a lake or pond)

Autochthonous Sediment that is derived from the plants growing at the point of deposition (e.g. on a bog or fen surface)

Awn A stiff bristle-like projection

Base-rich Soils or water containing relatively large amounts of basic ions, such as calcium and magnesium

Berry A fleshy fruit, usually many-seeded, without a stony layer surrounding the seeds

Bog A community that accumulates acid peat that is hydrologically isolated from groundwater

BP Before Present after a radiocarbon date. Present is taken as 1950 CE

Calcareous A sediment, soil or water that contains a high amount of calcium carbonate

Calciphile A plant found on soils containing free calcium carbonate

Calcifuge A plant not normally found on soils containing free calcium carbonate

Capsule A dry many-seeded dehiscent fruit formed from more than one fused carpel

Carpel The female part of a flower comprising the style, stigma, and ovary

H. H. Birks et al., *Seeds and Fruits of Aquatic Plants of Northern Europe - A Handbook*, Developments in Paleoenvironmental Research 23,
https://doi.org/10.1007/978-3-032-04170-8

Caryopsis A single seeded fruit with the pericarp fused to the testa, especially in Poaceae

Chalaza Where the integuments and nucellus are joined in an ovule. It is located opposite the micropyle

Cotyledon The first leaf contained in the seed

Diaspore A dispersal unit. It can be a seed or fruit, or a vegetative structure

Drupe A fleshy fruit with one (or more) seed surrounded by a stony layer

Ecosystem An integrated biotic community and its abiotic components (e.g. wetland, lake)

Emergent plant A plant that is rooted in the soil or sediment with its vegetative and reproductive parts above the water (telmatophyte)

Endemic Native and restricted to a small area or country

Elodeid Submerged macrophyte with long stems either rootless or rooted in the sediment, with no floating leaves, although it may produce flowering stems above the water.

Endocarp The inner, usually hardened part of a fruit

Epicarp/ectocarp/exocarp The outermost layer of a fruit, usually soft tissue which may be fleshy

Eutrophic An ecosystem with high amounts of nutrients, especially nitrogen and phosphorus

Eutrophication The addition of excess nutrients to an ecosystem, usually leading to an increase in primary productivity. Cultural eutrophication is caused by human activity

Fen A peat-accumulating wetland community on alkaline, neutral, or slightly acid wet peat connected to the groundwater

Follicle A dry fruit formed from one carpel and dehiscent along one side

Fruit The fleshy or dry ripened ovary of a flowering plant, enclosing the seed or seeds

Funicle The stalk of the ovule

Glacial period or ice age A period during the Quaternary with low global temperatures and the expansion of ice sheets and glaciers at high latitudes and altitudes

Gynoecium The female part of the flower, made of one or more ovaries with their styles and stigmas

Gyttja Organic sediment deposited in a lake; lake mud

Heterosporous A plant having female megaspores and male microspores that give rise to gametophytes

Hilum The scar left on a seed at its point of attachment to the stalk of the ovule (funicle)

Holocene The current interglacial period after the last glaciation, starting at about 11,700 years ago

Hydrology The depth, duration, frequency of fluctuations, and the chemical composition of water on a site, such as a wetland

Hydrophyte A plant that grows partly or entirely in water, either rooted in the substrate or free floating. Another word for macrophyte

Hydrosere The succession from open water to terrestrial vegetation enabled by the infilling of a lake basin by sediments and plant remains

Inflorescence Flowering branch, including its branchlets, flowers, and bracts
Interglacial period A warm period between glacial periods during the Quaternary
Introduced Not native, brought in purposely or accidentally by humans
Isoetid Stemless tufted plants with rosettes that grow submerged in shallow water, often forming a short turf, at lake margins, and characteristic of oligotrophic water.
Keel A sharp edge resembling the keel of a boat
Keystone species A species controlling the structure of an ecosystem
Late Glacial A period of warming (interstadial) followed by pronounced cooling (Younger Dryas) at the end of the last glacial period (Weichselian or Wisconsinan) between 14,500 and 11,700 years ago
Lemnaid Free-floating plant that has no connection with the sediment
Lenticular Two-sided, or lens shaped
Lumen-lumina The space(s) enclosed by wall in a reticulum
Macrofossil The remains of a plant in sediment that is visible to the naked eye and that can be manipulated by hand
Macrophyte An aquatic plant, large enough to see with the naked eye, that actively grows permanently or periodically submerged below, floating on, or up through the water surface
Marsh A herbaceous community on wet or periodically wet but not peaty soils
Meadow A grassy field cut for hay
Megaspore The female spore of a hetersporous plant
Mericarp A one-seeded segment formed by splitting of a schizocarp e.g. in Apiaceae
Mesotrophic With a medium amount of nutrients, especially nitrogen and phosphorus
Micropyle The point on a seed where the embryo exits during germination
Microspore The male spore (e.g. pollen grains) of a heterosporous plant
Multiproxy study A palaeolimnological study using more than one proxy
Native Not known to have been introduced by humans
Nut A relatively large indehiscent one-seeded fruit with a thick hard wall
Nutlet A dry, one-seeded fruit formed from a single indehiscent carpel with a woody wall
Nymphaeid A macrophyte rooted in the sediment with leaves floating on the water surface
Obovate Rounded and wider at the distal end than the proximal end
Oligotrophic Poor in nutrients such as nitrogen and phosphorus
Oospore Macroscopic fertilised spore of a Charophyte
Orbicular Circular, more or less equal in length and breadth
Ovary The part of the female part of a flower, the gynoecium, enclosing the ovules or seeds
Ovate Rounded and wider at the base than the distal end; egg shaped
Palaeolimnology The study of the past history of lakes and wetlands
Papilla A very small projection from the surface - papillate
Pasture A grazed grassy field

Pericarp The coat of a mature fruit derived from the ovary wall

Perigynium A globular, trigonous or lenticular sac surrounding the achene in *Carex* (utricle)

Perine The outer layer covering a fern spore

Proxy Biotic and abiotic remains in palaeolimnological sediment samples used to reconstruct past flora, fauna, communities, ecosystems, and environment

Quaternary The latest geological epoch, characterized by repeated glacial and interglacial periods over the last two million years

Representation The relationship of a macrofossil assemblage to the parent plant or vegetation

Reticulum A net like pattern with walls surrounding lumina - reticulate

Radicle The embryonic root

Rhizome An underground stem, usually swollen with food reserves, lasting more than one growing season

Scabrate Rough

Schizocarp A dry fruit that splits into two or more one-seeded portions (mericarps) at maturity (e.g. Apiaceae)

Seed A fertilized ovule containing the embryo and usually a store of food, either within the embryo or around it as endosperm

Seed bank Dormant seeds in the soil or substrate which germinate when favourable conditions return

SEM Scanning electron microscope

Spore A small haploid, usually unicellular, reproductive body produced in a sporangium that germinates to form a gametophyte

Sporangium The structure producing spores in pteridophytes and bryophytes

Stolon A short-lived thin creeping stem produced from a mature plant that roots along its length or at its apex

Stigma The receptive surface of the female part of the flower to which pollen grains adhere

Style Part of the female part of a flower connecting the stigma to the ovary

Taphonomy The processes resulting in the deposition of a fossil assemblage

Taxon (Taxa plural) A taxonomic entity, e.g. species, genus, family

Telmatophyte Emergent telmatophytes are rooted in the underwater sediment but the main body of the plant is above water

Tephra Volcanic ash dispersed by wind and deposited away from the volcano

Testa The outer coat of a seed

Trigonous Three-angled

Turion A detachable over-wintering bud

Utricle A globular, trigonous or lenticular sac surrounding the achene in *Carex* (Perigynium)

Verrucate Slightly rough with small wart-like outgrowths

Weichselian The last glacial period. Wisconsinan in USA. Wurmian in central Europe. Devensian in UK. Midlandian in Ireland

Wetland Transitional swamp between open water and dry ground. Where wetlands occur round a lake, the water body can be termed a 'wetland lake'

References

Aalto M (1970) Potamogetonaceae fruits. 1. Recent and subfossil endocarps of the Fennoscandian species. Acta Bot Fenn 88:85

Aalto MM, Coope GR, Gibbard PL (1984) Late Devensian river deposits beneath the floodplain terrace of the river Thames at Abingdon, Berkshire, England. Proc Geol Assoc 95:65–79

Abbasi S, Afsharzadeh S, Dinarvand M (2015) *Potamogeton friesii*, a new species for the flora of Iran. Iranian J Botany 21:39–42

Adamec L (2018a) Biological flora of Central Europe: *Aldrovanda vesiculosa* L. Perspect Plant Ecol Evol Syst 35:8–21

Adamec L (2018b) Ecophysiological characteristics of turions of aquatic plants: a review. Aquat Bot 148:64–77

Adamec L (2020) Biological flora of Central Europe: *Utricularia intermedia* Hayne, *U. ochroleuca* R.W. Hartm., *U. stygia* Thor and *U. bremii* Heer ex Kölliker. Perspect Plant Ecol Evol Syst 44:125520

Adams KJ (2017) Thin-leaved horned-pondweed new to the U.K. Essex Botany, Bryol & Mycol 6:8–9

Agami M, Waisel Y (1984) Germination of Najas marina L. Aquatic Botany 19:37–44

Agnew ADQ (1968) The interspecific relationships of *Juncus effusus* and *J. conglomeratus* in Britain. Watsonia 6:377–388

Akhani H (2014) *Groenlandia densa*. The IUCN red list of threatened species: Groenlandia densa (Opposite-leaved Pondweed). iucnredlist.org

Alderton E, Sayer CD, Davies R, Lambert SJ, Axmacher JC (2017) Buried alive: aquatic plants survive in 'ghost ponds' under agricultural fields. Biol Conserv 212:105–110

Alsos IG, Sjögren P, Edwards ME, Landvik JY, Gielly L, Forwick M, Coissac E, Brown AG, Jakobsen LV, Føreid MK, Pedersen MW (2016) Sedimentary ancient DNA from Lake Skartjørna, Svalbard: assessing the resilience of arctic flora to Holocene climate change. The Holocene 26:627–642

Arbor A (1920) Water plants. A study of aquatic angiosperms. Wheldon & Wesley Ltd, UK, reprinted by Cramer, Weinheim

Arts GHP (2002) Deterioration of Atlantic soft water macrophyte communities by acidification, eutrophication, and alkalinisation. Aquat Bot 73:373–393

H. H. Birks et al., *Seeds and Fruits of Aquatic Plants of Northern Europe - A Handbook*, Developments in Paleoenvironmental Research 23,
https://doi.org/10.1007/978-3-032-04170-8

Arts GHP, Heijden RAJM (1990) Germination ecology of *Littorella uniflora* (L.) Aschers. Aquat Bot 37:139–151

Artsdatabanken (2015). Norsk rødliste for arter (artsdatabanken.no)

Aykurt C, Deniz IG, Kaplan Z (2016) *Potamogeton acutifolius* (Potamogetonaceae)—a new species for the Flora of Turkey. Int J Bot 12:17–19

Ayres KR, Sayer CD, Skeate ER, Perrow MR (2008) Palaeolimnology as a tool to inform shallow lake management: an example from Upton Great Broad, Norfolk, UK. Biodivers Conserv 17:2153–2168

Baattrup-Pedersen A, Springe G, Riis T, Larsen SE, Sand-Jensen K, Kjellerup Larsen LM (2008) The search for reference conditions for stream vegetation in northern Europe. Freshw Biol 53:1890–1901

Backman AL (1941) *Najas marina* in Finnland wärend der Postglazialzeit. Acta Bot Fenn 30:1–38

Backman AL (1948) *Najas flexilis* in Europa während der Quartärzeit. Acta Bot Fenn 43:1–44

Backman AL (1950) *Najas tenuissima* (A. Br.) Magnus einst und jetzt. Societas Scientiarum Fennica, Commentationes Biologicae 10(19):1–36

Backman AL (1951) *Najas minor* All. in Europa einst und jetzt. Acta Bot Fenn 48:1–32

Bajzáth J (1996) Plant macrofossils from Hungarian Pleistocene II. Angiospermatophyta in Gyorújfalu, West Hungary. Annales Historico-Naturales Musei Nationalis Hungarici 88:5–20

Bansal S et al (2019) Typha (cattail) invasion in North American wetlands: biology, regional problems, impacts, ecosystem services, and management. Wetlands 39:645–684

Barry DH, Wade PM (1986) Biological Flora of the British Isles no. 162. *Callitriche truncata* Guss. J Ecol 74:289–294

Belyakov EA, Philippov DA (2018) The effect of changes in environmental conditions on the morphology of *Sparganium natans* L. (Typhaceae) in the taiga zone of European Russia. Ecosyst Transform 1:29–41

Belyakov EA, Nikolaenko SA, Glazunov VA, Lapirov AG (2022) Distribution, ecology and biology of Sparganium hyperboreum (Typhaceae) in Western Siberia. Ecosyst Transform 5:21–33

Bennett KD (1990) Postglacial history of alder (Alnus glutinosa (L.) Gaertn.) in the British Isles. J Quat Sci 5:123–133

Bennike O (1989) The Kap København formation: stratigraphy and palaeobotany of a Plio-Pleistocene sequence in Peary Land, North Greenland. Meddr Grønland. Geoscience 23:85. Copenhagen 1990-12–7

Bennike O (1995) Palaeoecology of two lake basins from Disko, West Greenland. J Quat Sci 10:149–155

Bennike O (2000) Palaeoecological studies of Holocene lake sediments from West Greenland. Palaeogeogr Palaeoclimatol Palaeoecol 155:285–304

Bennike O, Anderson J (2008) Potamogeton praelongus in West Greenland. Nord J Bot 18:499–501

Bennike O, Björck S (2000) Lake sediment coring in South Greenland in 1999. Geus 186:60–64

Bennike O, Hoek WZ (1999) Late glacial and early Holocene records of *Stratiotes aloides* L. from northwestern Europe. Rev Palaeobot Palynol 107:259–263

Bennike O, Houmark-Nielsen M, Wiberg-Larsen P (2006) A middle Weichselian interstadial lake deposit on Sejerø, Denmark: macrofossil studies and dating. J Quat Sci 22:647–651

Bennike O, Jensen JB (2013) A Baltic Ice Lake lowstand of latest Allerød age in the Arkona Basin, southern Baltic Sea. GEUS Bulletin 28:17–20

Bennike O, Jensen J, Lemke W (2001) Late Quaternary records of Najas spp. (Najadaceae) from the southwestern Baltic region. Rev Palaeobot Palynol 114:259–267

Bennike O, Pantmann P, Aarsleff E (2020) Lateglacial and Holocene floras and faunas from the Salpetermosen area, north-east Sjælland, Denmark. Bull Geol Soc Den 68:231–244

Bennion H, Sayer CD, Clarke SJ, Davidson TA, Rose N, Goldsmith B, Rawcliff R, Burgess A, Clarke G, Turner S, Wiik E (2018) Sedimentary macrofossil records reveal ecological change in English lakes: implications for conservation. J Paleolimnol 60:329–348

Berggren G (1964) Atlas of seeds. Part 2. Cyperaceae. Swedish Natural Science Research Council, Stockholm. 68 pp.

Berggren G (1974) Seed morphology of some Epilobium species in Scandinavia. Sven Bot Tidskr 68:164–168

Berglund BE, Sandgren P, Barnekow L, Hannon G, Jiang H, Skog G, Yu S-Y (2005) Early Holocene history of the Baltic Sea, as reflected in coastal sediments in Blekinge, southeastern Sweden. Quat Int 130:111–139

Beug HJ, Tonkov SB (2017) Evidence of Holocene fossil *Aldrovanda vesiculosa* (Droseraceae) seeds at Lake Arkutino, southeastern Bulgaria. Phytologia Balcanica 23:337–339

Bhandari S, Paudaya K, Momohara A (2011) Climate change on the basis of plant macrofossil assemblages from the late Quaternary sediments from the Mulpani section of the Gokarna formation, Kathmandu Valley. Nepal J Stratigraphic Assoc Nepal 7:47–58

Birkinshaw N, Kemp E, Clarke SJ (2013) The ecology of grass-wrack pondweed *Potamogeton compressus*. Report number: NECR130, Natural England

Birks HH (1973) Modern macrofossil assemblages in lake sediments in Minnesota. In: Birks HJB, West RG (eds) Quaternary Plant Ecology. Blackwell Scientific Publications, Oxford, pp 173–189

Birks HH (1980) Plant macrofossils in Quaternary lake sediments. Arch Hydrobiol 15:1–60

Birks HH (1991) Holocene vegetational history and climatic change in West Spitsbergen—plant macrofossils from Skardtjørna, an arctic lake. The Holocene 1:209–218

Birks HH (2000) Aquatic macrophyte vegetation development in Kråkenes Lake, western Norway, during the late-glacial and early-Holocene. J Paleolimnol 23:7–19

Birks HH (2001) Plant macrofossils. In: Smol JP, Birks HJB, Last WM (eds) Tracking Environmental Change Using Lake Sediments, Volume 3. Biological techniques and indicators. Kluwer Academic Publishers, Dordrecht, pp 49–74

Birks HH (2002) The recent extinction of Azolla nilotica in the Nile Delta, Egypt. Acta Palaeobotanica 42:203–213

Birks HH (2007) Plant macrofossils in palaeoecology—an overview. In: Elias SA (ed) Encyclopedia of Quaternary Science. Elsevier, pp 2266–2288

Birks HH (2008) The late-Quaternary history of arctic and alpine plants. Plant Ecol Diversity 1:135–146

Birks HH, Ammann B (2000) Two terrestrial records of rapid climatic change during the glacial–Holocene transition (14,000–9,000 calendar years B.P.) from Europe. Proc Natl Acad Sci 97:1390–1394

Birks HH, Birks HJB (2001) Recent ecosystem dynamics in nine North African lakes in the CASSARINA project. Aquat Ecol 35:461–478

Birks HH, Birks HJB (2006) Multi-proxy studies in palaeolimnology. Veg Hist Archaeobotany 15:235–251

Birks HH, Bjune AE (2010) Can we detect a west Norwegian tree line from modern samples of plant remains and pollen? Results from the DOORMAT project. Veg Hist Archaeobotany 19:325–340

Birks HH, Gaillard M-J (2007) Paleolimnological applications. In: Elias SA (ed) Encyclopedia of Quaternary Science. Elsevier BV

Birks HH, Mathewes RW (1978) Studies in the vegetational history of Scotland. V. Late Devensian and Early Flandrian pollen and macrofossil stratigraphy at Abernethy Forest, Inverness-Shire. New Phytologist 80:455–484

Birks HH, van Dinter M (2010) Lateglacial and early Holocene vegetation and climate gradients in the Nordfjord–Ålesund area, western Norway. Boreas 39:783–798

Birks HH, Wright HE (2000) Introduction to the reconstruction of the late-glacial and early-Holocene aquatic ecosystems at Kråkenes Lake, Norway. J Paleolimnol 23:1–5

Birks HH, Battarbee RW, Birks HJB (2000) The development of the aquatic ecosystem at Kråkenes Lake, western Norway, during the late-glacial and early-Holocene—a synthesis. J Paleolimnol 23:91–114

Birks HH, Jones VJ, Brooks SJ, Birks HJB, Telford RJ, Juggins S, Peglar SM (2012) From cold to cool in northernmost Norway: Lateglacial and early Holocene multi-proxy environmental and climate reconstructions from Jansvatnet, Hammerfest. Quat Sci Rev 33:100–120

Birks HH, Peglar SM, Boomer I, Flower RJ, Ramdani M (2001) Palaeolimnological responses of nine North African lakes in the CASSARINA project to recent environmental changes and human impact detected by plant macrofossil, pollen, and faunal analyses. Aquat Ecol 35:405–430

Birks HJB (2014) Challenges in the presentation and analysis of plant macrofossil stratigraphical data. Veg Hist Archaeobotany 23:309–330

Bishop IJ (2019) Understanding the habitat and decline of *Najas flexilis* (Willd.) Rostk. & W.L.E.Schmidt in the UK using ecology and palaeoecology. Dissertation, University College, London

Bishop IJ, Bennion H, Patmore IR, Sayer CD (2018) How effective are plant macrofossils as a proxy for macrophyte presence? The case of *Najas flexilis* in Scotland. J Paleolimnol 60:153–165

Bishop IJ, Bennion H, Sayer CD, Patmore IR, Yang H (2019) Filling the "data gap": using palaeoecology to investigate the decline of Najas flexilis (a rare aquatic plant). Geo: Geography Environ 6:e00081. https://doi.org/10.1002/geo2.81

Bittmann F (2007) Reconstruction of the Allerød vegetation of the Neuwied Basin, western Germany, and its surroundings at 12,900 cal BP. Veg Hist Archaeobotany 16:139–156

Blindow I, Werner K (1990) Bestämningsnyckel för svenska kransalger. Sven Bot Tidskr 84:119–160

Blockeel TL, Hodgetts HG, Pilkington SL, Prescott OL (2021) A census catalogue of British and Irish bryophytes 2021. British Bryological Society

Bobrov AA, Chemeris EV, Filippova VA, Maltseva SY (2018) European pondweed in East Siberia: evidence of *Potamogeton rutilus* (Potamogetonaceae) in Yakutia (Asian Russia) with evaluation of current distribution and conservation status. Phytotaxa 333:58–72

Bobrov AA, Volkova PA, Kopylov-Guskov YO, Mochalova OA, Kravchuk AE, Nekrasova DM (2022) Unknown sides of *Utricularia* (Lentibulariaceae) diversity in East Europe and North Asia or how hybridization explained old taxonomical puzzles. Perspect Plant Ecol Evol Syst 54:125649

Bobrov AA, Volkova PA, Mochalova OA, Chemeris EV (2023) High diversity of aquatic Sparganium (Xanthosparganium, Typhaceae) in North Eurasia is mostly explained by recurrent hybridization. Perspect Plant Ecol Evol Syst 60:125746

Böchelmann J, Tremetsberger K, Šumberová K, Grausgruber H, Bernhardt K-G (2017) Fitness and growth of the ephemeral mudflat species Cyperus fuscus in river and anthropogenic habitats in response to fluctuating water-levels. Flora 234:135–149

Bodner M (1994) Inorganic carbon source for photosynthesis in the aquatic macrophytes *Potamogeton natans* and *Ranunculus fluitans*. Aquat Bot 48:109–112

Bojňanský V, Fargašová A (2007) Atlas of seeds and fruits of central and east-European flora: the Carpathian Mountains region. Springer, Dordrecht

Borrill M (1956) A biosystematic study of some Glyceria species in Britain. Watsonia 3:291–298

Bos JAA, Urz R (2003) Late glacial and early Holocene environment in the middle Lahn river valley (Hessen, central-West Germany) and the local impact of early Mesolithic people—pollen and macrofossil evidence. Veg Hist Archaeobot 12:19–36

Bos JAA, Dickson JH, Coope GR, Jardine WG (2004) Flora, fauna and climate of Scotland during the Weichselian middle Pleniglacial—palynological, macrofossil and coleopteran investigations. Palaeogeogr Palaeoclimatol Palaeoecol 204:65–100

Boutahar L, Espinosa F, Richir J, Lepoint G, Gobert S, Maanan M, Bazairi H (2020) Deep-water *Zostera marina* meadows in the Mediterranean. Aquatic Botany 166:103269

Bradbury JP, Forester RM, Thompson RS (1989) Late Quaternary paleolimnology of Walker Lake, Nevada. J Paleolimnol 1:249–267

Bradshaw EG, Rasmussen P, Nielsen H, Anderson NJ (2005) Mid- to late-Holocene land-use change and lake development at Dallund Sø, Denmark: trends in lake primary production as reflected by algal and macrophyte remains. The Holocene 15:1130–1142

Brayshay BA, Dinnin M (1999) Integrated palaeoecological evidence for biodiversity at the floodplain-forest margin. J Biogeogr 26:115–131

Brinkhuis H et al (2006) Episodic fresh surface waters in the Eocene Arctic Ocean. Nature 441:606–609

Brinkkemper O, Weeda EJ, Bohncke SJP, Kuijper WJ (2008) The past and present occurrence of *Elatine* and implications for palaeoenvironmental reconstructions. Veg Hist Archaeobotany 17:15–24

Brouwer E, Bobbink R, Roelofs JGM (2002) Restoration of aquatic macrophyte vegetation in acidified and eutrophied softwater lakes: an overview. Aquat Bot 73:405–431

Brown A, Russell J, Scaife R, Tizzaard L, Whittaker J, Wyles SF (2018) Lateglacial/early Holocene palaeoenvironments in the southern North Sea Basin: new data from the Dudgeon offshore wind farm. J Quat Sci 33:597–610

Brux H, Todeskino D, Wiegleb G (1987) Growth and reproduction of *Potamogeton alpinus* Balbis growing in disturbed habitats. Arch Hydrobiol Beih 27:115–127

BSBI News (2018) BSBI News January 2018:61

Buckley S, Hardy K, Hallgren F, Kubiak-Martens L, Miliauskienė Z, Sheridan A, Sobkowiak-Tabaka I, Subirà ME (2023) Human consumption of seaweed and freshwater aquatic plants in ancient Europe. Nat Commun 14:6192

Buell MF (1935) Seed and seedling of *Acorus calamus*. Bot Gaz 96:758–765

Burrows CJ (1974) Plant macrofossils from late Devensian deposits at Nant Ffrancon, Caernarvonshire. New Phytol 73:1003–1033

Buurman J, van Geel B, van Reenen GBA (1995) Palaeoecological investigations of a late bronze age watering place at Bovenkarspel, The Netherlands. Meded Rijks Geol Dienst 52:249–270

CABI.org (n.d.) (Centre for Agriculture and Bioscience International) Trapa natans (water chestnut) (cabi.org)

Capo E, Barouillet C, Smol JP (2024) Tracking Environmental change using Lake sediments. Developments in Paleoenvironmental Research volume 6: Sedimentary DNA. Springer, Dordrecht

Cappers RTJ (1993a) The identification of Potamogetonaceae fruits found in The Netherlands. Acta Botanica Neerlandica 42(447):460

Cappers RTJ (1993b) Seed dispersal by water: a contribution to the interpretation of seed assemblages. Veg Hist Archaeobotany 2:173–186

Cappers RTJ, Bekke RM, Jans JEA (2012) Digitale Zadenatlas van Nederland—Digital Seed Atlas of The Netherlands, 2nd edn. Barkhuis & Groningen University Library. www.gas.ub.rug.nl

Carrion JS, van Geel B (1999) Fine-resolution upper Weichselian and Holocene palynological record from Navarrés (Valencia, Spain) and a discussion about factors of Mediterranean forest succession. Rev Palaeobot Palynol 106:209–236

Carrott J, Hall A, Issitt M, Jaques D, Kenward H, Large F (1996) Evaluation of biological remains from excavations at Welton Low Road, Elloughton, near Brough (site code: WLE96). In: Reports from the Environmental Archaeology Unit, York 96/17, p 5

Centre for Ecology and Hydrology (2004) Information sheet. *Lagarosiphon major*. Centre for Ecology and Hydrology, Wallingford

Chambers PA, Maberly SC (2024) Freshwater plants. Chapter 24 in Wetzel's Limnology (Fourth Edition), Lake and river ecosystems. Elsevier, pp 759–816. https://doi.org/10.1016/B978-0-12-822701-5.00024-0

Chambers PA, Lacoul P, Murphy KJ, Thomaz SM (2007) Global diversity of aquatic macrophytes in freshwater. In: Balian EV, Lévêque C, Segers H, Martens K (eds) Freshwater animal diversity assessment. Developments in Hydrobiology, vol 198. Springer, Dordrecht. https://doi.org/10.1007/978-1-4020-8259-7_2

Chandler MEJ (1922) The geological history of the genus Stratiotes: an account of the evolutionary changes which have occurred within the genus during Tertiary and Quaternary times. Q J Geol Soc Lond 79:117

Cholo F, Foden W (2006) Zannichellia palustris L. Red List of South African Plants version 2020.1. Threatened Species Programme. SANBI Red List of South African Plants

Chowdhury A, Chowdhury M, Das AP (2015) *Potamogeton gramineus* Linnaeus (Potamogetonaceae) a new record for the Indian sub-continent with a complete morphological description. Int J Curr Res 7:12362–12364

Clapham AR, Tutin TG, Warburg EF (1962) Flora of the British Isles, 2nd edn. Cambridge University Press, Cambridge

Clarke C, Alsos IG, Edwards M, Paus A, Gielly L, Haflidason H, Mangerud J, Regnéll C, Hughes PDM, Svendsen JI, Bjune AE (2020) A 24,000-year ancient DNA and pollen record from the polar Urals reveals temporal dynamics of arctic and boreal plant communities. Quat Sci Rev 247:106564

Clarke GH, Sayer CD, Turner S, Salgado J, Meis S, Patmore IR, Zhao Y (2014) Representation of aquatic vegetation change by plant macrofossils in a small and shallow freshwater lake. Veg Hist Archaeobotany 23:265–276

Collins PEF, Fenwick IM, Keith-Lucas M, Worsley P (1996) Late Devensian river and floodplain dynamics and related environmental change in Northwest Europe, with particular reference to a site at Woolhampton, Berkshire, England. J Quat Sci 11:357–375

Conway VM (1942) Biological Flora of the British Isles. *Cladium mariscus*. J Ecol 30:211–216

Cook CDK (1961) *Sparganium* in Britain. Watsonia 5:1–10

Cook CDK (1962) *Sparganium erectum* L. (*S. ramosum* Hudson, nom illeg). J Ecol 50:247–255

Cook CDK (1966) A monographic study of Ranunculus subgenus Batrachium (DC.) A. Gray. Mitteilungen der Botanischen Staatssammlung München 6:47–237

Cook CDK (1990) Seed dispersal of *Nymphoides peltata* (S.G.Gmelin) O. Kuntze (Menyanthaceae). Aquat Bot 37:325–340

Cook CDK, Lüönd R (1982) A revision of the genus *Hydrocharis* (Hydrocharitaceae). Aquat Bot 14:177–204

Cook CDK, Nicholls MS (1987) A monographic study of the genus Sparganium (Sparganiaceae). II: Subgenus Sparganium. Botanica Helvetica 97:1–44. https://api.semanticscholar.org/CorpusID:8871343

Cook CDK, Urmi-König K (1983) A revision of the genus *Stratiotes* (Hydrocharitaceae). Aquat Bot 19:213–249

Coops H, Van der Velde G (1995) Seed dispersal, germination and seedling growth of six helophyte species in relation to water-level zonation. Freshw Biol 34:13–20

Cope T, Gray A (2009) Grasses of the British Isles. BSBI handbook 13. Botanical Society of the British Isles

Courtney-Mustaphi CJ, Steiner E, von Fumetti S, Heiri O (2024) Aquatic invertebrate mandibles and sclerotized remains in Quaternary lake sediments. J Paleolimnol 71:45–83

Cox DD (2002) A naturalist's guide to wetland plants: An ecology for eastern North America. Syracuse University Press, Syracuse, NY

Cross A (2012) Aldrovanda: The waterwheel plant. Redfern Natural History Productions, Poole, Dorset

D'hondt B, D'hondt S, Bonte D, Brys R, Hoffmann M (2012) A data-driven simulation of endozoochory by ungulates illustrates directed dispersal. Ecol Model 230:114–122

Davidson T, Sayer CD, Bennion H, Rose N, David C, Wade M (2005) A 250 year comparison of historical, macrofossil and pollen records of aquatic plants in a shallow lake. Freshw Biol 50:1671–1686

Davidson TA, Sayer CD, Jeppesen E, Søndergaard M, Lauridsen TL, Johansson LS, Baker A, Graeber D (2023) Bimodality and alternative equilibria do not help explain long-term patterns in shallow lake chlorophyll-a. Nat Commun 14, 398. https://doi.org/10.1038/s41467-023-36043-9

Dawson FH (1988) Water flow and the vegetation of running waters. In: Symoens JJ (ed) Vegetation of inland waters. Handbook of vegetation science, vol 15-1. Springer, Dordrecht

Dendievel A-M, Riquier J, Mourier B, Winiarski T (2024) Tracking the effects of dam construction and restoration on side channel biodiversity using macrofossils in river sediment records. Sci Total Environ 957:177246. https://doi.org/10.1016/j.scitotenv.2024.177246

Dick J, Haynes D, Tibby J, Garcia A, Gel P (2011) A history of aquatic plants in the Coorong, a Ramsar-listed coastal wetland, South Australia. J Paleolimnol 46:623–635

Dickmann DI, Kuzovkina J (2014) Poplars and willows of the world, with emphasis on silviculturally important species. In: Richardson J, Isebrands JG (eds) Poplars and Willows: Trees for Society and the Environment, pp 8–91

Dickson CA (1970) The study of plant macrofossils in British Quaternary deposits. In: Walker D, West RG (eds) Studies in the vegetational history of the British Isles. Cambridge University Press, Cambridge, pp 233–254

Dickson JH (1973) Bryophytes of the Pleistocene. Cambridge University Press, Cambridge

Dieffenbacher-Krall AC, Halteman WA (2000) The relationship of modern plant remains to water depth in alkaline lakes in New England. J Paleolimnol 24:213–339

DiTamaso JM, Healy EA (2003) Aquatic and riparian weeds of the West. University of California Agriculture and Natural Resources, Oakland, pp 137–141

Donner J (1995) The Quaternary history of Scandinavia. Cambridge University Press, Cambridge

Douda J, Doudová J, Drašnarová A, Kuneš P, Hadincová V, Krak K, Zákravský P, Mandák B (2014) Migration patterns of subgenus *Alnus* in Europe since the Last glacial maximum: a systematic review. PLoS One 9(2):e88709. https://doi.org/10.1371/journal.pone.0088709

Dynowski P, Senetra A, Źróbek-Sokolnik A, Kozłowski J (2019) The impact of recreational activities on aquatic vegetation in Alpine lakes. Water 11:173

E-Flora BC (n.d.) Electronic Atlas of the Flora of British Columbia E-Flora BC Atlas Page (ubc.ca)

Eide W, Birks HH, Bigelow NH, Peglar SM, Birks HJB (2006) Holocene tree migrations in the Setesdal valley, southern Norway, reconstructed from macrofossil and pollen evidence. Veg Hist Archaeobotany 15:65–85

Elias SA (2001) Coleoptera and Trichoptera. In: Smol JP, Birks HJB, Last WM (eds) Tracking Environmental Change using Lake Sediments. Volume 4. Zoological Indicators. Kluwer Academic Publishers, Dordrecht, pp 67–80

Emson D, Sayer CD, Bennion H, Patmore IR, Rioual P (2018) Mission possible: diatoms can be used to infer past duckweed (lemnoid Araceae) dominance in ponds. J Paleolimnol 60:209–221

Fadne P (2023) Sjebnen til soleigro *Baldellia repens* i et gammelt vannverk i Leirvik, Stord, samt litt om artens økologi, systematikk og spredning. The occurrence and density of *Baldellia repens* in an old waterworks in Leirvik, western Norway and some notes on ecology, systematics and dispersal. Blyttia 81:163–170

Fajer J, Waga JM, Rzetala J, Szymczyk A, Nita M, Machowski R, Rzetala MA, Ruman M (2012) The late Vistulian and Holocene evolution of Jezioro Lake: a record of environmental change in southern Poland found in deposits and landforms. J Paleolimnol 48:651–667

Farmer AM, Spence DHN (1986) The growth strategies and distribution of isoetids in Scottish freshwater lochs. Aquat Bot 26:247–258

Feuchtmayr H, Moran R, Hatton K, Connor L, Heyes T, Moss B, Harvey I, Atkinson D (2009) Global warming and eutrophication: effects on water chemistry and autotrophic communities in experimental hypertrophic shallow lake mesocosms. J Appl Ecol 46:713–723

Field MH, Peglar SM (2010) A palaeobotanical investigation of the sediments from the West Runton mammoth site. Quat Int 228:38–45

Flenley JR, Maloney BK, Ford S, Hallam G (1975) *Trapa natans* in the British Flandrian. Nature 257:39–41

Flora of Svalbard (n.d.). http://svalbardflora.no/index.php?id=661

Flower RJ (2001) Change, stress, sustainability and aquatic ecosystem resilience in north African wetland lakes during the 20th century: An introduction to integrated biodiversity studies within the CASSARINA project. Aquat Ecol 35:261–280

Fluet-Chouinard E, Stocker BD, Zhang Z et al (2023) Extensive global wetland loss over the past three centuries. Nature 614:281–286

Fourounjian P, Slovin J, Messing J (2021) Flowering and seed production across the Lemnaceae. Int J Mol Sci 22:2733. https://doi.org/10.3390/ijms22052733

Francis DR (2001) Bryozoan statoblasts. In: Smol JP, Birks HJB, Last WM (eds) Tracking Environmental Change using Lake Sediments. Volume 4. Zoological indicators. Kluwer Academic Publishers, Dordrecht, pp 105–123

Fredskild B (1992) The Greenland limnophytes—their present distribution and Holocene history. Acta Bot Fenn 144:93–11

Frey DG (1964) Remains of animals in Quaternary lake and bog sediments and their interpretation. Ergebnisse der Limnologie 2:1–114

Fu L, Huang M, Han B, Sun X, Sree KS, Appenroth K-J, Zhang J (2017) Flower induction, microscope-aided cross-pollination, and seed production in the duckweed *Lemna gibba* with discovery of a male-sterile clone. Sci Rep 7:3047. https://doi.org/10.1038/s41598-017-03240-8

Gaillard M-J, Lemdahl G (1994) Early-Holocene coastal environments and climate in south-east Sweden: a reconstruction based on macrofossils from submarine deposits. The Holocene 4:53–68

Gałka M (2010) Subfossil seeds of Stratiotes aloides L. in northern and central Poland. Studia Quaternaria 27:11–15

Gałka M (2014) Pattern of plant succession from eutrophic lake to ombrotrophic bog in NE Poland over the last 9400 years based on high-resolution macrofossil analysis. Ann Bot Fenn 51:1–21

Gałka M, Kasper P (2011) Palaeobotanical peat analysis on contemporary Epipactis palustris (L.) Crontz. (marsh Helleborine) sites in northeastern part of Poland. Studia Limnol Telmatol 5:67–75

Galka M, Sznel M (2013) Late glacial and Early Holocene development of lakes in northeastern Poland in view of plant macrofossil analyses. Quat Int 292:124–135

Gałka M, Apolinarska K, Bubak I, Dezsö J, Feurdean A, Salem A, Słowik M, Zawisza E (2024) Long-term ecological studies on the oxbow ecosystems development and fire history in the Drava river valley (Central Europe): implications for ecological restoration. Progress Phys Geography Earth Environ 48:1. https://doi.org/10.1177/03091333241280073

Gałka M, Bergonzini L, Williamson D, Majule A, Masao C, Huguet A (2015) Macrofossil evidence of late Holocene presence of Aldrovanda vesiculosa L. in Central-Eastern Europe (Poland) and East Africa (Tanzania). Quat Int 386:186–190

Gałka M, Lewandowska A, Niedzielski P, Sim TG, Swindle GT, Szczurek G (2020) Late glacial and early Holocene development of an oxbow lake in Central Europe (Poland) based on plant macrofossil and geochemical data. The Holocene 30:178–189

Gałka M, Tobolski K (2013) Macrofossil evidence of early Holocene presence of Picea abies (Norway spruce) in NE Poland. Ann Bot Fenn 50:129–141

Gałka M, Tobolski K, Kołaczek P (2012) The Holocene decline of slender naiad (*Najas flexilis* (Willd.). Rostk. & W.L.E. Schmidt) in NE Poland in the light of new palaeobotanical data. Acta Palaeobotanica 52:127–138

Gałka M, Tobolski K, Zawisza E, Goslar T (2014) Postglacial history of vegetation, human activity and lake-level changes at Jezioro Linówek in northeast Poland, based on multi-proxy data. Veg Hist Archaeobotany 23:123–152

Garcés-Pastor S, Nota K, Rijal DP, Liu S, Jia W, Leunda M, Schwörer C, Crump SE, Parducci L, Alsos IG (2023) Terrestrial plant DNA from lake sediments. In: Capo E, Barouillet C, Smol JP (eds) Developments in Paleoenvironmental Research, Volume 21. Springer, Cham

Garnock-Jones PJ (1978) *Rorippa* (Cruciferace Arabideae) in New Zealand. NZ J Bot 16:119–122

Gaskin JF, Andreas J, Grewel BJ, Haefliger P, Harms HE (2021) Diversity and origins of *Butomus umbellatus* (flowering rush) invasion in North America. Aquat Bot 173:103400

Geertsema W (2005) Spatial dynamics of plant species in an agricultural landscape in The Netherlands. Plant Ecol 178:237–247

Gibbard PL, Aalto MM, Beales PW (1977) A Hoxnian interglacial site at Fishers Green, Stevenage, Hertfordshire. New Phytol 78:505–523

Godwin H (1923) Dispersal of pond floras. Journal of Ecology 11:160–164

Godwin H (1975) History of the British Flora. A factual basis for phytogeography. Second edition. Cambridge University Press, Cambridge

Goetcheus VG, Birks HH (2001) Full-glacial upland tundra vegetation preserved under tephra in the Beringia National Park, Seward Peninsula, Alaska. Quat Sci Rev 20:135–147

Green AJ, Lovas-Kiss A, Chevonne Reynolds C, Sebastián-González E, Silva GG, van Leeuwen CHA, Wilkinson DM (2022) Dispersal of aquatic and terrestrial organisms by waterbirds: a review of current knowledge and future priorities. Freshw Biol 68:175–190

Green AJ, Soons M, Brochet A-L, Kleyheeg E (2016) Dispersal of plants by waterbirds. In: Sekercioglu ÇH, Wenny DG, Whelan CJ (eds) Why birds matter. Avian ecological function and ecosystem services. The University of Chicago Press, Chicago, pp 147–195

Grimm EC (1990) TILIA and TILIA.GRAPH, PC spreadsheet and graphics software for pollen data. INQUA Working Group on Data Handling Methods Newsletter 4:5–7

Grimm EC (2004) TGView 2.02. Illinois State Museum, Springfield, IL

Grizett B, Poikane S (2024) The importance of inland waters. In: Wetzel's Limnology 4th edn, Elsevier, pp 7–13

Groenteman R, Heenan P, Barton J (2020) Feasibility for biological control of water celery (*Helosciadium nodiflorum).* Contract report LC3783. Landcare research New Zealand Ltd and Nelson City Council 2020

Grosse-Brauckmann G (1972) Über pflanzliche Makrofossilien mitteleuropäischer Torfe—I. Gewebereste krautiger Pflanzen und ihre Merkmale. Telma 2:19–55

Grüger J (1973) Studies on the Late Quaternary vegetation history of Northeastern Kansas. Geol Soc Am Bull 84:239–250

Gundersen H, Bryan T, Chen W, Moy FE, Sandman AN, Sundblad G, Schneider S, Andersen JH, Langaas S, Walday MG (2017) Ecosystem services in the coastal zone of the Nordic countries. Technical Report, Nordic Council of Ministers

Guppy HB (1897) On the postponement of the germination of the seeds of aquatic plants. Proc Royal Phys Soc Edin 13:344–359

Gurnell AM, O'Hare MT, O'Hare JM, Scarlett P, Liffen TMR (2013) The geomorphological context and impact of the linear emergent macrophyte Sparganium erectum L.: a statistical analysis of observations from British rivers Earth Surf Process Landf 38(15), 1869–1880. https://doi.org/10.1002/esp.3473

Haas JN (1994) First identification key for charophyte oospores from central Europe. Eur J Phycol 29:227–235

Hájková P, Jamrichová E, Horsák M, Hájek M (2013) Holocene history of a *Cladium mariscus*-dominated calcareous fen in Slovakia: vegetation stability and landscape development. Preslia 85:289–315

Hall A, Kenward H, Large F (1995) Biological remains from a medieval 'pond' at Higher Lane, Fazakerley, Merseyside (site code FAZ94). Rep Environ Archaeol Unit York 96/5:29

Handley RJ, Davy AJ (2000) Discovery of male plants of *Najas marina* L. (Hydrocharitaceae) in Britain. Watsonia 23:331–334

Handley RJ, Davy AJ (2002) Seedling root establishment may limit *Najas marina* L. to sediments of low cohesive strength. Aquat Bot 73:129–136

Hannon GE, Gaillard M-J (1997) The plant-macrofossil record of past lake-level changes. J Paleolimnol 18:15–28

Hart KH, Cox PA (1995) Dispersal ecology of *Nuphar lutea* (L.) Sibthorp & Smith: abiotic seed dispersal mechanisms. Bot J Linn Soc 119:87–100

Haslam SM (1972) Biological Flora of the British Isles no. 128. *Phragmites communis* Trin. J Ecol 60:585–610

Hauser TP, Loeschcke V (1994) Inbreeding depression and mating-distance dependent offspring fitness in large and small populations of *Lychnis flos-cuculi* (Caryophyllaceae). J Evol Biol 7:609–622

Havinga AJ, van den Berg RM, van Saparoea RM (1992) Pollen and macrofossil analyses of a unique peat remnant of the former extensive peat bogs in West Friesland, The Netherlands: a detailed record of coastal raised bog formation. Veg Hist Archaeobotany 1:185–188

Heiri O, Lotter AF, Lemcke G (2001) Loss on ignition as a method for estimating organic and carbonate content in sediments: reproducibility and comparability of results. J Paleolimnol 25:101–110

Hill MO (1992) Sphagnum: a field guide. Updated by NG Hodgetts and AG Payne. The UK Joint Nature Conservation Committee, Peterborough

Hilt S, Alirangues Nuñez MM, Bakke ES, Blindow I, Davidson TA, Gillefalk M, Hansson L-A, Jansen JH, Janssen ABG, Jeppesen E, Kabus T, Kelly A, Köhler J, Lauridsen TL, Mooij WM, Noordhuis R, Phillips G, Rücker J, Schuster H-H, Søndergaard M, Teurlincx S, van de Weyer K, van Donk E, Waterstraat A, Willby N, Sayer CD (2018) Response of submerged macrophyte communities to external and internal restoration measures in north temperate shallow lakes. Front Plant Sci 9:194

Hilton J, O'Hare M, Bowes MJ, Jones IJ (2006) How green is my river? A new paradigm of eutrophication in rivers. Sci Total Environ 365:66–83

Hinojosa-Garro D, Mason CF, Underwood GJ (2008) Macrophyte assemblages in ditches of coastal marshes in relation to land-use, salinity and water quality. Fundam Appl Limnol 172:325–327

Hofstra E, Schoelynch J, Ferreoll J, Coetzee J, de Winton M, Bickel TO, Champion P, Madsen J, Bakker ES, Hilt S, Matheson R, Niteherland M, Gross EM (2020) On the move: new insights on the ecology and management of native and alien macrophytes. Aquat Bot 162:103190

Holyoak DT (1983) A Late Pleistocene interglacial flora and molluscan fauna from Thatcham, Berkshire, with notes on Mollusca from the interglacial deposits at Aveley, Essex. Geol Mag120: 623–629

Howard HW, Lyon AG (1952) Biological Flora of the British Isles: *Nasturtium officinale* R.Br. (*Rorippa nasturtium-aquaticum* (L.) Hayek). J Ecol 40:228–245

Hroudová Z, Hrouda L, Zákravský P, Ostrý I (1988) Ecobiology and distribution of Sagittaria sagittifolia L. in Czechoslovakia. Folia geobotanica et phytotaxonomica 23:337–374

Hroudová Z, Krahulcová A, Zákravský P, Jarolímová V (1996) The biology of *Butomus umbellatus* in shallow waters with fluctuating water levels. Hydrobiologia 340:27–30

Hroudová Z, Zákravský P, Ducháček M, Marhold K (2007) Taxonomy, distribution and ecology of *Bolboschoenus* in Europe. Ann Bot Fenn 44:81–102

Hroudová Z, Zákravsky P, Hrouda L, Ostrý I (1992) *Oenanthe aquatica* (L.) Poir.: seed reproduction, population structure, habitat conditions and distribution in Czechoslovakia. Folia Geobot 27:301–335

Hubbard CE (1954) Grasses. 3nd edition revised by JCE Hubbard (1982). Penguin books

Hulme PD, Durno SE (1980) A contribution to the phytogeography of Shetland. New Phytol 84:165–169

Hultén E, Fries M (1986) Atlas of north European vascular plants I-III. Koeltz Scientific Books, Königstein, p 498

Hunter K (2012) Specialist report 19: plant macrofossils. In: Biddulph E, Foreman S, Stafford E, Stansbie D, Nicholson R (eds) London gateway: Iron age and Roman salt making in the Thames Estuary, excavation at Stanford Wharf Nature Reserve, Essex. (Oxford archaeology monograph). Oxford Archaeology, Oxford

Hutchinson GE (1970) The chemical ecology of three species of *Myriophyllum* (Angiospermae Haloragaceae). Limnol Oceanography 15:1–5

Imam H, Riaz Z, Azhar M, Sofi G, Hussain A (2013) Sweet flag (*Acorus calamus* Linn.): An incredible medicinal herb. Int J Green Pharm 7:288–296

Ingólfsson Ó, Rognvaldsson F, Bergsten H, Hedenäs L, Lemdahl G, Lirio JM, Sejrup HP (1995) Late Quaternary glacial and environmental history of Kongsøya, Svalbard. Polar Res 14:123–139

Iversen J (1954) The late-glacial flora of Denmark and its relation to climate and soil. Danm Geol Unders Series 2, 80: 87–119

Jackson ST, Booth RK (2007) Validation of pollen studies. In: Elias SA (ed) Encyclopedia of Quaternary Science. Elsevier BV, pp 2413–2422

Jackson ST, Futyma RP, Wilcox DA (1988) A paleoecological test of a classical hydrosere in the Lake Michigan dunes. Ecology 69:928–936

Jacobson HA (1994) Interaction of Holocene climate, water balance, vegetation, fire, and cultural land-use in the Swedish borderland. LUNDQUA Thesis 30. Department of Quaternary Geology, Lund University

Jacomet S (2007) Use of macrofossils in environmental archaeology. In: Elias SA (ed) Encyclopedia of Quaternary Science. Elsevier BV, pp 2384–2412

Janssens JA (1983) Past and extant distribution of *Drepanocladus* in North America, with notes on the differentiation of fossil fragments. J Hattori Bot Lab 54:251–298

Jenkins J (2019) Sedges of the Northern Forest. A photographic guide. Cornell University Press, Ithaca

Jensen HA (1987) Macrofossils and their contribution to the history of the spermatophyte flora in southern Scandinavia from 13000 BP to 1536 AD. Biol Skrifter 29:1–74

Jeppesen E, Lauridsen TL, Kairesalo T, Perrow M (1998) Impact of submerged macrophytes on fish-zooplankton interactions in lakes. In: Jeppesen E, Ma S, Mo S, Christoffersen K (eds) The structuring role of submerged macrophytes in lakes. Ecological Studies, vol 131. Springer, pp 91–114

Jessen K (1949) Studies in Late Quaternary deposits and flora-history of Ireland. Proc R Ir Acad B 52:85–290

Jessen K (1955) Key to subfossil *Potamogeton*. Botanisk Tidsskrift 52:1–7

Jiménez-Majías P, Martinetto E (2013) Toward an accurate taxonomic interpretation of *Carex* fossil fruits (Cyperaceae): a case study in section *Phacocystis* in the western palaeartic. Am J Bot 100:1580–1603

Johnson WC (2000) Tree recruitment and survival in rivers: influence of hydrological processes. Hydrol Process 47:3051–3074

Jonsell B, Karlsson T (eds) (2010) Flora Nordica. Royal Swedish Academy of Sciences, Stockholm

Jonsgard B, Birks HH (1995) Late-glacial mosses and environmental reconstructions at Kråkenes, Western Norway. Lindbergia 20:64–82

Jowsey PC (1966) An improved peat sampler. New Phytol 65:245–224

Juggins S (2022) C2: software for ecological and palaeoecological data analysis and visualisation (Version 1.8.0)

Juggins S (2023) riojaPlot: stratigraphic diagrams in R, package version (0.1-20)

Juggins S (2024) "rioja: Analysis of Quaternary Science Data." R package version 1.0-7. https://cran.r-project.org/package=rioja

Kamayama Y, Toyama M, Ohara M (2005) Hybrid origins and F_1 dominance in the free-floating, sterile bladderwort, *Utricularia australis* f. *australis* (Lentibulariaceae). Am J Bot 92:469–476

Kaneko K, Jinguji H (2012) Effects of environmental factors on *Sparganium emersum* and *Sparganium erectum* colonization in two drainage ditches with different maintenance. Agric Sci 3:538–544

Kaplan Z (2008) A taxonomic revision of *Stuckenia* (Potamogetonaceae) in Asia, with notes on the diversity and variation of the genus on a worldwide scale. Folia Geobot 43:159–234

Kaplan Z (2010) Potamogetonaceae Dumort. In: Chrtek JJ SJJ, Kaplan ZZ (eds) Flora of The Czech Republic, vol 8. Academia, Praha, pp 329–384

Kaplan Z, Marhold K (2012) Multivariate morphometric analysis of the *Potamogeton compressus* group (Potamogetonaceae). Bot J Linn Soc 170:112–130

Kaplan Z, Štěpánek J (2003) Genetic variation within and between populations of *Potamogeton pusillus* agg. Plant Syst Evol 239:95–112

Kaplan Z, Symoens J-J (2005) Taxonomy, distribution and nomenclature of three confused broad-leaved *Potamogeton* species occurring in Africa and on surrounding islands. Bot J Linn Soc 148:329–357

Kaplan Z, Jarolímová V, Fehrer J (2013) Revision of chromosome numbers of Potamogetonaceae: a new basis for taxonomic and evolutionary implications. Preslia 85:421–482

Kaplan Z, Danihelka J, Štěpánková J, Bureš P, Zázvorka J, Hroudová Z, Ducháček M, Grulich V, Řepka R, Dančák M, Prančl J, Šumberová K, Wild J, Trávníček B (2015) Distributions of vascular plants in The Czech Republic. Part 1. Preslia 87:417–500

Kaplan Z, Šumberova K, Formanova I, Duchaček M (2014) Re-establishment of an extinct population of the endangered aquatic plant *Potamogeton coloratus*. Aquat Bot 119:91–99

Karhulec F, Lepš J (1993) The migration of vascular plants to a new water reservoir: geographic relationships. Preslia 65:147–162

Karrenberg S, Edwards PJ, Kollmann J (2002) The life history of Salicaceae living in the active zone of floodplains. Freshw Biol 47:733–748

Katz NJ, Katz SV, Kipiani MG (1965) Atlas and keys of fruits and seeds occurring in the Quaternary deposits of the USSR. Nauka, Moscow. (in Russian)

Katz NJ, Katz SV, Skobejeva EI (1977) Atlas of plant remains in peat-soil. "Nedra", Moscow. (In Russian)

Keddy PA (2024) Wetland ecology. Principles and conservation. 3rd edn. Cambridge University Press, Cambridge

Kelly MR (1964) The Middle Pleistocene of North Birmingham. Philos Trans R Soc Lond B 247:533–559

Kew Plants of the World (2023) Juncus conglomeratus L. | Plants of the World Online. Kew Science

Kim SH, Nam J.M, Kim JG (2017) Establishment strategy of a rare wetland species *Sparganium erectum* in Korea. J Ecol Environ 41: 27

Kirschner J (2002) Juncaceae 3: Juncus subgen. Agathryon—species Plantarum. Flora of the World 8:192

Klimešová J, Martínkova J, Kočvarová M (2004) Biological flora of Central Europe: *Rorippa palustris* (L.) Besse. Flora 199:453–463

Kołaczek P, Gałka M, Karpinska-Kołaczek M, Lutynska M (2015) Late Pleniglacial and Late Glacial lake-mire transformations in south-eastern Poland reflected in aquatic and wetland vegetation changes. Quat Int 388:39–50

Körber-Grohne U (1964) Bestimmungsschlussel für subfossile *Juncus*-Samen und Gramineen-Früchte. Probl Küstenforsch südl Nordseegebiet 7:33–37. August Lax Verlagsbuchhandlung, Hildesheim

Körber-Grohne U (1967) Geobotanische Untersuchungen auf der Feddersen Wierde. Feddersen Wierde 1. Steiner, Wiesbaden

Körber-Grohne U (1991) Bestimmungsschluüssel für subfossile *Juncus*-Samen und Gramineen-Früchte. Probl Küstenforsch südl Nordseegebiet 18:169–234. August Lax Verlagsbuchhandlung, Hildesheim

Korhola A, Rautio M (2001) Cladocera and other branchiopod crustaceans. In: Smol JP, Birks HJB, Last WM (eds) Tracking Environmental Change Using Lake Sediments. Volume 4. Zoological indicators, Kluwer Academic Publishers, Dordrecht, pp 5–42

Koshkarova VL, Koshkarov AD, Kol'tsova VG (2006) Climatic dependence of the late Holocene dynamics of tree species cenotypes in the intrazonal ecotone of eastern Sayan forests. Ekologiya 5:350–359

Kozelková Z, Prausová R, Tomášová Z, Šafářová L (2021) Differences in *Potamogeton praelongus* morphology and habitats in Europe. Acta Soc Bot Pol 90:1–19

Kozlowski G, Jones RA, Nicholls-Vuille F-L (2008) Biological Flora of Central Europe: *Baldellia ranunculoides* (Alismataceae). Perspect Plant Ecol Evol Syst 10:109–142

Krause W (1986) Zur Bestimmungsmöglichkeit subfossiler Characeen-Oosporen an Beispeilen aus Schweizer Seen. Viertljahrsschrift der Naturforschenden Gesellschaft in Zürich 131:295–313

Krause W (1997) Charales (Charophyceae). Susswasserflora von Mitteleuropa 18. G. Fischer, Jena

Lamentowicz M, Milecka K (2004) Lobelia dortmanna L. seeds in lake sediments from the Tuchola Forest (Pomerania, northern Poland). Acta Palaeobotanica 44:281–285

Lamentowicz M, Gałka M, Draga M, Jassey VEJ, Fritz C, Glatzel S, Robroek B, Meyer H, Lehmann J, Juszczak R, Chojnick BH, Knorr K-H (2024) Investigating ecological baselines and critical thresholds in ombrotrophic nemoral peatlands: implications for ecological restoration. In: EGU General Assembly, Vienna, Austria, 14–19 Apr 2024, EGU24–3047. https://doi.org/10.5194/egusphere-egu24-3047

Lampinen J, Metsäranta N (2020) *Hottonia palustris* L. (Primulaceae)—a new vascular plant for Finland found in Turku. Memoranda Soc Fauna Flora Fennica 96:32–37

Landolt E (1975) Morphological differentiation and geographical distribution of the *Lemna gibba—Lemna minor* group. Aquat Bot 1:345–363

Landolt E (1986) The family of Lemnaceae—a monographic study, vol 1. Veröffentlichungen des Geobotanischen Institutes der Eidg. Techn, Hochschule, Stiftung Rübel., in Zürich 71, p 566

Lang G (1994) Quartäre Vegetationsgeschichte Europas. G. Fischer, Jena, Stuttgart, New York

Langangen A (1974) Ecology and distribution of Norwegian charophytes. Nor J Bot 21:31–52

Langangen A (2007) Charophytes of the Nordic countries. Saeculum ANS, 1903 Gan, Norway

Lansdown RV (2008) Water-Starworts *Callitriche* of Europe. BSBI handbook no. 11. Botanical Society of the British Isles, London

Lansdown RV (2011a) *Callitriche cophocarpa*. The IUCN red list of threatened species 2013: Callitriche cophocarpa (Long-styled water-starwort) (iucnredlist.org)

Lansdown RV (2011b) *Hypericum elodes*. The IUCN red list of threatened species 2011, e.T167868A6401761. https://doi.org/10.2305/IUCN.UK.2011-1.RLTS.T167868A6401761.en

Lansdown RV (2022) The genus *Callitriche* (Plantaginaceae, Callitricheae) in Australasia and Oceania. Phytotaxa 547:243–284

Lansdown R, Pankhurst T (2023) *Caltha palustris* var. *radicans*: overlooked and under-recorded? BSBI News (Botanical Society of Britain and Ireland) 152:31–33

Lansdown RV, Wade PM (2003) Ecology of the floating water-plantain *Luronium natans*, Conserving Natura 2000 Rivers, Ecology Series no. 9. English Nature, Peterborough

Lansdown RV, Kitchener G, Jones E (2022) Wolffia columbiana and W. globosa (Araceae) new to Britain. British and Irish Botany 4:14–26

Lapateva EG, Zaretskaya NE, Lychagina EL, Trofimova SS, Demakov DA, Kopytov SV, Chernov AV (2023) Holocene vegetation dynamics, river valley evolution and human settlement of the upper Kama valley, Ural region, Russia. Veg Hist Archaeobotany 32:361–385

Large MF, Braggins JE (1993) Spore morphology of New Zealand *Azolla filiculoides* Lam. (Salviniaceae). NZ J Bot 31:419–423

Last WM, Smol JP (2001) Physical and Geochemical Methods. Tracking Environmental Change using Lake Sediments. Volume 2. Kluwer Academic Publishers

Law A, Baker A, Sayer CD, Foster G, Gunn I, MacAdam C, Willby N (2024) Repeatable patterns in the distribution of freshwater biodiversity indicators across contrasting landscapes? Landsc Ecol 39:195. https://doi.org/10.1007/s10980-024-01992-z

Les DH, Peredo EL, Tippery HP, Benoit LK, Razifard H, King UM, Na HR, Choi HK, Chen L, Shannon RK, Sheldon SP (2015) *Najas minor* (Hydrocharitaceae) in North America: a reappraisal. Aquat Bot 126:60–72

Levi EE, Cakiroĝlu AI, Bucak T, Odgaard BV, Davidson TA, Jeppesen E, Beklioĝlu M (2014) Similarity between contemporary vegetation and plant remains in the surface sediment in Mediterranean lakes. Freshw Biol 59:724–736

Li Y, Wang L, Chao C, Yu H, Yu D, Liu C (2021) Submerged macrophytes successfully restored a subtropical aquacultural lake by controlling its internal phosphorus loading. Environ Pollut 268:115949

Light JJ (1975) Clear lakes and aquatic bryophytes in the mountains of Scotland. J Ecol 63:937–943

Light JJ, Smith RIL (1976) Deep-water bryophytes from the highest Scottish lochs. J Bryol 9:55–62

Lindstrøm E-A, Brettum P, Johansen SW, Mjelde M (2004) Vannvegetasjon i norske vassdrag. Kritiske grenseverdier for forsuring. Effekter av kalking. NIVA-rapport nr. 4821–2004

Livingstone D (1957) On the sigmoid growth phase in the history of Linsley Pond. Am J Sci 255:364–373

Lodwick L (2014) Condiments before Claudius: new plant foods at the late iron age oppidum at Silchester, UK. Veg Hist Archaeobotany 23:543–549

Lodwick LA (2017) Agricultural innovations at a Late Iron Age oppidum: Archeobotanical evidence for flax, food, and fodder from Calleva Atrebatum, UK. Quat Int 460:198–219

Loque G, Meinesz A (1988) Flowering and fruiting of *Zostera noltiii* in Golfe Juan (French Mediterranean). Aquat Bot 32:341–352

Lucassen ECHET, Spierenberg P, Fraaije RGA, Smolders AJP, Roelofs JGM (2009) Alkalinity generation and sediment CO_2 uptake influence establishment of *Sparganium angustifolium* in softwater lakes. Freshw Biol 54:2300–2314

Lukács AB, Molnár A, Mészáros A, Lovas-Kiss A, Vincze O, Süveges K, Fekete R, Mesterházy A (2020) The decline and recovery of populations of *Potamogeton coloratus* in Hungary. Preslia 92:73–86

Maberly SC (1993) Morphological and photosynthetic characteristics of *Potamogeton obtusifolius* from different depths. J Aquat Plant Manag 31:34–39

Maberly S (2023) Aquatic plants of China, Q. Wang, W. Li, G. Wang, K. Liao, Changjiang Publishing & Media, China (2021), p. 552, ISBN 978-7-5706-1150-8. Aquatic Botany 188:103684

Maberly SC, Madsen TV (2002) Freshwater angiosperm carbon concentrating mechanisms: processes and patterns. Funct Plant Biol 29:393–405

Maberly SC, Spence DHN (1983) Photosynthetic inorganic carbon use by freshwater plants. J Ecol 71:705–724

MacDonald GM, Beukens RP, Kieserm WE (1991) Radiocarbon dating of limnic sediments: a comparative analysis and discussion. Ecology 72:1150–1155

Madalsky J (1949) Cechy morfologiczne pestek Europejskich gatunków Potamogeton (Tourn.) L. i klucz do osnaczania ich szczątków dyluwialnych. Travaux de la Societé des Sciences et des Lettres de Wroclaw. Seria B Nr24: 5-25. In Czech. (Not seen by us)

Madgwick G, Emson D, Sayer CD, Willby NJ, Rose N, Jackson MJ, Kelly A (2011) Centennial-scale changes to the aquatic vegetation structure of a shallow eutrophic lake and implications for restoration. Freshw Biol 56:2620–2636

Marrotte RG, Chmura GA, Stone PA (2012) The utility of Nymphaeaceae sclereids in paleoenvironmental research. Rev Palaeobot Palynol 160:29–37

Martin AC (1951) Identifying pondweed seeds eaten by ducks. J Wildl Manag 15:253–258

Matthews JV (1974) Wisconsin environment of interior Alaska: pollen and macrofossil analysis of a 27 meter core from the Isabella Basin (Fairbanks, Alaska). Can J Earth Sci 11:828–841

Matthews JV (1975) Insects and plant macrofossils from two Quaternary exposures in the Old Crow - Porcupine region, Yukon Territory, Canada. Arct Alp Res 7:249–259

Maul LC, Stebich M, Frenzel P, Hambach U, Henkel T, Katzschmann L, Kienast F, Meng S, Penkman K, Rolf C, Thomas M, Kahlke R-D (2013) Age and palaeoenvironment of the enigmatic Arternian interglacial—evidence from the Muschelton at Voigtstedt/Hackelsberg (Thuringia, Central Germany). Palaeogeogr Palaeoclimatol Palaeoecol 386:68–85

Mauquoy D, van Geel B (2007) Mire and peat macros. In: Elias SA (ed) Encyclopedia of Quaternary Science. Elsevier, pp 2315–2336

Mazur M, Marcysiak K, Dunajska A, Gawlak M, Kałuski T (2022) Taxonomic Significance of Seed Morphology in Veronica L. (Plantaginaceae) Species from Central Europe. Plants 2022, 11: 88. https://doi.org/10.3390/plants11010088

McConnachie AJ, Hill MP, Byrne JH (2004) Field assessment of a frond-feeding weevil, a successful biological control agent of red waterfern, *Azolla filiculoides*, in southern Africa. Biol Control 29:326–331

McCrackin ML, Jones HP, Jones PC, Moreno-Mateos D (2017) Recovery of lakes and coastal marine ecosystems from eutrophication: a global meta-analysis. Limnol Oceanogr 62:507–518

McVean DN (1953) Biological Flora of the British Isles, *Alnus* Mill. J Ecol 41:447–466

Meredith TC, Grubb PJ (1993) *Peucedanum palustre* (L.) Moench. J Ecol 81:813–826

Miotk-Szpiganowicz G, Gałka, M (2009) A new site of Holocene fossil Trapa natans L. at the Kaszuby Lakeland (Poland). Limnological Review 9:165–173

Mitchell GF (1952) Further identifications of macroscopic plant fossils from Irish Quaternary deposits, especially from a Late-Glacial deposit at Mapastown, Co. Louth. (Studies in Irish Quaternary deposits: no. 8). Proc Royal Irish Acad Sect B 55(1952/1953):225–281

Mittal N, Varshney VK, Song BH, Ginwall HS (2015) High levels of diversity in the phytochemistry, ploidy and genetics of the medicinal plant *Acorus calamus L.* Med Aromat Plants S1:002

Mochalova OA, Efimov DY (2022) Environmental patterns of distribution of *Sparganium emersum* and *S. hyperboreum* (Typhaceae) in Northeast Asia. Inland Water Biol 15:784–793

Moody ML, Les DH (2007) Geographic distribution and genotypic composition of invasive hybrid watermilfoil (*Myriophyllum spicatum* x *M. sibiricum*) populations in North America. Biol Invasions 9:559–570

Moody ML, Les DH (2010) Systematics of the aquatic angiosperm genus *Myriophyllum* (Haloragaceae). Syst Bot 35:121–113

Moon H-K, Hong S-P (2006) Nutlet morphology and anatomy of the genus Lycopus (Lamiaceae: Mentheae). J Plant Res 119:633–644

Moore JA (1986) Charophytes of Great Britain and Ireland. BSBI handbook no 5. British Museum, London

Moore KA, Short FT (2006) *Zostera*: biology, ecology, and management. In: Larkum AWD, Orth RJ, Duarte CM (eds) Seagrasses: biology, ecology, and conservation. Springer, Dordrecht, pp 361–386

Moravcová A, Tichá A, Carter VA, Vondrák D, Čtvrtlíková M, van Leeuwen JFN, Heurich M, Tinner W, Kuneš P (2021) Mountain aquatic *Isoëtes* populations reflect millennial-scale environmental changes in the Bohemian Forest Ecosystem, Central Europe. The Holocene 31:746–759

Mortensen MF, Birks HH, Christensen C, Holm J, Noe-Nygaard N, Odgaard BV, Olsen J, Rasmussen KL (2011) Lateglacial vegetation development in Denmark – new evidence based on macrofossils and pollen from Slotseng, a small-scale site in southern Jutland. Quaternary Science Reviews 30:2534–2550

Mossberg B, Stenberg L (2018) Nordens Flora. Bonnier Fakta, Sweden. (in Swedish)

Murphy KJ (2002) Plant communities and plant diversity in softwater lakes of northern Europe. Aquat Bot 73:287–324

Murray DR (1987) Seed dispersal by water. In: Murray DR (ed) Seed Dispersal. Academic Press, Marrickville, pp 49–86

Nagasaka M, Yoshizawa K, Ariizumi K, Hirabayashi K (2002) Temporal changes and vertical distribution of macrophytes in Lake Kawaguchi. Limnology 3:107–114

Naul M, Mikulyuk A (2009) Yellow Floating Heart (*Nymphoides peltata*): A technical review of distribution, ecology, impacts, and management. Wisconsin Department of Natural Resources Bureau of Science Services, PUB-SS-1051 2009, Madison, Wisconsin

Navarro-Ramos MJ, Green AJ, de Vries R, van Leeuwen CHA (2024) Float, fly, then sink: wetland plant seed buoyancy is lost after internal dispersal by waterbirds. Hydrobiologia 851:4033–4048

Neid SL (2006) *Utricularia minor* L. (lesser bladderwort): a technical conservation assessment. USDA Forest Service, Rocky Mountain region. http://www.cnhp.colostate.edu/download/documents/Spp_assessments/utriculariaminor.pdf

Newbold C (2003) The ecology and status of *Potamogeton acutifolius* Link. (sharp-leaved pondweed) in 2003: A resurvey of BSBI records for the period 1970–1999. Report to English Nature

Nielsen SR, Martinsen KT, Pedersen O, Baastrup-Spohr L (2023) Reasons for the dramatic loss of *Lobelia dortmanna*, a keystone plant species of softwater lakes in the Northern Hemisphere. Freshw Biol 68:1673–1684

Nilsson T (1961) Kvartärpaleontologi. II Planscher. Lund University, Sweden

Nilsson Ö, Hjelmqvist H (1967) Studies on the nutlet structure of south Scandinavia species of *Carex*. Bot Notiser 120:460–485

Novenko EY, Volkova EM, Glasko MP, Zuganova IS (2012) Palaeoecological evidence for the middle and late Holocene vegetation, climate and land use in the upper Don River basin (Russia). Veg Hist Archaeobotany 21:337–352

Novenko EY, Volkova EM, Nosova NB, Zuganova IS (2009) Late Glacial and Holocene landscape dynamics in the southern taiga zone of East European plain according to pollen and macrofossil records from the Central Forest State Reserve (Valdai Hills, Russia). Quat Int 207:93–103

Nowak RS, Nowak CL, Tausch RJ (2000) Probability that a fossil absent from a sample is also absent from the paleolandscape. Quat Res 54:144–154

Odgaard BV, Rasmussen P (2001) The occurrence of egg-cocoons of the leech *Piscicola geometra* (L.) in recent lake sediments and their relationships with remains of submerged macrophytes. Arch Hydrobiol 152:671–686

Odland A (2001) *Elatine orthosperma* new to Voss, Western Norway, and comments on its ecology and management potentials (Nordlig krossevjeblom *Elatine orthosperma*: nyfunn på Voss og kommentarer til artens økologi og vurdering av mulige skjøtselstiltak). Blyttia 59:15–21

Olesen B, Krause-Jensen D, Marbà N, Christensen PB (2015) Eelgrass *Zostera marina* in subarctic Greenland: dense meadows with slow biomass turnover in cold waters. Marine Ecol Progress Ser 518:107–121

Ota J, Koike N, Asaeda T, Fujino T (2016) Annual growth pattern of Sparganium erectum and its effects on the stream current and substrate materials. Proc Hydraulic Eng 50:1183–1188

Ozimek T, Prejs A, Prejs K (1976) Biomass and distribution of underground parts of *Potamogeton perfoliatus* L. and *P. lucens* L. in MikoIajskie Lake, Poland. Aquat Bot 2:309–316

Palm N, Franzaring J, Schweiger AH (2024) International Biological Flora: Trapa natans. Journal of Ecology 112:2386–2420.

Partridge JW (2001) *Persicaria amphibia* (L.) Gray (*Polygonum amphibium* L.). J Ecol 89:487–501

Patmore IR, Sayer CD, Goldsmith B, Davidson TA, Rawcliffe R, Salgado J (2014) Big Ben: a new wide-bore piston corer for multi-proxy palaeolimnology. J Paleolimnol 51:79–86

Pedersen O, Andersen T, Ikejima K, Hossain MDZ, Andersen FØ (2006) A multidisciplinary approach to understanding the recent and historical occurrence of the freshwater plant, *Littorella uniflora*. Freshw Biol 51:865–877

Peglar SM, Fritz SC, Alapieti T, Saarnisto M, Birks HJB (1984) Composition and formation of laminated sediments in Diss Mere, Norfolk, England. Boreas 13:13–28

Perego R, Furlanetto G, Rapi M, Ravazzi C (2025) Plant-human interactions: from pristine forest to bronze age farming—vegetation history and depositional processes offshore the lake dwelling of Lavagnone, N-Italy. Veget History Archaeobot 34:459. https://doi.org/10.1007/s00334-024-01027-2

Petr L, Žáčková P, Grygar TM, Píšková A, Křížek M, Treml V (2013) Šúr, a former late-glacial and Holocene lake at the westernmost margin of the Carpathians. Preslia 85:239–263

Philbrick CT, Aakjar RA, Stuckey RL (1998) Invasion and spread of *Callitriche stagnalis* (Callitrichaceae) in North America. Rhodora 100:25–38

Philippsen B (2013) The freshwater reservoir effect in radiocarbon dating. Herit Sci 1:24. http://www.heritagesciencejournal.com/content/1/1/24

Pietryka M, Richter D, Podlaska M (2023) Distribution and ecology of Wolffia arrhiza (L.) Horkel ex Wimm. in the lowland part of Lower Silesia (Poland). Biologia 78:971–978

Pip E (1987) The ecology of *Potamogeton* species in central North America. Hydrobiologia 153:203–216

Plant Atlas 2020 (2023) BSBI Online Plant Atlas 2020. Botanical Society of Britain and Ireland

Pokorny P, Klimešová J, Klimeš L (2000) Late Holocene history and vegetation dynamics of a floodplain alder carr: a case study from eastern Bohemia, Czech Republic. Folia Geobot 35:43–58

Popiela A, Łysko A, Białecka B, Marta M, Sramkó G, Staroń W, Wieczorek A, Molnár A (2017) Seed morphometric characteristics of European species of *Elatine* (Elatinaceae). PeerJ 5:e3399. https://doi.org/10.7717/peerj.3399

Popiela A, Łysko A, Molnár A (2013) Recent distribution of the euro-Siberian-sub-Mediterranean species *Elatine alsinastrum* L. (Elatinaceae). Acta Bot Croatia 72:375–386

Popiela A, Łysko A, Molnár A, Kącki Z, Lukács BA (2015) Distribution, morphology and habitats of *Elatine triandra* (Elatinaceae) in Europe, with particular reference to the central part of the continent. Acta Bot Gall Botany lett 162:325–337

Poschlod P, Rosbakh S (2018) Mudflat species: threatened or hidden? An extensive seed bank survey of 108 fish ponds in southern Germany. Biol Conserv 225:154–163

Praeger RL (1913) The buoyancy of the seeds of some Britannic plants. Sci Proc Roy Dublin Soc 14:1

Prausová R, Holzbauerová H, Špringrová I, Jará N, Šafářová L, Cross AT, Adamec L (2022) Seed gemination ecology of common bladderwort (*Utricularia vulgaris* L.). Aquatic Botany 182:103545

Preston CD (1995) Pondweeds of Great Britain and Ireland. BSBI handbook no. 8. BSBI, London

Preston CD (1999) RDB Species Account: Lythrum hyssopifolia. In: Wigginton, M.J. (ed.). 1999, British Red Data Books. 1. Vascular Plants (3rd edn). JNCC.

Preston CD, Croft JM (1997) Aquatic Plants in Britain and Ireland. Harley Books

Rachal DM, Dello-Russo R, Cuba M (2024) The Pleistocene footprints are younger than we thought: correcting the radiocarbon dates of *Ruppia* seeds, Tularosa Basin, New Mexico. Quat Res 117:67–78

Raimondo FM (1996) *Potamogeton gramineus* and *P. angustifolius*, new to the Sicilian flora. Flora Mediterranea 6:81–84

Ralska-Jasiewiczowa M, Rzetkowska A (1987) Pollen and macrofossil stratigraphy of fossil lake sediments at Niechorze I, W. Baltic coast. Acta Palaeobotanica 27:153–178

Rasmussen P, Pantopoulos G, Jensen JB, Olsen J, Røy H, Bennike O (2020) Holocene sedimentary and environmental development of Aarhus Bay, Denmark—a multi-proxy study. Boreas 49:108–128

Raunkiær C (1895-1899) De Danske blomsterplanters naturhistorie, vol 1. Gyldendanske Boghandels Forlag, p 60

Renberg I, Wallin J-E (1985) The history of the acidification of Lake Gårdsjön as deduced from diatoms and Sphagnum leaves in the sediment. Ecol Bull 37:47–52

Revéret A, Rijal DP, Heintzman PD, Brown AG, Stoof-Leichsenring K-R, Alsos IG (2023) Environmental DNA of aquatic macrophytes: The potential for reconstructing past and present vegetation and environments. Freshwater Biol 68:1929–1950

Richards PW, Clapham AR (1941a) *Juncus effusus* L. (*Juncus communis β effusus* E. Mey). J Ecol 29:375–380

Richards PW, Clapham AR (1941b) *Juncus conglomeratus* L. (*J. communis β conglomeratus* E. Mey.; J. Leersii Marsson). J Ecol 29:381–384

Richardson DC, Holgerson MA, Farragher MJ, Hoffman KK, King KBS, Alfonso MB, Andersen MR, Cheruveil KS, Coleman KA, Farrugia MJ, Fernandes RL, Hondola KL, López Moreira Mazacotte GA, Paul K, Peierls BL, Rabaey JS, Sadro S, Sánchez ML, Smyth RL, Sweetman JN (2022) A functional definition to distinguish ponds from lakes and wetlands. Sci Rep 12:10472

Ridley HN (1930) The dispersal of plants throughout the world. Reeve, Kent, UK. Reprinted 1990 by Singh & Singh, Dehra Dun, India

Roberts J (2023) *Callitriche palustris* (narrow-fruited water-starwort): a Cumbrian *annus mirabilis*. BSBI News (Botanical Society of Britain and Ireland) 152:18–23

Robionek AE, Banaś K, Chmara R, Szmeja J (2018) Periodic events of *Potamogeton alpinus* in NW Poland (Pomerania region). Oceanol Hydrobiol Stud 47:41–49

Robson NKB (1981) Studies in the genus *Hypericum* L. (Guttiferae). 2. Characters of the genus. Bulletin of the British Museum (Natural History). Botany Series 8:53–226

Roelofs JGM (1983) Impact of acidification and eutrophication on macrophyte communities in soft waters in The Netherlands. I. Field observations. Aquat Bot 17:139–155

Rørslett B (2022) *Callitriche platycarpa* and its hybrid *C.* x *vigens* in Norway. Blyttia 80:225–233

Rørslett B, Brettum P (1989) The genus *Isoëtes* in Scandinavia: an ecological review and perspectives. Aquat Bot 35:223–261

Rørslett B, Mjelde M (2021a) Faktaark: Sparganium angustifolium Flótgras. Versjon 1. Fotoflora vannplanter. Norsk institutt for vannforskning. https://www.niva.no/omradesider/fotoflora-for-norske-vannplanter

Rørslett B, Mjelde M (2021b) Faktaark: Sparganium emersum Rankpiggknopp. Versjon 1. Fotoflora vannplanter. Norsk institutt for vannforskning

Rørslett B, Mjelde M (2021c) Faktaark: Sparganium glomeratum Nøstepiggknopp. Versjon 1. Fotoflora vannplanter. Norsk institutt for vannforskning

Rørslett B, Mjelde M (2021d) Faktaark: Sparganium gramineum Sjøpiggknopp. Versjon 1. Fotoflora vannplanter. Norsk institutt for vannforskning

Rose F (1989) Colour identification guide to grasses, sedges, rushes, and ferns of the British Isles and Northwest Europe. Viking/Penguin

Rose N (2015) Spheroidal carbonaceous fly ash particles provide a globally synchronous stratigraphic marker for the Anthropocene. Environ Sci Technol 49:4155–4162

Rosset V, Angélibert S, Arthaud F, Bornette G, Robin J, Wezel A, Vallod D, Oertl B (2014) Is eutrophication a major impairment for small waterbody biodiversity? J Appl Ecol 51:415–425

Rosset V, Simaika JP, Arthaud F, Bornette G, Vallod D, Samways MJ, Oertli B (2013) Comparative assessment of scoring methods to evaluate the conservation value of pond and small lake biodiversity. Aquat Conserv-Mar Freshw Ecosyst 23(1), 23–36. https://doi.org/10.1002/aqc.2287

Sadori L, Mercuri AM, Lippi MM (2010) Reconstructing past cultural landscape and human impact using pollen and plant macroremains. Plant Biosyst 144:940–951

Salgado J, Sayer CD, Carvalho L, Davidson TA, Gunn I (2010) Assessing aquatic macrophyte community change through the integration of palaeolimnological and historical data at Loch Leven, Scotland. J Paleolimnol 43:191–204

Salisbury EJ (1967) On the reproduction and biology of *Elatine hexandra* (Lapierre) DC. (Elatinaceae); a typical species of exposed mud. Kew Bull 21:139–149

Salmina L (2004) Factors influencing distribution of *Cladium mariscus* in Latvia. Ann Bot Fenn 41:367–371

Samuelsson G (1934) Die Verbreitung der höheren Wasserpflanzen in Nordeuropa (Fennoskandien und Dänemark). Acta Phytogeographica Suecica 6:1–21

Sand-Jensen K, Pedersen NL, Thorsgaard I, Moeslund B, Borum J, Brodersen KP (2008) 100 years of vegetation decline and recovery in Lake Fure, Denmark. J Ecol 96:260–271

Sand-Jensen K, Riis T, Vestergaard O, Larsen SE (2000) Macrophyte decline in Danish lakes and streams over the past 100 years. J Ecol 88:1030–1040

Sapelko T, Kalinska E, Kuznetsov D, Naumenko J, Galka M (2022) Holocene history of the lake and forest island ecosystem at and around lake Seliger, Valdai Hills (East European plain, Russia). Int J Earth Sci 111:1947–1960

Sayer CD (2014) Conservation of aquatic landscapes: ponds, lakes, and rivers as integrated systems. Wiley Interdiscip Rev Water 1:573–585

Sayer CD, Bennion H, Davidson TA, Burgess A, Clarke G, Hoare D, Frings P, Hatton-Ellis T (2012) The application of palaeolimnology to practical lake management and conservation: examples from UK lakes. Aquatic Conserv Mar Freshwater Ecosyst 22:165–180

Sayer CD, Burgess A, Kari K, Davidson TA, Peglar SM, Yang H, Rose N (2010a) Long-term dynamics of submerged macrophytes and algae in a small and shallow, eutrophic lake: implications for the stability of macrophyte-dominance. Freshw Biol 55:565–583

Sayer CD, Davidson TA, Jones JI (2010b) Seasonal dynamics of macrophytes and phytoplankton in shallow lakes: a eutrophication-driven pathway from plants to plankton? Freshw Biol 55:500–513

Sayer CD, Davidson TA, Rawcliffe R, Langdon PG, Leavitt PR, Cockerton G, Rose N, Croft T (2016) Consequences of fish kills for long-term trophic structure in shallow lakes: implications for theory and restoration. Ecosystems 19:1289–1309

Sayer CD, Greaves H (2020) Making an impact on UK farmland pond conservation. Aquat Conserv Mar Freshwat Ecosyst 30:1821–1826

Sayer CD, Parmenter J (2021) Resurrection of a Norfolk pond gem: grass-poly *Lythrum hyssopifolia*. BSBI NEWS 146:11–14

Schepers M, Cappers RTJ, Bekker RM (2013) A review of prehistoric and early historic mainland salt marsh vegetation in the northern-Netherlands based on the analysis of plant macrofossils. J Coast Conserv 17:755–773

Schlütz F, Enters D, Bittmann F (2021) From dust till drowned: the Holocene landscape development at Norderney, East Frisian Islands. Netherlands J Geosci 100:E7

Schmitz U, Kelm H (2017) First discovery of flowering *Wolffia arrhiza* in Central Europe. Aquat Bot 143:33–35

Schneider SC, Moe TF, Hessen DO, Kaste Ø (2013) *Juncus bulbosus* nuisance growth in oligotrophic freshwater ecosystems: different triggers for the same phenomenon in rivers and lakes? Aquat Bot 104:15–24

Schneider-Binder E (2018) Aquatic macrophyte communities of the Gorgova-Isac-Uzlina area (Danube Delta, Romania). Transylv Rev Syst Ecol Res 20:39–56

Schnittler M, Gunther KF (1999) Central European vascular plants requiring priority conservation measures—an analysis from national red list distribution maps. Biodivers Conserv 8:891–925

Schoch WH, Pawlik B, Schweingruber FH (1988) Botanische Makroreste, botanical macroremains, Macrorestes botaniques. Paul Haupt Publishers, Bern and Stuttgart

Schou JC, Moeslund B, van de Weuer K, Lansdown RV, Weigleb G, Holm P, Baastrup-Spohr L, Sand-Jensen K (2023) Aquatic plants of northern and central Europe including Britain and Ireland. Princetown University Press, New Jersey

Schubert H, Blindow I, Nat E, Korsch H, Gregor T, Denys L, Stewart N, van de Weyer K, Romanov R, Casanova MT (eds) (2024) Charophytes of Europe. Springer

Scribailo RW, Posluszny U (1985) The reproductive biology of *Hydrocharis morsus-ranae*. II. Seed and seedling morphology. Can J Bot 63:492–496

Sculthorpe CD (1967) The Biology of Aquatic Vascular Plants. Edward Arnold, London

Seddon B (1972) Aquatic macrophytes as limnological indicators. Freshw Biol 2:107–130

Šegota V, Zlatković B, Vukov D, Alegro A, Koletić N, Vuković N, Rimac A (2019) Status assessment of the rare aquatic plant *Groenlandia densa* (L.) Fourr. (Potamogetonaceae) in the Western Balkans. Bot Lett 166:125–133

Sell PD, Murrell G (1996–2018) Flora of Great Britain and Ireland. 5 volumes. Cambridge University Press, Cambridge

Shin H, Choi H-K (1998) Taxonomy and distribution of *Zostera* (Zosteraceae) in eastern Asia, with special reference to Korea. Aquat Bot 60:49–66

Shin CJ, Nam JM, Kim JG (2013) Comparison of environmental characteristics at *Cicuta virosa* habitats, and endangered species in South Korea. J Ecol Environ 36:19–29

Short F, Carruther T, Dennison W, Waycott M (2007) Global seagrass distribution and diversity: a bioregional model. J Exp Marine Biol Ecol 350:3–120

Short FT, Carruthers TJR, Waycott M, Kendrick GA, Fourqurean JW, Callabine A, Kenworthy WJ, Dennison WC (2010) "Zostera noltii". IUCN Red List of Threatened Species 2010: e.T173361A6999224. Zostera noltii (Dwarf Eelgrass) (iucnredlist.org)

Shutoh K, Usuba M, Yamagishi H, Fujita Y, Hiramatsu S, Tsujimura O, Ishidoya Y, Kasa M, Kasai N, Matsumoto A, Norita T, Yokoyama A, Kaneko S, Shiga T (2018) A new record of the critically endangered pondweed, *Potamogeton lucens* (Potamogetonaceae) from Aomori prefecture, Japan. J Japanese Botany 93:240–252

Shutoh K, Usuba M, Yamagishi H, Fujita Y, Shiga T (2020) A new record of Potamogeton × angustifolius J. Presl (Potamogetonaceae) in Japan. Acta Phytotax Geobot 71:33–44

Siggery B, Bennion H, Herd J, Kodeeswaran S, Murphy R, Morse S, Waite M (2025) Talking the same language: co-production of a palaeoecological investigation to inform heathland management. J Environ Manag 377:124652

Smith AJE (1978) The Moss Flora of Britain and Ireland. Cambridge University Press, Cambridge

Smith DN, Roseff R, Butler S (2001) The sediments, pollen, plant macrofossils and insects from a Bronze Age channel fill at Yoxall Bridge, Staffordshire. Environ Archaeol 6:1–12

Smits AJM, Van Ruremonde R, van der Velde G (1989) Seed dispersal of three nymphaeid macrophytes. Aquat Bot 35:167–180

Smol JP (2008) Pollution of lakes and Rivers. A Paleoenvironmental perspective, 2nd edn. Blackwell Publishing

Smol JP (2023) Lakes in the Anthropocene: reflections on tracking ecosystem change in the Arctic. Excellence in Ecology 30. International Ecology Institute, Oldendorf/Luhe. ISBN: 978-3-946729-30-3

Smol JP, Birks HJB, Last WM (2001a) Tracking Environmental Change Using Lake Sediments. Volume 3. Terrestrial, Algal, and Silicious indicators. Kluwer Academic Publishers, Dordrecht

Smol JP, Birks HJB, Last WM (2001b) Tracking Environmental Change Using Lake Sediments. Volume 4. Zoological Indicators, Kluwer Academic Publishers, Dordrecht

Smolders AJP, den Hartog C, Roelofs JGM (1995) Germination and seedling development in Stratiotes aloides L. Aquat Bot 51:269–279

Smolders AJP, Lucassen ECHET, Roelofs JGM (2002) The isoetid environment: Biogeochemistry and threats. Aquat Bot 73:325–350

Solhøy T (2001) Oribatid mites. In: Smol JP, Birks HJB, Last WM (eds) Tracking Environmental Change Using Lake Sediments. Volume 4. Zoological Indicators. Kluwer Academic Publishers, Dordrecht, pp 81–104

Søndergaard M, Jeppesen E, Lauridsen TL, Skov C, van Nes EH, Roijackers R, Lammens E, Portielje R (2007) Lake restoration: successes, failures and long-term effects. J Appl Ecol 44:1095–1105

Soomers H, Karssenberg D, Soons MB, Verweij PA, Verhoeven JTA, Wasse MJ (2013) Wind and water dispersal of wetland plants across fragmented landscapes. Ecosystems 16:434–451

Soons MB, van Der Vlugt C, van Lith B, Heil GW, Klaassen M (2008) Small seed size increases the potential for dispersal of wetland plants by ducks. J Ecol 96:619–627

Šoštarić R, Küster H (2001) Roman plant remains from Veli Brijun (Island of Brioni), Croatia. Veg Hist Archaeobotany 10:227–233

Spencer DF, Anderson LWJ, Ksander GG (1994) Field and greenhouse investigations on winter bud production by *Potamogeton gramineus* L. Aquat Bot 48:285–295

Stace CA (2009) Eleven new combinations in the British flora. Watsonia 27:246

Stace C (2019) New Flora of the British Isles, 4th edn. C & M Floristics, Suffolk

Stachowicz-Rybka R (2015) Environmental and climate changes reflected in the Domuraty 2 section (NE Poland) based on analysis of plant macroremains. Acta Palaeobotanica 55:213–231

Stachowicz-Rybka R, Gałka M, Alexandrowicz WP, Alexandrowicz SW (2009) Plant macrofossils and malacocoenoses of Quaternary mineral-organic sediments at Starunia palaeontological site and vicinity (Carpathia region, Ukraine). Ann Soc Geol Pol 79:297–313

Stafford PJ (1995) Marsileaceae. The Northwest European Pollen Flora, 52. Rev Palaeobot Palynol 88:3–24

Stančikaitė M, Milkevičius M, Kisielienė D (2003) Palaeoenvironmental changes in the environs of Žadeikiai Bog, NW Lithuania, during the late glacial and the Holocene according to palaeobotanical and ^{14}C data. Geologija 43:47–60

Steiner C-MCJE, von Fumetti S, Heiri O (2024) Aquatic invertebrate mandibles and sclerotized remains in Quaternary lake sediments. J Paleolimnol 71:45–83

Stewart RIA, Andersson GKS, Brönmark C, Klatt BK, Hansson L-A, Zülsdorff V, Smith HG (2017) Ecosystem services across the aquatic-terrestrial boundary: linking ponds to pollination. Basic Appl Ecol 18:13–20

Stift M, Luttikhuizen PC, Wisser EJW, van Tienderen PH (2008) Different flooding responses of *Rorippa amphibia* and *Rorippa sylvestris*, and their modes of expression in F_1 hybrids. New Phytol 180:229–239

Stoof-Leichsenring KR, Huang S, Liu S, Jia W, Li K, Liu X, Pestryakova LA, Herzschuh U (2022) Sedimentary DNA identifies modern and past macrophyte diversity and its environmental drivers in high-latitude and high-elevation lakes in Siberia and China. Limnol Oceanogr 67: 1126–1141

Štorchová H, Krüger M, Small I, Adamec L, Nevill PG, Howell KA, Dixon KW, Derbyshire M, Zhong X, Cross AT (2023) Global mitochondrial and chloroplast genome diversity in the threatened aquatic carnivorous plant *Aldrovanda vesiculosa*. Aquat Bot 191:103742

Strazisar T, Koch MS, Dutra E, Madden CJ (2013) *Ruppia maritima* L. seed bank viability at the Everglades-Florida Bay ecotone. Aquat Bot 11:126–134

Stroh PA (2015) *Oenanthe fistulosa* L. Tubular Water Dropwort. Species Account. Botanical Society of Britain and Ireland, London

Strykstra RJ, Bekker RM, Andel JV (2002) Dispersal and life span spectra in plant communities: a key to safe site dynamics, species coexistence and conservation. Ecography 25:145–160

Stuckey RL (1985) Distribution history of *Najas marina* (spiny naiad) in North America. Bartonia 51:2–16

Sun J, Hunter PD, Tyler AN, Willby NJ (2018) The influence of hydrological and land use indicators on macrophyte richness in lakes—a comparison of catchment and landscape buffers across multiple scales. Ecol Indic 89:227–239

Sutherland WJ (1990) Biological Flora of the British Isles: *Iris pseudacorus* L. no. 169. J Ecol 78:833–848

Święta-Musznicka J, Latałowa M, Szmeja J, Badura M (2011) *Salvinia natans* in medieval wetland deposits in Gdańsk, northern Poland: evidence for the early medieval climate warming. J Paleolimnol 45:369–383

Tallantire PA (1972) The palaeohistory of the grey alder (*Alnus incana* (L.) Moench.) and black alder (*A. glutinosa* (L.) Gaertn.) in Fennoscandia. New Phytol 73:529–546

Tallantire PA (1973) Some data on the history of alder in Trøndelag, Norway. Grana 13:18–24

Tallantire PA (1992) The alder [*Alnus glutinosa* (L.) Gaertn.] problem in the British Isles: a third approach to its palaeohistory. New Phytol 122:717–731

Tanaka N, Ito Y, Uehara K (2006) Distribution and new localities of Zannichellia palustris L. (Zannichelliaceae) in Japan. Annal Tsukuba Botanical Garden 25:1–6

Taylor K, Rowland P (2011) Biological Flora of the British Isles: *Stachys palustris* L. J Ecol 99:1081–1090

The Flora of Svalbard Ranunculus hyperboreus ssp. (n.d.) arnellii - The Flora of Svalbard (svalbardflora.no)

Thiébaut G, Giamberini L, Muller S (2002) Can *Potamogeton polygonifolius* be considered as a bioindicator species of eutrophication and of acidification in low mineral waters? Verhandlungen des Internationalen Verein Limnologie 28:1–4

Thomaz SM (2023) Ecosystem services provided by freshwater macrophytes. Hydrobiologia 850:2757–2777

Thor G (1988) The genus *Utricularia* in the Nordic countries, with special emphasis on *U. stygia* and *U. ochroleuca*. Nord J Bot 8:213–225

Timson J (1966) Polygonum hydropiper L. J Ecol 54:815–821

Tonkov S, Bozilova E, Marinova E, Jüngner H (2008) History of vegetation and landscape during the last 4000 years in the area of Straldzha mire (SE Bulgaria). Phytol Balcanica 14:185–191

Tralau H (1962) Najas tenuissima (A. Br.) Magnus during the late Cainozoic period in Europe. Botaniska Notiser 115:421–428

Tralau H (1963) The recent and fossil distribution of some boreal and arctic montane species in Europe. Arkiv für Botanik Series 2(5):533–582

Triest L (1988) A revision of the genus *Najas* L. (Najadaceae) in the Old World. Mém Acad Roy Sci Outre-Mer Cl Sci Nat Méd 22:1–172

Troels-Smith J (1955) Karakterisering af løse jordarter. Characterisation of unconsolidated sediments. Danmarks Geologiske Undersøgelse Ser IV 3(10):73

U.S. Fish & Wildlife Service Web Version (2019) 1/31/2019 Water Chestnut (Trapa natans) ERSS (fws.gov)

Väliranta M, Birks HH, Helmens K, Engels S, Piirainen M (2009) Early glacial interstadial summer temperatures were higher than today in north Scandinavia. Quat Sci Rev 28:777–782

Van den Broek T, van Diggelen R, Bobbink R (2005) Variation in seed buoyancy of species in wetland ecosystems with different flooding dynamics. J Veg Sci 16:579–586

Van der Ham RWJM, Kuijper WJ, Kortselius MJH, van der Burgh J, Stone GN, Brewer JG (2008) Plant remains from the Kreftenheye Formation (Eemian) at Raalte, The Netherlands. Veg Hist Archaeobotany 17:127–144

Van Geel B, Aptroot A, Baittinger C, Birks HH, Bull ID, Cross HB, Evershed RP, Gravendeel B, Kompanje EJO, Kuperus P, Mol D, Nierop KGJ, Pals JP, Tikhonov AN, van Reenen G, van Tienderen PH (2008) The ecological implications of a Yakutian mammoth's last meal. Quat Res 69:361–376

Van Geel B, Brinkkemper O, van Reenen GBA, van der Putten NNL, Sybenga JE, Soonius C, Kooijman AM, Hakbijl T, Gosling WD (2020) Multicore study of upper Holocene mire development in West-Frisia, Northern Netherlands: ecological and archaeological aspects. Quaternary 3:12. https://doi.org/10.3390/quat3020012

Van Wijk RJ (1989) Ecological studies on *Potamogeton pectinatus* L. III. Reproductive strategies and germination ecology. Aquat Bot 33:271–299

Van Wijk RJ, Trompenaars HJAJ (1985) On the germination of turions and the life cycle of Potamogeton trichoides Cham. et Schld. Aquatic Botany 22:165–172

Vargot EV, Shcherbakov AV, Bolotova YV, Uotila PJ (2016) Current distribution and conservation of Najas tenuissima (Hydrocharitaceae). Nat Conserv Res Zapovednaâ nauka 1:2–10

Verhoeven JTA (1979) The ecology of *Ruppia*-dominated communities in western Europe. 1. Distribution of *Ruppia* representatives in relation to the autecology. Aquat Bot 6:197–268

Verhofstad MJJM, Núñes MMA, Riechman E, van Donk E, Lamers LPM, Bakker ES (2017) Mass development of monospecific submerged macrophyte vegetation after the restoration of shallow lakes: roles of light, sediment nutrient levels, and propagule density. Aquat Bot 141:29–38

Vermaat JE, Verhagen FCA (1996) Seasonal variation in the intertidal seagrass *Zostera noltii* Hornem: coupling demographic and physiological patterns. Aquat Bot 52:259–281

Vogt K, Rasran L, Jensen K (2004) Water-borne seed transport and seed deposition during flooding in a small river-valley in northern Germany. Flora 199:377–388

Von Post L (1925) Gotlands-agen (Cladium Mariscus R.Br.) i Sveriges postarktikum. Ymer. Tidskr utg av Sv Sallskap f Antrop o Geogr

Wade PM, Edwards RW (1980). The effect of channel maintenance on the aquatic macrophytes of the drainage channels of the Monmouthshire Levels, South Wales, 1840–1976. Aquatic Botany 8:307–322

Walker D (1966) The Late Quaternary history of the Cumberland Lowland. Philos Trans R Soc Lond B 251:1–210

Wallace G (2005) The functional ecology of Potamogeton rutilus Wolfg. PhD thesis, University of Glasgow, http://theses.gla.ac.uk/4018/

Waller M (2010) Ashtead Common, the evolution of a cultural landscape: a spatially precise vegetation record for the last 2000 years from southeast England. The Holocene 20:733–746

Waller M, Early R (2015) Vegetation dynamics from a coastal peatland: insights from combined plant macrofossil and pollen data. J Quat Sci 30:779–789

Walton RE, Sayer CD, Bennion H, Axmacher JC (2021a) Open-canopy ponds benefit diurnal pollinator communities in an agricultural landscape: Implications for farmland pond management. Insect Conservation and Diversity 14:307–324

Walton RE, Sayer CD, Bennion H, Axmacher JC (2021b) Once a pond in time: employing palaeoecology to inform farmland pond restoration. Restor Ecol 29:e13301. https://doi.org/10.1111/rec.13301

Wang X, Byun E, Turner S, Yang H, Baker A, Roberts C, Bennion H (2024) Shallow urban lakes record industrial history over the last century: a paleolimnological study of three lochs in Greater Glasgow, Scotland, United Kingdom. J Paleolimnol 72:381–399

Warner B (1988) Methods in Quaternary Ecology #3. Plant Macrofossils. Geosci Canada 15:121–129

Wasylikowa K (1967) Late Quaternary plant macrofossils from Lake Zeribar, western Iran. Rev Palaeobot Palynol 2:313–318

Wasylikowa K, Maczyńska M, Ralska-Jasiewiczowa M, Tomczyńska Z, Mueller-Bieniek A (2012) The impact of the Przeworsk culture settlement (200-500 AD) on the vegetation in the Liswarta River basin, south-Central Poland: combined pollen and plant macrofossil evidence. Acta Palaeobotanica 52:33–58

Wasylikowa K, Witkowski A, Walanus A, Hutorowicz A, Alexandrowicz SW, Langer JL (2006) Palaeolimnology of Lake Zeribar, Iran, and its climatic implications. Quat Res 66:477–493

Watts WA (1959) Interglacial deposits at Kilbeg and Newtown. Proc R Ir Acad 60:79–134

Watts WA (1978) Plant macrofossils and Quaternary paleoecology. In: Walker D, Guppy JC (eds) Biology and Quaternary environments. Australian Academy of Science, Canberra, pp 53–67

Watts WA, Bright RC (1968) Pollen, seed, and mollusk analysis of a sediment Core from Pickerel Lake, northeastern South Dakota. Geol Soc Am Bull 79:855–876

Watts WA, Winter TC (1966) Plant macrofossils from Kirchner Marsh, Minnesota - a paleoecological study. Geol Soc Am Bull 77:1339–1360

Weber JA, Noodén LD (2005) The causes of sinking and floating in turions of *Myriophyllum verticillatum*. Aquat Bot 83:219–226

West RG (1953) The occurrence of *Azolla* in British interglacial deposits. New Phytol 52:267–272

Weyl PSR, Thum RA, Moody ML, Newman RL, Coetzee JA (2016) Was *Myriophyllum spicatum* L. (Haloragaceae) recently introduced to South Africa from Eurasia? Aquat Bot 128:7–12

Wiegleb G (1989) On *Potamogeton coloratus* (Potamogetonaceae) in Turkey. Willdenowia 19:121–125

Wiegleb G, Kaplan Z (1998) An account of the species of *Potamogeton* L. (Potamogetonaceae). Folia Geobot 33:241–316

Wiegleb G, Todeskino D (1983) Habitat conditions of *Potamogeton alpinus* Balbis stands and relations to the plants characters. Proceedings of the International Symposium on Aquatic Macrophytes, Nijmegen, pp 311–316

Wiegleb G, Bobro AA, Zalewdka-Gałosz J (2017) A taxonomic account of Ranunculus section Batrachium (Ranunculaceae). Phytotaxa 319:1–55

Wiegleb G, Bru H, Herr W (1991) Human impact on the ecological performance of *Potamogeton* species in northwestern Germany. Vegetatio 97:161–172

Wiersema JH (1988) Reproductive biology of *Nymphaea* (Nymphaeaceae). Ann Mo Bot Gard 75:795–804

Wilkinson DM (2023) Plants on the wing: waterbirds and plant dispersal in Britain. British Wildlife 35:167–171

Willis KJ (1992) The late Quaternary vegetational history of Northwest Greece. New Phytol 121:101–117

Wohlfarth B, Hannon G, Feurdean A, Ghergari L, Onac BP, Possnert G (2001) Reconstruction of climatic and environmental changes in NW Romania during the early part of the last deglaciation (~15,000-13,600 cal yr BP). Quat Sci Rev 20:1897–1914

Wood JR, Walker S (2008) Macrofossil evidence for pre-settlement vegetation of Central Otago's basin floors and gorges. NZ J Bot 46:239–255

Wright HE (1992) Patterns of Holocene climatic change in the Midwestern United States. Quat Res 38:129–134

Wright HE, Mann DH, Glaser PH (1984) Piston corers for peat and lake sediments. Ecology 65:657–659

Xie D, Zhou H, Zhu H, Ji H, Li N, An S (2015) Differences in the regeneration traits of *Potamogeton crispus* turions from macrophyte- and phytoplankton-dominated lakes. Sci Rep 5:12907

Yansa CH, Basinger JF (1999) A postglacial plant macrofossil record of vegetation and climate change in southern Saskatchewan. Holocene climate and environmental change in the Palliser triangle: a geoscientific context for evaluating the impacts of climate change on the southern Canadian prairies. Geological Survey of Canada, Ottawa. Bulletin 534(295):139–172

Young DA, Green CP, Batchelor CR, Austin P, Elias SA, Athersuch J, Lincoln P (2021) Macrofossil evidence of alder (*Alnus* sp.) in Britain early in the late glacial interstadial: implications for the northern cryptic refugia debate. J Quat Sci 36:40–55

Yu D, Dong W, Li Z-Y, Funston AM (2002) Taxonomic revision of the genus *Myriophyllum* (Haloragaceae) in China. Rhodora 104:396–421

Yu Q, Wang H-Z, Chi X, Li Y, Ma S-N, Liang X-M, Jeppesen E, Wang HJ (2018) Higher tolerance of canopy-forming *Potamogeton crispus* than rosette-forming *Vallsneria natans* to high nitrogen concentration as evidenced from experiments in 10 ponds with contrasting nitrogen levels. Front Plant Sci 9:1845

Zalat AA (1996) Charophyte gyrogonites from Holocene lacustrine sediments of the Fayoum depression. Egypt Neues Jahrbuch für Geologie und Paläontologie 1996:502–516

Zalewska-Galosz J, Nowak A, Dajdok Z (2012) Ecological variation between marginal and central populations of Potamogeton polygonifolius, a rare and endangered species in Central Europe. J Nat Conserv 20:76–84

Zhang Q, Wu Y, Yang L, Ze L, Zo L, Yang Y, Chen S, Liu E (2025) Tracing aquatic macrophyte development in Nansi Lake, northern China's largest freshwater lake: plant macrofossils from 1855 to present. Ecol Evol 15:e70878. https://doi.org/10.1002/ece3.70878

Zhang S, Li Z, Wang Q, Chen C, Xue-pin J, Sun H (2009) A newly recorded species *Rorippa amphibia* (L.) Besser in China. J Subtrop Trop Biol 17:176–178

Zhao Y, Sayer CD, Birks HH, Hughes M, Peglar SM (2006) Spatial representation of aquatic vegetation by macrofossils and pollen in a small and shallow lake. J Paleolimnol 35:335–350

Zona S, Davis P, Gunathilake LAAH, Prince J, Horn JW (2012) Seeds of Eriocaulaceae of the United States and Canada. Castanea 77:37–45

Index

A
Acid rain, 5, 329
Acorus calamus, 38
Aldrovanda vesiculosa, 139
Alisma gramineum, 39
Alisma lanceolatum, 40
Alisma plantago-aquatica, 41
Alismataceae, 39–47
Alnus glutinosa, 77–79
Alnus incana, 78, 79
Alopecurus aequalis, 222–224
Alopecurus geniculatus, 222–224
Amblystegiaceae, 65
Apiaceae, 18, 48–57
Aquatic ecosystems, 1, 2, 4, 7, 8, 10–12, 15, 20–22, 36
Araceae, 58–63
Araliaceae, 64
Aspleniaceae, 67
Asteraceae, 18, 34, 68–75
Azolla filiculoides, 12, 23, 312–313

B
Baldellia ranunculoides, 42–43, 46
Balsaminaceae, 76
Berula erecta, 8, 48
Betulaceae, 77–79
Bidens cernua, 68
Bidens radiata, 69
Bidens tripartita, 70
Biodiversity, 2, 4, 7, 8, 10, 12, 21, 37, 93, 189
Bolboschoenus maritimus, 101
Boraginaceae, 80
Brassicaceae, 18, 81–86
Butomaceae, 87
Butomus umbellatus, 87

C
Calamagrostis canescens, 225
Calamagrostis epigejos, 226
Calamogrostis purpurea, 227
Calamogrostis stricta, 228
Calla palustris, 58
Callitrichaceae, 204–214
Callitriche, 204–214, 246
Callitriche brutia, 205–207
Callitriche cophocarpa, 204, 208
Callitriche hermaphroditica, 205, 209
Callitriche obtusangula, 210
Callitriche palustris, 211
Callitriche platycarpa, 212, 213
Callitriche stagnalis, 213
Callitriche truncata, 204, 205, 214
Caltha palustris, 291
Campanulaceae, 88
Caprifoliaceae, 89–90
Carex, 102–120
Carex acuta, 104–105, 115
Carex appropinquata, 106
Carex aquatilis, 107, 115
Carex chordorrhiza, 108
Carex demissa, 109
Carex diandra, 110
Carex dioica, 111
Carex elata, 112, 115
Carex hispida, 113

H. H. Birks et al., *Seeds and Fruits of Aquatic Plants of Northern Europe - A Handbook*, Developments in Paleoenvironmental Research 23,
https://doi.org/10.1007/978-3-032-04170-8

Carex lasiocarpa, 114
Carex nigra, 115
Carex rostrata, 102, 119, 190
Carex vesicaria, 102, 120
Caryophyllaceae, 91, 193
Catabrosa aquatica, 222, 229
Ceratophyllaceae, 92–93
Ceratophyllum, 8, 18, 21, 23, 25, 34, 92, 93
Ceratophyllum submersum, 92–93
Ceratopyllum demersum, 8, 23, 25, 34, 92, 93
Chara, 23, 95–98
Chara virgata, 97, 98
Characeae, 18, 26, 34, 94–99
Cicuta virosa, 49
Cirsium palustre, 71, 177
Cladium mariscus, 18, 121–122
Comarum palustre, 303
Conservation, 10, 11, 15, 17, 23, 44
Crassula aquatica, 100
Crassulaceae, 100
Cyperaceae, 10, 17, 101–138, 336
Cyperus fuscus, 123
Cyperus longus, 124

D
Damasonium alisma, 44
Diaspore(s), 1, 18, 25, 27, 32–36, 342
Dispersal, 2, 18, 26, 27, 30, 33–35, 39, 50, 53, 55, 58–60, 90, 102, 149, 152, 164, 165, 186, 189, 222, 223, 225–229, 231, 236, 238, 240, 241, 244, 267, 281, 286, 303, 305, 311, 326, 330, 334
Drepanocladus, 20, 65–66
Droseraceae, 139

E
Ecosystem services, 4, 10–12
Elatinaceae, 140
Elatine, 140–142
Elatine alsinastrum, 140–143
Elatine hexandra, 140–143
Elatine hydropiper, 140–143
Elatine orthosperma, 140, 141, 143
Elatine triandra, 140–143
Eleocharis acicularis, 125, 133, 142
Eleocharis mamillata, 126
Eleocharis multicaulis, 127
Eleocharis palustris, 128
Eleocharis quinqueflora, 129
Eleocharis uniglumis, 130
Elodea, 8, 12, 34, 153
Elodea canadensis, 8, 153
Elodea nuttallii, 153
Elodeid(s), 3, 8, 11, 93, 94, 98, 99, 132, 149, 150, 152, 153, 155, 157, 160–162, 181, 183, 184, 205, 211–213, 216, 251, 260, 265, 279, 342
Epilobium hirsutum, 199, 201
Epilobium palustre, 199–201
Epilobium parviflorum, 199, 201
Equisetum, 144
Equisetum fluviatile, 144
Equisetum limosum, 144
Equisetum palustre, 144
Eriocaulaceae, 145
Eriocaulon aquaticum, 18, 145
Eupatorium cannabinum, 72
Eutrophication, 5, 10–12, 19, 22–23, 36, 43, 46, 87, 88, 93, 139, 142, 143, 150, 158, 162, 164, 168, 184, 195, 197, 209, 216, 217, 246, 253, 255, 258, 259, 261, 269, 274, 277, 280, 281, 286, 313, 319, 326, 333, 342

F
Fabaceae, 146
Filipendula ulmaria, 304
Fontinalaceae, 147
Fontinalis antipyretica, 147

G
Galium palustre, 306
Galium uliginosum, 307
Geum rivale, 305
Glyceria declinata, 230, 235
Glyceria fluitans, 222, 230–233, 235
Glyceria lithuanica, 233
Glyceria maxima, 222, 234
Glyceria notata, 235
Gratiola officinalis, 34, 215
Groenlandia densa, 245, 246, 248, 250

H
Haloragaceae, 148, 150, 152
Helosciadium inundatum, 50
Helosciadium nodiflorum, 51
Hippuridaceae, 216
Hippuris vulgaris, 216
Hottonia palustris, 288
Hydrocharis morsus-ranae, 3, 154
Hydrocharitaceae, 153–156, 159, 161–163

Hydrocotyle vulgaris, 64
Hydrosere, 11, 15, 21, 22, 197, 238
Hypericaceae, 165
Hypericum elodes, 165

I
Impatiens glandulifera, 76
Invasive, 4, 12, 47, 72, 75, 87, 153, 166, 178, 183, 187, 189, 191, 199, 234, 236, 263, 304, 313, 316, 328
Invasive species, 4, 34–35, 151, 183
Iridaceae, 166
Iris pseudacorus, 166
Isoetaceae, 167–168
Isoetes echinospora, 167–168
Isoetes lacustris, 167–168
Isoetid(s), 3, 8, 11, 21, 86, 88, 99, 142, 145, 168, 217, 314, 343
Isolepis cernua, 131
Isolepis fluitans, 132
Isolepis setacea, 125, 133

J
Jacobaea aquatica, 73
Juncaceae, 169–173, 176, 177
Juncus, 125, 169, 176, 327
Juncus acutiflorus, 170, 171
Juncus articulatus, 171
Juncus bufonius, 172
Juncus bulbosus, 173
Juncus conglomeratus, 174–175
Juncus effusus, 174–175
Juncus filiformis, 176
Juncus subnodulosus, 177

K
Keystone species, 10

L
Lagarosiphon major, 155
Lake levels, 8, 15
Lamiaceae, 178–180
Lathyrus palustris, 146
Lemna aequinoctialis, 59
Lemnaceae, 11, 18, 59–63
Lemnaids, 3
Lemna gibba, 59–61, 63
Lemna minor, 59–63
Lemna minuta, 59
Lemna trisulca, 59, 62, 63
Lemna turionifera, 59
Lemna valdiviana, 59
Lentibulariaceae, 181–184
Limosella aquatica, 142, 314
Littorella uniflora, 18, 88, 168, 217–218
Lobelia dortmanna, 88, 142, 168
Luronium natans, 45–46
Lysimachia thyrsiflora, 289
Lysimachia vulgaris, 290
Lythrum hyssopifolia, 185
Lythrum portula, 186
Lythrum salicaria, 187

M
Macrofossil analysis, method, 28–30
Management, 23, 36, 142, 246, 253, 260
Marsileaceae, 192
Mentha aquatica, 179
Menyanthaceae, 190, 191
Menyanthes trifoliata, 18, 190
Mimulus guttatus, 202
Mitigation, 12–13
Montia fontana, 193
Montiaceae, 193
Multiproxy studies, 1, 17, 20, 23, 28, 36
Myriophyllum alterniflorum, 21, 148–149, 152
Myriophyllum aquaticum, 151
Myriophyllum spicatum, 148–152
Myriophyllum verticillatum, 148, 149, 151, 152

N
Najadaceae, 156–162
Najas flexilis, 32, 156–158
Najas marina, 23–25, 159–160
Najas minor, 161
Najas tenuissima, 162
Nasturtium microphyllum, 81, 82
Nasturtium officinale, 81, 82
Nitella, 95–99
Nitella flexilis, 99
Nitella opaca group, 99
Nuphar lutea, 8, 194, 195
Nuphar pumila, 195
Nymphaea alba, 21, 194, 196–198
Nymphaea tetragona, 198
Nymphaeid(s), 3, 154, 191, 198, 271, 343
Nymphoides peltata, 34, 191

O
Oenanthe crocata, 52
Oenanthe fistulosa, 53
Oenanthe lachenalii, 54
Onagraceae, 199–201
Orobanchaceae, 203

P
Palaeolimnology, 1, 5, 15–24, 343
Pedicularis palustris, 203
Persicaria amphibia, 240
Persicaria hydropiper, 241
Petasites hybridus, 74
Peucedanum ostruthium, 55
Peucedanum palustre, 56
Phalaris arundinacea, 222, 236
Phrymaceae, 202
Pilularia globulifera, 192
Plantaginaceae, 204–220
Poaceae, 18, 221–239
Poa palustris, 222, 239
Pollen, 1, 16–18, 36, 59, 64, 93, 122, 149, 151, 194, 197, 205, 218, 304, 328, 344
Polygonaceae, 240–244
Potamogeton, 8, 18, 21, 25, 26, 32, 197, 247–287
Potamogetonaceae, 245–287
Potamogeton acutifolius, 249–253
Potamogeton alpinus, 248–250, 254–255
Potamogeton berchtoldii, 249, 251, 257–258, 265, 273, 278, 279
Potamogeton coloratus, 250, 259, 276
Potamogeton compressus, 249, 250, 260–261
Potamogeton crispus, 8, 248–250, 262–263
Potamogeton epihydrus, 264
Potamogeton friesii, 249–251, 265–266, 274, 278, 279
Potamogeton gramineus, 249, 250, 256, 267–268
Potamogeton lucens, 8, 248–251, 256, 269
Potamogeton natans, 248, 250, 251, 270–271
Potamogeton nodosus, 272
Potamogeton obtusifolius, 248–251, 273–274
Potamogeton perfoliatus, 247, 248, 250, 275
Potamogeton polygonifolius, 249, 250, 259, 271, 276–277
Potamogeton pusillus, 249–251, 258, 265, 278–279
Potamogeton rutilus, 249–251, 278–280
Potamogeton trichodes, 249–251, 278, 281–282
Potamogeton x angustifolius, 249, 250, 256
Primulaceae, 288–290

R
Radiocarbon, 5, 20, 29, 36, 66, 194, 235, 337–341
Ranunculaceae, 291–302
Ranunculus flammula, 43, 298
Ranunculus hyperboreus, 299, 302
Ranunculus lingua, 300
Ranunculus reptans, 298, 301
Ranunculus sceleratus, 302
Ranunculus Sect. *Batrachium*, 292–297
Representation, 18, 19, 29, 30, 32, 33, 35, 36
Restoration, 4, 12–13, 15, 23, 36, 185
Rorippa amphibia, 83
Rorippa islandica, 84, 85
Rorippa palustris, 84, 85
Rorippa sylvestris, 85
Rosaceae, 303–305
Rubiaceae, 306, 307
Rumex aquaticus, 242, 243
Rumex hydrolapathum, 242, 243
Rumex palustris, 244
Ruppiaceae, 308–309
Ruppia cirrhosa, 308–309
Ruppia maritima, 308–309

S
Sagittaria sagittifolia, 8, 47
Salicaceae, 310–311
Salix, 9, 21, 25, 106, 107, 112, 239, 310–311
Salviniaceae, 312–313
Schoenoplectus lacustris, 8, 136, 137
Schoenoplectus tabernaemontani, 137
Schoenoplectus triqueter, 138
Scirpoides holoschoenus, 134
Scirpus sylvaticus, 135
Scrophulariaceae, 202, 314, 315
Scrophularia auriculata, 315
Scrophularia nodosa, 315
Seed bank(s), 2, 9, 12, 36, 42, 44, 82, 84, 88, 123, 139, 143, 160, 208, 217, 298, 302, 309, 314, 317, 319, 344
Silene flos-cuculi, 91
Sium latifolium, 57
Solanaceae, 316
Solanum dulcamara, 316
Sonchus palustris, 75
Sparganiaceae, 317–326
Sparganium angustifolium, 317
Sparganium emersum, 8, 318–319
Sparganium erectum, 10, 319–321
Sparganium glomeratum, 322
Sparganium gramineum, 323
Sparganium hyperboreum, 324

Sparganium natans, 325–326
Sphagnaceae, 329
Sphagnum, 10, 12, 18, 25, 114, 127, 145, 149, 175, 190, 276, 303, 329, 330
Sphagnum palustre, 329
Sphagnum recurvum, 329
Spirodela polyrhiza, 4, 59, 63
Stachys palustris, 180
Stratiotes aloides, 3, 4, 163–164
Stuckenia filiformis, 249, 250, 283–284, 287
Stuckenia pectinata, 249, 250, 285–286
Stuckenia vaginata, 249, 250, 287
Subularia aquatica, 86, 88, 142, 168

T
Taphonomy, 28, 32–34, 344
Telmatophytes, 3, 21, 34, 324, 342, 344
Trapa natans, 10, 188–189
Typha, 10, 18, 23, 34, 40, 137, 169, 327, 328
Typha angustifolia, 327–328
Typhaceae, 317–324, 327–328
Typha domingensis, 327–328
Typha latifolia, 18, 40, 327–328

U
Utricularia australis, 181, 184
Utricularia bremii, 181, 183
Utricularia intermedia, 181–182
Utricularia minor, 181, 183
Utricularia ochroleuca, 181
Utricularia stygia, 181
Utricularia vulgaris, 181, 183, 184

V
Valerianaceae, 89–90
Valeriana dioica, 89–90
Valeriana excelsa ssp. *sambucifolia*, 89–90
Valeriana officinalis, 89–90
Veronica, 219, 220
Veronica anagallis-aquatica, 219, 220
Veronica beccabunga, 219, 220
Veronica catenata, 219, 220
Veronica scutellata, 220
Viola palustris, 330
Violaceae, 330

W
Water level(s), 5, 7, 8, 11, 12, 22, 27, 33, 36, 43, 46, 47, 68, 83, 87, 113, 125, 139, 142, 164, 189, 211, 217, 238, 259, 288, 305, 317, 319, 324, 326
Wolffia arrhiza, 59, 63

Z
Zannichellia obtusifolia, 331
Zannichellia palustris, 32, 331
Zannichelliaceae, 331
Zosteraceae, 332–334
Zostera marina, 332–334
Zostera noltii, 334

MIX
Papier aus verantwortungsvollen Quellen
Paper from responsible sources
FSC® C105338

If you have any concerns about our products,
you can contact us on
ProductSafety@springernature.com

In case Publisher is established outside the EU,
the EU authorized representative is:
Springer Nature Customer Service Center GmbH
Europaplatz 3, 69115 Heidelberg, Germany

Printed by Libri Plureos GmbH
in Hamburg, Germany